AF380110

Practical Biostatistics for Medical and Health Sciences

Seyed Hassan Saneii · Hassan Doosti

Practical Biostatistics for Medical and Health Sciences

 Springer

Seyed Hassan Saneii
School of Rehabilitation Science
Iran University of Medical Sciences
Tehran, Iran

Hassan Doosti
Macquarie University
Arncliffe, NSW, Australia

ISBN 978-981-97-3082-7   ISBN 978-981-97-3083-4 (eBook)
https://doi.org/10.1007/978-981-97-3083-4

This Springer imprint is published by the registered company Springer Nature Singapore Pte Ltd.
The registered company address is: 152 Beach Road, #21-01/04 Gateway East, Singapore 189721, Singapore

If disposing of this product, please recycle the paper.

Preface

Statistics permeates nearly every aspect of our lives. Its significance has particularly grown in the biological, medical, and health sciences, spanning from opinion polls to clinical trials in medicine, as well as the analysis of big data from health applications. The influence of statistics shapes the world around us. In *Practical Biostatistics for Medical and Health Sciences*, we, as authors, establish a strong connection between statistics and our reality by employing a diverse range of real-life applications that bring theory and methods to life.

This book is specifically designed for students majoring in the medical, biological, and health sciences, catering to a wide variety of learners who are taking their first biostatistics course. Recognizing that some students may not have completed advanced mathematical and algebraic courses beforehand, we have minimized the use of algebra and calculus. As a result, the book does not demand a higher level of mathematical rigor than what students have typically covered in high school, where their mathematical background is limited to basic algebra.

The primary audience for this biostatistics book comprises medical and health science researchers, whether in clinical or non-clinical fields. Additionally, students and clinicians seeking an understanding of research design and analysis, as well as those attending postgraduate courses in medical and health science statistics, will find it useful. Our intention, as authors, is to start from a beginner level and progress to a research level, with a strong focus on the interpretation and presentation of results. Throughout the book, we adopt a non-theoretical approach, omitting formal proofs and instead explaining concepts intuitively, supported by ample examples.

Our motivation for writing this book stemmed from the belief that any writing style can be understood by a portion of its audience. However, we observed that most introductory texts fail to sufficiently explain the underlying concepts of statistics and often detach from the reality of conducting and evaluating medical research.

The book comprises 13 chapters, each accompanied by detailed solved examples. In Chap. 1, we elucidate the measurement scales, types of data, and the application of statistics in medical and health sciences. Chapter 2 offers guidance on tabulating raw data and creating various plots. Chapter 3 focuses on data summarization, including measures of central tendency, measures of dispersion, and the impact of constant changes on these measures. Chapter 4 delves into the principles of probability, conditional probability, Bayes' theorem, epidemiology, odds and probability, and counting principles. Chapter 5 explores discrete distributions such as binomial, multinomial, geometric, hypergeometric, and Poisson distributions. Chapter 6 is dedicated to normal and standard normal distributions, as well as t and Chi-squared distributions. Chapter 7 primarily discusses sampling techniques, emphasizing probabilistic approaches and sample size calculations. Chapter 8 delves into estimation (point and interval) and statistical inference. Chapter 9 focuses on health and disease measures, particularly relevant to epidemiological studies. Chapter 10 introduces the foreground toolbox for hypothesis testing, while Chap. 11 delves further into hypothesis testing. Chapter 12 covers correlations, including Pearson, Spearman, and point-biserial correlations, as well as linear correlation. Finally, Chap. 13 centers around one-way ANOVA and post-hoc tests. All the procedures are meticulously calculated step by step and have been validated using SPSS.

<table>
<tr><td>Tehran, Iran</td><td>Seyed Hassan Saneii</td></tr>
<tr><td>Arncliffe, Australia</td><td>Hassan Doosti</td></tr>
</table>

Contents

Chapter 1
Measuring Scales

1.1 Statistics and Its Applications

In the realm of theoretical and pure sciences, such as mathematics and physics, the proof of a theorem or principle leads to an unequivocal conclusion. For instance, in a right triangle, the square of the hypotenuse is always equal to the sum of the squares of the other two sides. In physics, the velocity of an object is determined by dividing the object's displacement by the time taken. These events in the pure sciences always occur with 100% certainty and are not subject to occasional or partial occurrences. If even a single case were to be found where such a relationship does not hold true, it would challenge the validity of the principle or theorem.

However, in the biological sciences, events do not exhibit such absolute certainty. For example, despite the emphasis on handwashing, maintaining social distance, and wearing masks to prevent Covid-19, there are instances where individuals who follow these guidelines still contract the virus. Conversely, there are individuals who do not adhere to the health protocols but do not contract the coronavirus.

When evaluating the efficacy of a new Covid-19 vaccine, people vary in their response to the vaccine and their exposure to the disease. This means that some unvaccinated individuals may not contract Covid-19, while some vaccinated individuals might. In such cases, if the percentage of cases is significantly lower among the vaccinated group compared to the unvaccinated, can we conclude that the vaccine was not effective? Or could it be attributed to chance, or even differences between the two groups in terms of social class, age, gender, or occupation?

Research has demonstrated that smoking and alcohol consumption pose health risks and can lead to heart disease and lung cancer. However, one can find elderly individuals who have been long-time smokers and drinkers but show no symptoms of these diseases and maintain good physical health. Conversely, there are young individuals who have never used alcohol or tobacco but still suffer from heart disease or lung cancer. Moreover, many drugs used to treat a specific disease are effective for the majority of patients, but there are cases where these drugs do not work.

S. H. Saneii and H. Doosti, *Practical Biostatistics for Medical and Health Sciences*, https://doi.org/10.1007/978-981-97-3083-4_1

In the biological sciences, observations often vary from one situation to another, and exact replication of results is not always possible. These variations arise due to genuine and inherent differences or due to sampling variability. It is the responsibility of statisticians to identify these differences and determine the extent to which variations can be attributed to chance and how much is related to inherent differences in the studied phenomena.

Statistics encompasses the scientific processes of planning, collecting, organizing, summarizing, presenting, analyzing, and interpreting data, and it finds application in various disciplines and fields. In the realm of medical and biological sciences, statistics serves numerous purposes, including:

- Assessing the accuracy of measurements.
- Comparing different methods and techniques of measurement.
- Evaluating the performance of diagnostic tests.
- Establishing normal clinical and laboratory values.
- Estimating disease prognosis.
- Monitoring patients' progress and response to treatment.
- Determining the appropriate sample size for research studies.
- Determining the incidence, prevalence, and mortality rates of diseases.

These applications highlight the vital role of statistics in guiding decision-making, drawing meaningful conclusions, and gaining valuable insights in the medical and biological sciences. By employing statistical techniques, researchers and practitioners can make informed decisions, improve healthcare outcomes, and advance scientific knowledge.

1.2   Type of Data

Observation refers to the act of noting a fact or event, often involving the measurement of a quantity using appropriate instruments, as defined by Merriam Webster. Not all observations are of the same kind and possess varying characteristics. Each phenomenon and observation differs in nature from others, leading to distinct approaches in measuring and recording their properties. Such diverse nature of observations necessitates different analytical methods. Observations can be recorded through measurement, counting, sorting, or simply naming. In general, observations are broadly categorized into two groups: "categorical or qualitative" and "numerical or quantitative". These categories serve as fundamental distinctions in understanding and analyzing different types of data.

Categorical or Qualitative Data

Categorical or qualitative observations encompass observations that cannot be quantitatively measured or assigned numerical values. Instead, they are classified based on their kind or similarity to one another. These observations can be represented using codes or letters, which serve as symbols rather than actual quantities or values. Examples of categorical data include pain levels in patients (e.g., no pain, slight, mild, moderate, severe), the gender of patients (male, female), the types of diseases (e.g., cardiac, cancer, infectious, physical), and blood types (A, B, O, AB).

The number of levels or categories assigned to each variable in categorical data is limited and can be predetermined. When a variable has only two possible responses, such as male or female for gender, it is referred to as a dichotomous variable. On the other hand, if a variable has more than two possible responses, it is called a polytomous variable. For example, the gender of individuals is a dichotomous variable, while the opinions of individuals about a health program, with responses ranging from "very poor" to "very good", represent a polytomous variable.

Qualitative observations can further be classified into the following two groups.

Nominal Data

Some observations cannot be subjected to a meaningful scale of measurement, making it impossible to compare the answers in terms of better or worse, bigger or smaller. In these cases, the only approach is to assign distinct names or categories to the various possible answers or modes of the variable. Consequently, the data is primarily distinguished by their separate categories. Examples of such nominal data include the gender of a baby (boy, girl), blood types (A, B, O, and AB), marital status (single, married, divorced, widowed), and colors of objects (green, blue, yellow, black, red, etc.).

In nominal data, the focus is on identifying the distinct categories rather than establishing any inherent order or hierarchy. These categories are not amenable to numerical comparison or quantitative analysis. Instead, they serve as labels or identifiers for different groups or characteristics within the dataset.

Ordinal Data

Certain types of data can be arranged in a specific order, either from small to large or from bad to good (and vice versa). In this type of data, the focus is on the relative positions and the order of items, without making any assumptions about the precise differences between two states or stages. This type of measurement, based on rankings, is known as ordinal scales.

For instance, the severity of a disease can be categorized into "mild, moderate, severe" as a way of indicating its intensity. However, on an ordinal scale, it is not possible to determine if the difference between a severity level of 3 and 4 is the same as that between 7 and 8. Furthermore, two babies with the same Apgar score of 6 can still exhibit differences between them.

Similarly, let's consider a customer satisfaction survey where respondents rate their experience on a scale of "very dissatisfied, dissatisfied, neutral, satisfied, very satisfied". While we can determine that "very satisfied" is ranked higher than "satisfied", we cannot conclude that the difference between "neutral" and "satisfied" is the same as the difference between "dissatisfied" and "neutral". The distances between these levels are not consistently measurable in terms of the underlying variable being measured.

However, as scientific understanding progresses, new scales may be developed to transform qualitative data into measurable and quantitative data. For instance, in the past, intelligence was often categorized as "very clever, clever, moderate", etc. But now, intelligence levels can be measured and expressed numerically through intelligence quotient (IQ) scores.

Tip

If the ordinal scale has many levels (e.g., summed score of a questionnaire), the data may have the statistical properties of interval scale.

As new measurement techniques and tools emerge, some qualitative data can be translated into quantitative data, enabling a more precise and numerical representation.

Numerical or Quantitative Data

Quantitative or numerical data refers to observations that can be measured or counted and represented by numbers. Examples of quantitative data include the number of family members, the prevalence of a specific disease, body temperature, height, age, weight, white blood cell (WBC) count per unit of blood, and blood pressure. Numerical measurements provide a higher level of precision and consistency compared to qualitative data.

For instance, when we compare weights, we can confidently state that 500 g is exactly double the weight of 250 g. The difference between 200 and 500 g carries the same significance as the difference between 600 and 900 g. It is important to note that not all quantitative data are alike, and they can be categorized into two distinct groups: discrete and continuous.

Discrete Variables

Discrete data refers to variables that can only take specific, distinct values and cannot assume any values in between. These values can be listed, even if the list is extensive or infinite. Examples of discrete variables include family size, number of decayed teeth, number of health personnel, number of white blood cells (WBC) in a unit of blood, eyeglass size, and shoe size.

Count data, which are characterized by specific values such as 1, 2, 3, and so on, are part of the discrete scale. There are no conceivable values between two successive values, and they exhibit a distinct separation or "step" in their continuity. For instance, family size can only be expressed as whole numbers such as 3, 4, 5, and so forth, whereas stating the family size as 3.36 persons would not be meaningful. However, certain articles may provide such numbers as an average of family sizes in a population. These numbers represent estimates of the mean or average based on observations of family sizes, obtained by dividing the sum of family sizes by the total number of families. Typically, these average values are not whole numbers.

Similarly, eyeglass lenses (e.g., 1.00, 1.25, 1.50, 1.75, 2.00, 2.25, and 2.50) and shoe sizes (e.g., 37.5, 38, 38.5, and 39) are examples of discrete data. Each value is distinct, with no other values existing between them, creating a noticeable gap.

Age is also considered a discrete variable since it is measured and expressed in separate units such as months and years, without decimal units of time. To illustrate this concept, imagine that a person's age corresponds to the number of candles lit on their most recent birthday cake (assuming everyone celebrates their birthday).

> **Tips**
> Age is also a discrete variable because it is expressed in separate units like month and year, with no decimal unit of time.
> Age is equal to the number of candles lit at their last birthday party.

Continuous Variable

If an observation has the potential to take on any conceivable value, it is considered to have a continuous scale. This implies that there can exist additional values between any two successive values, even at infinitesimally small intervals. In other words, the range of possible values for a continuous variable is not restricted to a finite list. Concrete examples of continuous data include length, volume, and weight. When someone states their weight as 75 kg, it does not necessarily mean that their exact weight is precisely 75,000 g. Rather, the actual weight could be slightly higher or lower if measured with greater precision. In settings such as pharmaceutical factories or jewelry workshops, goods and raw materials are often weighed to several decimal places for enhanced accuracy. This allows for the conceptualization of a different value between any two consecutive measurements with greater precision. The following illustration demonstrates the distinction between discrete and continuous scales.

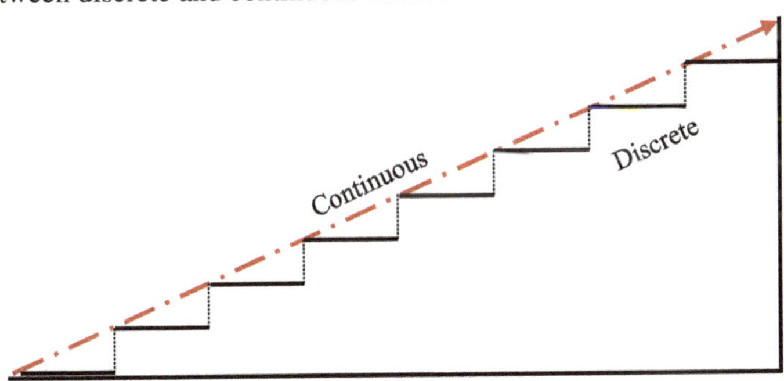

Unlike discrete data, continuous data does not have distinct separations or gaps between values. It allows for infinite possibilities within a given range or interval. For example, between any two heights, there can be an infinite number of values.

It is important to note that continuous data is not limited to physical measurements. Time can also be considered a continuous variable, where it can be measured with precision down to the fraction of a second.

> **Tips**
> Continuous data must be measured, but not counted.
>   Next value for any particular value of discrete data is known, but not for continuous data.

In summary, continuous data encompasses variables that can assume any value within a specified range or interval, allowing for decimal precision and infinite possibilities between two points.

In the realm of quantitative data, there are two subcategories: interval and ratio scales. The distinction between these scales lies in the arbitrariness or substantive difference of their origin (i.e., the zero value). Nevertheless, in most cases, the methods employed for their analysis remain the same. Software programs like SPSS commonly classify both nominal and ordinal observations as either nominal or ordinal scales. However, the statistical analysis approach remains consistent for both types, and SPSS treats them as scales.

Interval Data

Observations that have a conventional and consistent origin, where the absence of something does not imply "nothing" or "lack of", are categorized as interval data. In this case, negative values are also possible, as the starting point is arbitrary and does not represent an absence. For instance, the freezing point of water serves as the origin (zero point) for temperature, indicating the absence of a specific temperature level. Thus, a temperature of $-12°$ signifies being $12°$ below zero (the starting point). The fundamental principle of the interval scale is that "equal intervals" correspond to "equal variations". However, there is no fixed relationship between the ratios of the observed quantities. For example, if the temperatures of areas A and B are 40 and 10 °C, respectively, it does not imply that the temperature of area A is four times that of area B. It simply means that the temperature of area B is 30 °C higher and warmer than that of area A.

Another illustration of an interval scale is the calendar date. Typically, the birth year of a prominent figure or the occurrence of a significant event is considered the reference point for each country's calendar date. For instance, Islamic countries often designate the migration of Prophet Muhammad as the starting date (zero), while European countries use the birth year of Jesus. Suppose event A took place in the year 2000 AD and event B occurred in the year 500 AD. This does not imply that event A transpired four times later than event B, but it signifies that event A occurred 1500 years after event B. On an interval scale, the temperature difference between $8°$ and $11°$ holds the same meaning as the difference between $4°$ and $7°$. Conversely, on an ordinal scale, the distinction between ranks 8 and 11 is not equivalent to the difference between ranks 4 and 7.

In some cases, if an ordinal scale exhibits a large number of levels (e.g., with a summed score ranging from 0 to 100, unlike a five-level rating on an opinion sheet), the data may possess statistical properties similar to interval scales. Summated scales, where scores are derived by summing responses to multiple questions, are often treated as interval scales as well.

For instance, in a questionnaire consisting of 20 ordinal questions (coded from 0 to 4), the total score can range from 0 to 80. In this context, the score of the questionnaire can be regarded as an interval scale. Similarly, IQ is determined through a set of multiple-choice questions and is thus considered an interval scale. When comparing two individuals with IQ scores of 150 and 100, it is incorrect to state that the first person's score is 1.5 times higher than the latter's. The only valid assertion is that the difference in IQ between the two individuals is 50. This is because the IQ test lacks an actual zero point; a score of zero does not signify a complete absence of intelligence but rather reflects the individual's performance in that specific test. If the person were evaluated using different criteria and questions, their IQ score would inevitably change.

When converting the measurement system for interval observations, the value of zero in the new system no longer aligns with zero in the original system. For example, if temperature is converted from 0 °C, the corresponding value in Fahrenheit would be 15°. Similarly, when transitioning from the solar calendar to the Gregorian calendar, the value of solar year zero changes to 624 AD.

The key distinction between the interval scale and the ordinal scale lies in the fact that the intervals between points on the interval scale are equal, whereas, on the ordinal scale, the intervals between points are not equal.

Ratio Data

The ratio scale is characterized by having an actual value of zero, which represents nothingness and absence. On this scale, negative values are not possible. For example, a value of zero in the ratio scale indicates no weight or complete absence of the measured attribute. In this scale, an object weighing 150 g is precisely three times heavier than an object weighing 50 g. If we combine three objects weighing 50 g each, their total weight will equal that of a single object weighing 150 g. Units of measurement such as volume, length, weight, distance, area, speed, and time exemplify the ratio scale.

By defining fundamental mathematical operations and associating them with different scales and data, we can understand their characteristics as follows:

- The nominal scale is not associated with any mathematical operations. It does not allow for mathematical comparisons or operations of any kind.
- The ordinal scale corresponds to the symbols "=", "<", and ">". It allows for comparisons between observations, but other mathematical operations such as addition or subtraction are not applicable.
- The interval scale corresponds to the symbols "+" and "−". In addition to comparisons, it enables us to determine the magnitude of the difference between values.
- The ratio scale corresponds to the symbols "×" and "/". It encompasses comparisons, measures the difference between values, and provides information about the magnitude of one value relative to another in terms of multiplication or division (Fig. 1.1).

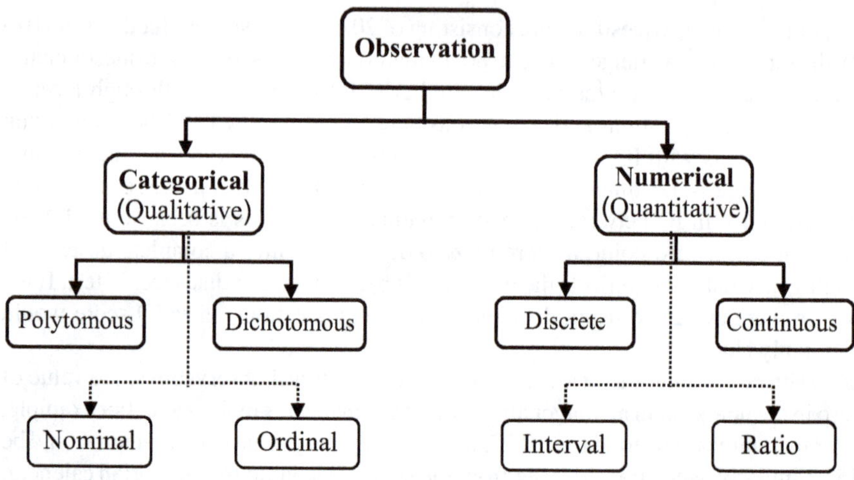

Fig. 1.1 Types of observations

When analyzing data with statistical software, the type of input data and its measurement scale must be specified. In SPSS, these scales must be introduced as "nominal" (for polytomous and dichotomous data), "ordinal" (for ordinal variables), and "scale" (for quantitative data including interval and ratio) via the variable view tab and then the "measure" column.

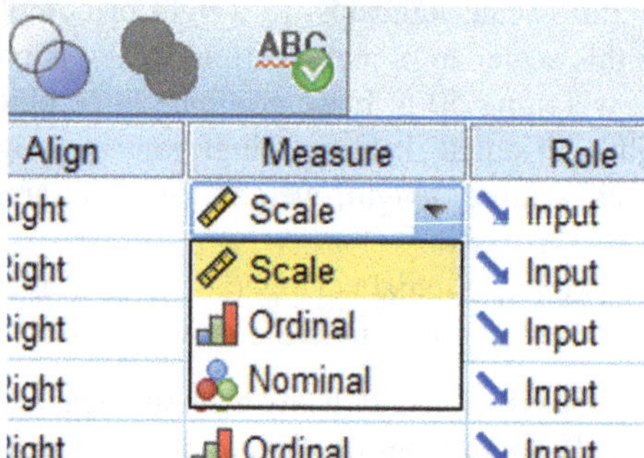

1.3   Exercises

1. Determining the structure of data

- A shoe store wants to analyze its sales data to identify the most popular shoe sizes. What type of data structure and measurement scale would be most appropriate for this analysis?

- A vision clinic is conducting a study on the effectiveness of different eyeglass prescriptions for correcting refractive errors. Which data structure and measurement scale would be suitable for recording and analyzing the data?
- A nutritionist is investigating the relationship between calcium intake and bone health. What data structure and measurement scale should be used to collect and analyze data on the amount of calcium available in different nutrition sources?
- A pharmaceutical company is conducting a clinical trial to determine the optimal dosage of a new medication. Which data structure and measurement scale would be most suitable for recording and analyzing the dosage data?

2. Data Collection Methods

For each item listed below, determine the most appropriate data collection method:

A. Shoes and clothes size.
B. Refractive error (glasses).
C. The amount of calcium available in nutrition.
D. Drug dosage.
E. Treatment duration of a specific disease.
F. Infant weight.
G. The score of a student on biostatistics.
H. The volume of a gallon.
I. Workload of a clerk in a health center.
J. The infant gender.
K. The height of an infant.
L. Blood type.
M. Race of people.
N. The interval between two deliveries of a mother.
O. Number of decayed teeth.
P. Relative humidity.
Q. Systolic blood pressure.
R. Pulse rate.
S. Education level.
T. The price of toothpaste.
U. The opinions of TV viewers.
V. The amount of lead in the air at gas stations.
W. The amount of chlorine in one cubic meter of water.
X. Amount of doctor's visits.
Y. Carat of a piece of gold.
Z. Age of a student.

Note: In this exercise, you can suggest appropriate data collection methods based on the context of each item.

Chapter 2
Tables and Charts

2.1 Introduction

Presenting data and observed values in the best possible way is another task of statistics. Without organizing the observations measured and collected, they can neither be analyzed, nor interpreted, nor used. Therefore, classifying and organizing observations in the form of frequency distribution tables and graphs is one of the basic requirements for statistical methods.

2.2 How to Construct a Frequency Table

The construction of one- and multidimensional tables for qualitative data (nominal or ordinal, dichotomous or polytomous) is straightforward, but for quantitative data (discrete or continuous), it requires more experience and precision. First, the number of classes and their limits are determined, then one person reads the raw data and another person enters tally marks in the appropriate row. The tally marks should be diagonal or vertical, and the fifth tally mark should always intersect with the first four tally marks to make counting tally marks easier. After reading all the raw data and ensuring that the data has been correctly transferred to the table, the number of strokes in each row is counted and written as the absolute frequency (f_i). The index i indicates the class number.

There are different points of views about the number of classes in a table, which all are based on experience. The recommended number of classes is between (5–25) or (8–15) or (10–25) or … I recommend the number of classes between 5 and 15 in relation to the total number of observations. For qualitative data such as blood types (A, B, AB, O), gender (male, female), geographical area (urban, rural), the same number of responses of each qualitative variable can be used.

S. H. Saneii and H. Doosti, *Practical Biostatistics for Medical and Health Sciences*,
https://doi.org/10.1007/978-981-97-3083-4_2

Table 2.1 Cumulative frequency for risk rate

Risk	f	F
Low	21	21
Moderate	17	38
High	32	70
Total	70	–

For ordinal data, the order of the classes must be considered. To interpret this type of data, calculating the cumulative frequency (the sum of the frequency of each class with its predecessor frequencies) is useful.

Example 1 The risk in 70 patients with Covid-19 is (mild 21, moderate 17, and severe 32). By constructing Table 2.1 and calculating the cumulative frequency, it can be determined that 38 individuals are in the moderate stage and less than.

Classification is simple even for ordinal data. Even in most cases, the classes are the same as the observed data or the fusion of two or more adjacent categories.

In data tabulation, the number of classes affects the accuracy of the calculations. The larger the number of classes in a dataset, the shorter the width of the classes and the closer the results of the calculations are to the raw data. The smaller the number of classes, the wider the width of the classes and the further away the results of the tabulation calculations are from the raw data.

In the schematic table below, each cell is identified as a class and the midpoint of each class is identified by a dot in the cell. The number of classes in the first row is high, and the difference between the midpoint of the class and the farthest value on the same class is less than in the second row, where the number of classes is reduced. This difference and deviation in the second row is less than in the third row, where the number of classes has decreased again, and so on.

On the other hand, because of the large number of classes in the first series, the calculations are longer and more difficult than in the second series. Likewise, the second series from the third series, the third series from the fourth series, are more difficult and longer, and so on.

However, when classifying the data is classified, some information of the raw data is lost. This is because in the calculations, instead of using the initial value of each observation, the midpoint of the class to which it belongs is used.

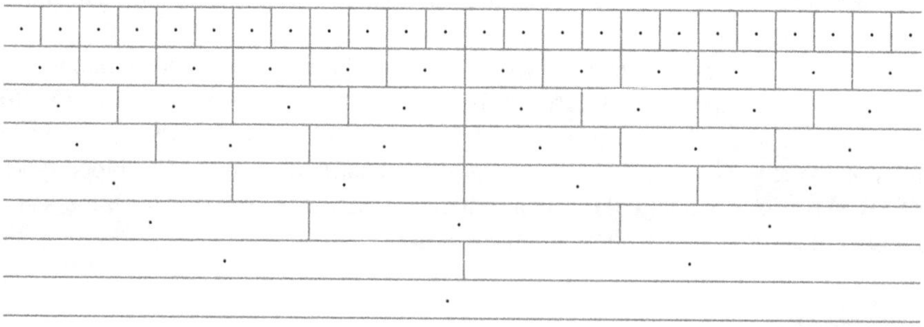

Classification of continuous data requires some experience and expertise. In general, the classification of quantitative data should be done as follows:

1. Specify the number of data or sample size (n).
2. Calculate the range ($R = X_{\max} - X_{\min}$), where $X_{\max}$ and $X_{\min}$ are the maximum and minimum values of observed data, respectively.
3. Decide on the number of classes (K) depending on the range, variety and nature of data, and the number of data observed. If the range is wide and the number of data is large, the number of classes should be larger. Conversely, if the range and the number of data are small, you should opt for fewer classes. Sturges suggests that you get the number of classes using the formula $K = 1 + 3.32 \log n$ with an additional approximation (upward), where n is the number of data.
4. Determine the class width (C): The class width is the difference between the lower limit of two consecutive classes or the upper limit difference of two consecutive classes. In another words, the difference between the upper and lower bounds of each class is called the width of the classes. When the width of the classes is chosen to be equal, drawing graphs and statistical calculations becomes easier than when the classes are unequal in width. However, unequal widths of classes are quite common in epidemiological studies. Fisher suggests that the class width for normally distributed data is equal to one-quarter of the standard deviation of the observed data $\left(C \leq \frac{Sd}{4}\right)$.

> **Tip**
> In discrete and sequential data, the actual limits are the same values as those shown in the table. But in continuous data, the actual limits are the average low limit of each category and the upper limit of the previous category. For example, if the weight for one category is about 70–74 and for the other category 75–80, 74.5 will be the actual upper limit of the first category and the actual lower limit of the next category. Because the first category includes up to 74.4999999 and the next category will start from 74.5000001.

There is always a relationship between the number of classes (K), the class interval (C) and the range (R).

$$C = \frac{R}{K} = \frac{X_{\mathrm{Max}} - X_{\mathrm{Min}}}{K}$$

In practice, steps 2 and 3 of data classification can be interchanged. That is, one first decides on the width of the classes and then calculates the number of classes. In any case, the determination of the values of C and K depends on the subject of study.

Tips

Range: It is the difference between the maximum and minimum data entries.

Class limits: There are two limits for each class. Class limits have the same accuracy as the data values; the same number of decimal places as the data values. Each class will have a "lower-class limit" and an "upper-class limit" which are the lowest and highest numbers that can be in each class.

Lower-class limit: It is the smallest data value that can belong to each class.

Upper-class limit: It is the largest data value that can belong to each class.

Class boundaries: There are two limits for each class. They are a halfway point that separates the classes but includes the gaps created by class limits.

Lower-class boundary: It is obtained by averaging the upper limit of the previous class and the lower limit of the given class.

Upper-class boundary: It is obtained by averaging the upper limit of the class and the lower limit of the next class.

Class width: It is the difference between the upper and lower boundaries of any class (or two consecutive lower-class boundaries or two consecutive upper-class boundaries) in a frequency table.

Class midpoint: It is the value in the middle of each class and equals to the average of the upper- and lower-class limits of each class.

5. Determining the lower limit of the first class: First, determine the lower limit of the first class, that is, the beginning of the classes. Then add the width of the classes (C) to it to get the lower limit of the next class. Usually, the lower limit of the first class starts with the smallest observed value (X_{min}) or any other suitable value smaller than it. This suitable value can be started so that the smallest observed value is the midpoint of the first class.

   That is, the limits of the first class are from $(X_{Min} - \frac{C}{2})$ to $(X_{Min} + \frac{C}{2})$.

6. Read the observed values and tally mark the corresponding class in the table.

Example 2 The ages of 90 patient admitted in a hospital are: 59, 29, 72, 80, 40, 65, 37, 66, 68, 51, 53, 32, 88, 48, 61, 60, 61, 48, 36, 84, 69, 82, 47, 80, 84, 80, 83, 45, 59, 79, 42, 49, 68, 31, 69, 55, 86, 36, 67, 58, 75, 45, 35, 84, 41, 81, 38, 63, 62, 85, 45, 27, 53, 16, 64, 34, 46, 90, 21, 61, 50, 52, 61, 71, 81, 22, 50, 26, 73, 54, 72, 27, 38, 62, 56, 37, 65, 42, 76, 80, 57, 47, 32, 66, 61, 89, 40, 48, 63, 42.

Table 2.2 Frequency distribution for example 2

Class width	Tally mark	Frequency
15–24	///	3
25–34	//// //	7
35–44	//// //// //	12
45–54	//// //// //// //	17
55–64	//// //// //// //	17
65–74	//// //// ///	13
75–84	//// //// ////	14
85–94	////	5
Total	–	90

To construct the frequency table, we have to do the following steps.

$$X_{\max} = 90, \quad X_{\min} = 16 \quad \text{and}$$
$$R = X_{\max} - X_{\min} = 90 - 16 = 74.$$

Using suggested Sturges guideline, the number of classes is (Table 2.2):

$$K = 1 + 3.32 \times \log n = 1 + 3.32 \times 1.95$$
$$= 7.48 \uparrow = 8 \text{ and } C = \frac{R}{K} = \frac{74}{8}$$
$$= 9.25 \approx 10.$$

Example 3 The weight of 100 casting workers is as follows: 67, 72, 80, 74, 65, 53, 59, 83, 69, 67, 69, 79, 76, 65, 64, 60, 57, 71, 73, 68, 87, 70, 72, 67, 62, 59, 66, 58, 75, 79, 87, 79, 77, 65, 60, 75, 64, 59, 60, 63, 72, 65, 70, 81, 79, 68, 73, 70, 84, 76, 58, 94, 86, 83, 69, 67, 75, 64, 66, 74, 97, 62, 60, 68, 71, 78, 71, 75, 67, 64, 69, 77, 72, 67, 73, 69, 78, 69, 70, 67, 66, 74, 68, 73, 79, 73, 69, 99, 70, 75, 68, 73, 71, 58, 65, 71, 78, 68, 97, 89.

Solution: $X_{\max} = 100$, $X_{\min} = 53$ and $R = X_{\max} - X_{\min} = 100 - 53 = 47$
Using suggested Sturges guideline the number of classes is:

$$K = 1 + 3.32 \times \log n = 1 + 3.32 \times 2 = 7.64 \uparrow = 8$$

and $C = \frac{R}{K} = \frac{47}{8} = 5.88 \approx 6$. But we decided to classify the weight data with class width of 5 and the first class begins with 51 kg. Then we have (Table 2.3):

Table 2.3 Frequency distribution for example 3

Class width	Tally mark	Frequency
51–55	/	1
56–60	//// //// /	11
61–65	//// //// //	12
66–70	//// //// //// //// //// ///	28
71–75	//// //// //// //// ///	23
76–80	//// //// ///	13
81–85	////	4
86–90	////	4
91–95	/	1
96–100	///	3
Total	–	100

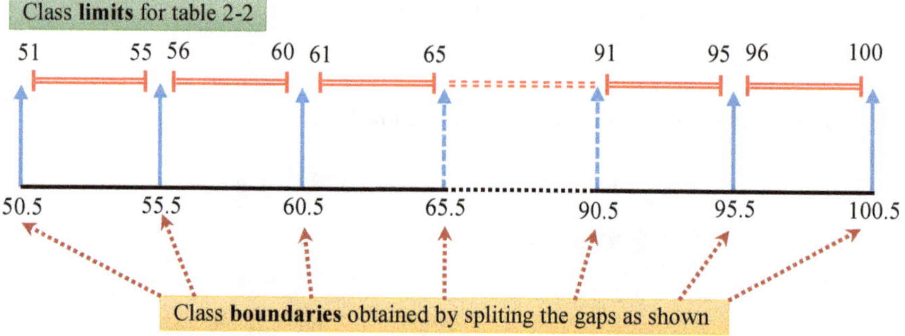

Example 4 The education level of 150 retirees is: primary 35, diploma 96, Associate Degree 11, BSc 6, MSc, and above 2. Find and interpret the cumulative frequency and the relative cumulative frequency.

Solution: Since it is an ordinal variable, we write them in Table 2.4 in the same order and add each class to the previous classes.

Table 2.4 Frequency distribution hemoglobin levels in 70 women

Hemoglobin level	Tally mark	f_i
8.5–9.4	///	3
9.5–10.4	//// /	6
10.5–11.4	//// //// //// //	19
11.5–12.4	//// //// //// /	16
12.5–13.4	//// //// ////	15
13.5–14.4	//// /	6
14.5–15.4	////	5
Total	–	70

Table 2.5 Frequency distribution hemoglobin levels in 70 women

Hemoglobin level	Tally mark	f_i
8.5–9.4	///	3
9.5–10.4	//// /	6
10.5–11.4	//// //// //// //	19
11.5–12.4	//// //// //// /	16
12.5–13.4	//// //// ////	15
13.5–14.4	//// /	6
14.5–15.4	////	5
Total	–	70

From the fourth column of this table, it can be seen that the literacy of 23.3% of retirees is at the primary level and 87.3% of them have a diploma and less.

Example 5 The following data are the hemoglobin level of 70 women g/100 ml:11.9, 13.4, 12.0, 9.3, 12.1, 10.6, 13.3, 10.2, 10.8, 12.9, 13.5, 12.9, 10.5, 12.9, 13.7, 14.6, 11.2, 10.4, 11.4, 13.7, 12.1, 10.4, 11.1, 15.1, 11.4, 12.1, 12.2, 9.4, 14.9, 13.5, 10.7, 12.5, 10.9, 11.1, 14.6, 11.0, 10.9, 12.9, 13.0, 11.3, 8.8, 11.2, 12.7, 11.8, 9.7, 11.6, 14.7, 10.2, 11.7, 10.6, 14.1, 13.2, 11.5, 13.3, 11.6, 10.9, 11.4, 10.3, 10.0, 12.8, 13.1, 13.4, 12.5, 11.9, 13.6, 11.7, 11.0, 11.8, 12.3, 13.1.

Solution: First we find the range, the number of classes, and the class width.

$$X_{\text{Max}} = 15.1 \quad X_{\text{Min}} = 8.8$$
$$R = X_{\text{Max}} - X_{\text{Min}} = 15.1 - 8.8 = 6.3$$
$$K = 1 + 3.32 \log 50 = 1 + 3.32 * 1.845 = 7.12 \approx 8$$
$$C = \frac{R}{K} = \frac{6.3}{8} = 0.788 \approx 0.8.$$

For easier classification, we consider the class width of 1 instead of 0.8 and the number of classes will be $K = 7$ and we start the first class with 8.5. After reading the values and putting tally marks for each observation in our corresponding category, we have Table 2.5:

2.3 Types of Frequencies

In the frequency distribution table, other frequencies can be calculated in addition to the absolute frequency (f_i).

Cumulative frequency: The cumulative frequency of a class is the sum of the absolute frequencies of this class and all preceding classes. For example, the cumulative frequency of the class is:

$$F_4 = f_1 + f_2 + f_3 + f_4.$$

Cumulative frequency and absolute frequency are always equal in the first class $(F_1 = f_1)$, and the cumulative frequency of the last class is equal to the total frequency. The cumulative frequency has a non-decreasing trend. That is, it always increases except when the frequency of a class is zero. In this case, the cumulative frequency remains constant and equal to the previous class.

Relative frequency: The relative frequency of a class is given by dividing the absolute frequency of each class (f_i) by the sum of all frequencies (N) and denoting it by f_p.

$$f_p = \frac{\text{frequency for a class}}{\text{sum of all frequencies}} = \frac{f_i}{N}$$

When the relative frequency (f_p) is multiplied by 100, the percentage of frequency (or proportion) is obtained and is denoted by p.

$$p = 100 \times f_p.$$

Cumulative relative frequency (F_p): The cumulative relative frequency of a class is the cumulative frequency divided by the sum of all frequencies or the sum of the relative frequencies of each class with its predecessors.

$$F_p = \frac{\text{Cumulative frequency}}{\text{Sum of all frequencies}} = \frac{F_i}{N}$$

The relative cumulative frequency of the last class is always equal to 1. For cumulative frequency and relative cumulative frequency, the sum has no meaning and should be left blank. The calculations for the previous example are shown in Table 2.6.

In SPSS, tables can be formed from two paths:

Table 2.6 Different frequencies for hemoglobin levels in 70 women

Hemoglobin level	f_i	F_i	f_p	p	F_p
8.5–9.4	3	3	0.043	4.3	0.043
9.5–10.4	6	9	0.086	8.6	0.129
10.5–11.4	19	28	0.271	27.1	0.400
11.5–12.4	16	44	0.229	22.9	0.629
12.5–13.4	15	59	0.214	21.4	0.843
13.5–14.4	6	65	0.086	8.6	0.929
14.5–15.4	5	70	0.071	7.1	1.00
Total	70	–	1	100	–

- For two-dimensional tables (only one variable in a row and one variable in a column), select the analysis\descriptive statistics\crosstabs path to open the following window.

Then enter one categorical variable in the Row(s) field and one in the Column(s) field. If another categorical variable is entered in the layer field, it will cause the row variable to be divided into smaller layers. But the columns are not divided into smaller groups.

- For multidimensional tables where rows and columns are divided into other subgroups, select the path: Analyze\tables\custom tables, to open a window like the following. Then select the name of the categorical variable and drag it to the column or row. In this way, it is possible to create combined and layered tables. You can even split a given subset into smaller categories and layers. But the interpretation of multidimensional combined tables is difficult for inexperienced users.

To transform quantitative data (continuous or discrete) into tables, you must first define the necessary classes for them using the transform\recode command into different variables. For example, the age variable is first classified into age groups of 5 or 10 years under the new name age group. The new age group variable is then used to extract the tables.

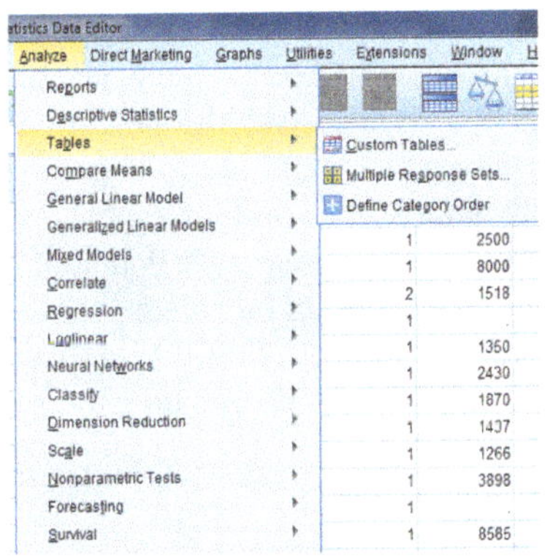
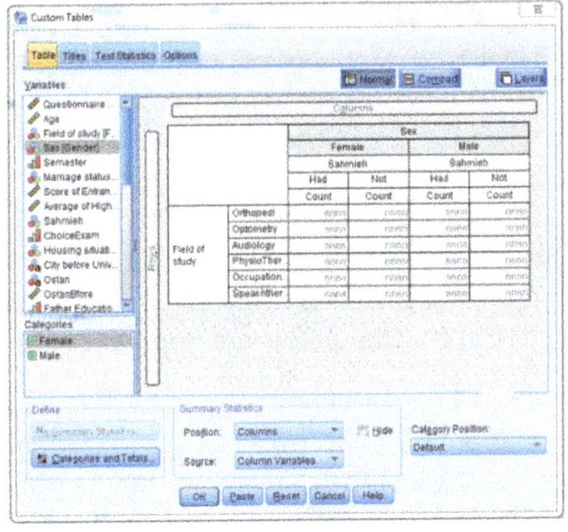

2.4 Features of Table Classification

Table classification should have the following two features:

> **Tips**
>
> - Classes should not overlap at all, even those with a frequency of zero. That is, each original observation can be assigned to only and exclusively one class.
>
> $$S_i \cap S_j = \emptyset : \forall i \neq j$$
>
> - The set of these classes should be comprehensive, covering all original observations. That is, no observation is without a class.
>
> $$\bigcup_{i=1}^{k} S_i = S_1 \cup S_2 \cup \cdots \cup S_{k-1} \cup S_k = S.$$

For this reason, the boundaries of the classes must be precisely defined. The boundaries of classes depend on the nature (discrete or continuous) of the variable. For a discrete variable (such as age and number of people), the class limits and the class boundaries are the same. But for a continuous variable (such as height and weight), the lower boundary of a class is equal to the average of the lower-class limit and the upper limit of the previous class. Then by adding the width of the class to the lower boundary, the upper boundary is obtained. For greater ease of calculations, it is recommended to add a column for the class boundaries and a column for the midpoint of the class (average of upper and lower limits $= x_i$).

The following points are important in the table of frequency distribution:

- Including a row and a column for horizontal and vertical total with the sigma symbol (Σ) improves the understanding of the table.
- It is advised to decide on the number of classes and class width before looking at the data.
- The number of classes and class width depend on the nature of the subject and the number of observations. Therefore, decide on these in consultation with an expert.
- Although equal width is common in frequency tables, they are not necessary. In the epidemiology of diseases, unequal width is sometimes necessary.
- Always put a number and title for the table so that you can refer the reader to it.
- An explicit and plain table is better understood by non-specialists. Therefore, presenting a simple two-dimensional table is better than presenting a complex multidimensional table.

- Identify the source of the information on table so that it can gain the reader's confidence.
- Define the unit of measurement and the symbols used in it.
- The time, place, and subject of the table should be defined so that the question about when? Where? And what? It does not happen to the reader.

Example 6 Meningitis, measles, and rubella are more common in early life and adolescence, and elderlies are less likely to get them. But cancer, rheumatism, shingles and Covid-19 usually occur at an older age. Therefore, the age group of patients with meningitis, measles and rubella cannot be classified as the age group of patients with cancer, rheumatism, shingles, and Covid-19 with equal width of 5 or 10 years. Because the first group of 10 years, data on meningitis is condensed, but elderly groups may be accompanied by a lack of observation. As a result, their classification is even useless. Conversely, considering equal class width for cancer, rheumatism, shingles, and Covid-19 results in a very low frequency in the lower classes and a higher data density in the last classes of the table.

Example 7 The birth weight of 30 infants in Kg and their mothers' smoking status during pregnancy is given in Table 2.7.

In this study, smoking of pregnant mothers during pregnancy and neonatal birth weight in kg were collected. Therefore, two-dimensional tables show both attributes simultaneously. The number of classes of the first variable, which is binomial, is the same as being "smoker" and "non-smoker". But for the second variable which is continuous, the minimum and maximum observations are equal to 2.34 and 4.13, respectively. Since the total number of observations is 30, the number of classes cannot be large. Therefore, 5 classes of "less than 2.50", "2.50–2.99", "3.00–3.49", "3.50–3.99", and "more than 4.00 kg" will be suitable. so (Table 2.8):

In this table, the number 2 indicates that there are 2 infants weighing less than 2.5 kg whose mothers smoked during pregnancy. The number 10 also indicates that

Table 2.7 Birth weight of 30 babies and their mothers' smoking habits

Smk-3.59, Non-3.60, Non-3.99, Smk-3.63, Non-4.08, Non-3.79, Non-4.13, Smk-3.52, Non-3.60, Non-3.26, Smk-3.23, Smk-3.18, Non-3.54, Smk-2.76, Smk-2.84, Smk-2.38, Non-3.61, Non-3.73, Smk-2.34, Smk-3.60, Smk-2.90, Non-3.51, Smk-3.75, Smk-3.27, Non-2.71, Non-3.83, Smk-3.85, Non-3.69, Non-3.31, Non-3.21

Table 2.8 Frequency table of birth weight of infants (kg) by mothers' smoking habit

Mothers' habit	Weight					
	<2.50	2.50–2.99	3.00–3.49	3.50–3.99	>4.00	Total
Smoker	2	3	3	6	0	14
Non-smoker	0	1	3	10	2	16
Total	2	4	6	16	2	30

there are 10 infants weighing between 3.50 and 3.99 kg whose mothers did not smoke at all during pregnancy.

2.5 Graphs

For many people, graphical presentation is more tangible and easier than presenting tables and raw numbers and data. This is because they may not have the statistical knowledge or patience to calculate percent and …. Therefore, the information obtained by looking at a chart is much easier than that obtained by reading and explaining. For this reason, it will be useful to present a graph of the study results.

Before explaining in detail the pie chart, the bar chart, the histogram, the polygon, the stem-and-leaf diagram, the boxplot, and the scattergram, we will outline the common features to be observed in these diagrams.

1. Each diagram must have a title.
2. Include the source of the information.
3. A guideline should be included on the top corner.
4. Unit of measurement and scale should also be given.
5. The beauty and neatness of the chart are important for the attention of the viewer.
6. If the distance of the smallest value observed in each of the axes from the origin of coordinates (zero) is large, it is necessary to put a cut sign (⊣⊦) in front of it.

Pie Chart

A frequently used graphical representation for illustrating the proportions within categorical data is the pie chart, wherein a circle is segmented into sectors, each representing a distinct category. The size of each sector corresponds to the frequency count of its respective category, or its relative frequency. For instance, if a category possesses a relative frequency of 0.18, its sector would occupy 18% of the circle. If the data is on qualitative scale (nominal and ordinal), the observations themselves can be used directly to draw a pie chart. But if the data is quantitative, they should be categorized first and then the categorized data used for the pie chart.

To draw a pie chart, the area of the circle ($360°$ of central angle) is divided according to the relative frequency of each group. To do this, we multiply the relative frequency of each category by $360°$ and separate the corresponding sectors. For the above example that has 18% of the total frequencies, the sector should have $0.18 \times 360 = 64.8°$.

Different sectors can be prominent by different colors or patterns. Software usually places a color guide in the upper corner. This chart can be simple or three-dimensional and prominent. Sometimes, one of its sectors is raised a little to be shown separately.

The steps of drawing pie chart with SPSS software for 200 people with blood type (A = 59, B = 49, AB = 16, O = 76) are shown in the figure below. To do this, choose path graphs\legacy dialogs\pie to open another sub-window (Fig. 2.1).

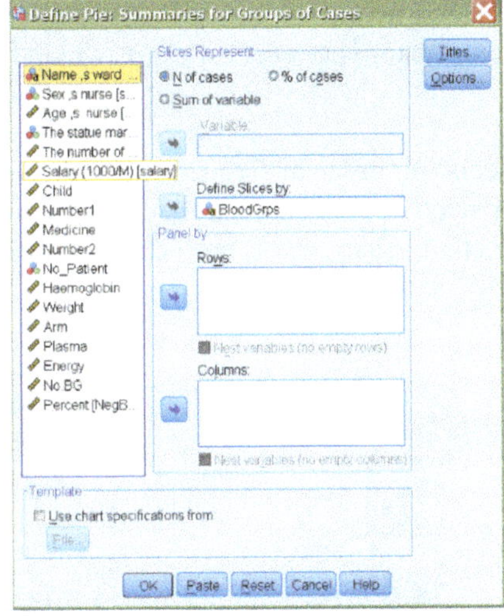

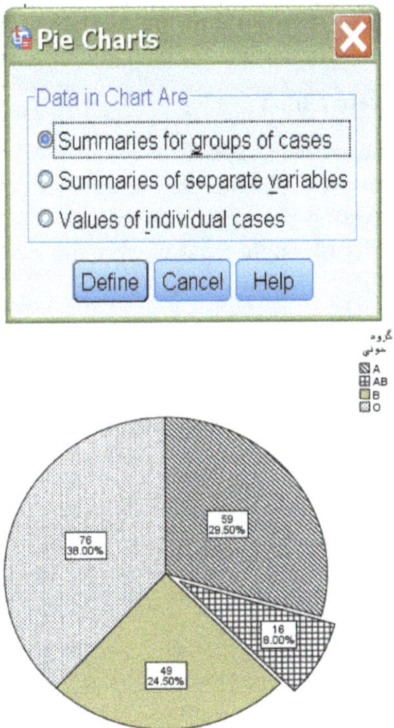

Fig. 2.1 Pie chart of blood types for a sample of 200

Bar Graph/Bar Chart

A bar chart serves as a visual depiction of the frequency distribution of categorical data, offering an alternative to the pie chart. Comprised of equally wide bars (rectangles), each corresponding to a category, the bar graph illustrates the frequencies of categorical data. The height of each rectangle reflects the frequency (or relative frequency) of its respective category. To prevent visual errors and ambiguity and misinterpretation, the width of the bars should be equal and it is recommended that the distance between the bars be equal.

> **Tip**
>
> It is easier to compare different categories with this chart than with a pie chart.

To draw this graph, the Y-axis is divided according to the absolute frequencies (or relative frequency or percentage). The X-axis should also be according to the different classes of the variable under study. The height of the bars should also be proportional to the frequency of each category. The use of different colors and patterns such as hatching and shading makes the graph more beautiful.

The steps for drawing a bar chart with SPSS software are shown in Fig. 2.2. To do this, first select the path graphs\legacy dialogs\bar to open the bar charts window.

Pareto Chart

When a bar graph arranges its bars in descending order of frequency or relative frequency, placing the highest frequency or relative frequency on the left and the lowest on the right, it is termed as a Pareto chart. These charts are valuable when clarity is sought regarding the most commonly occurring categories.

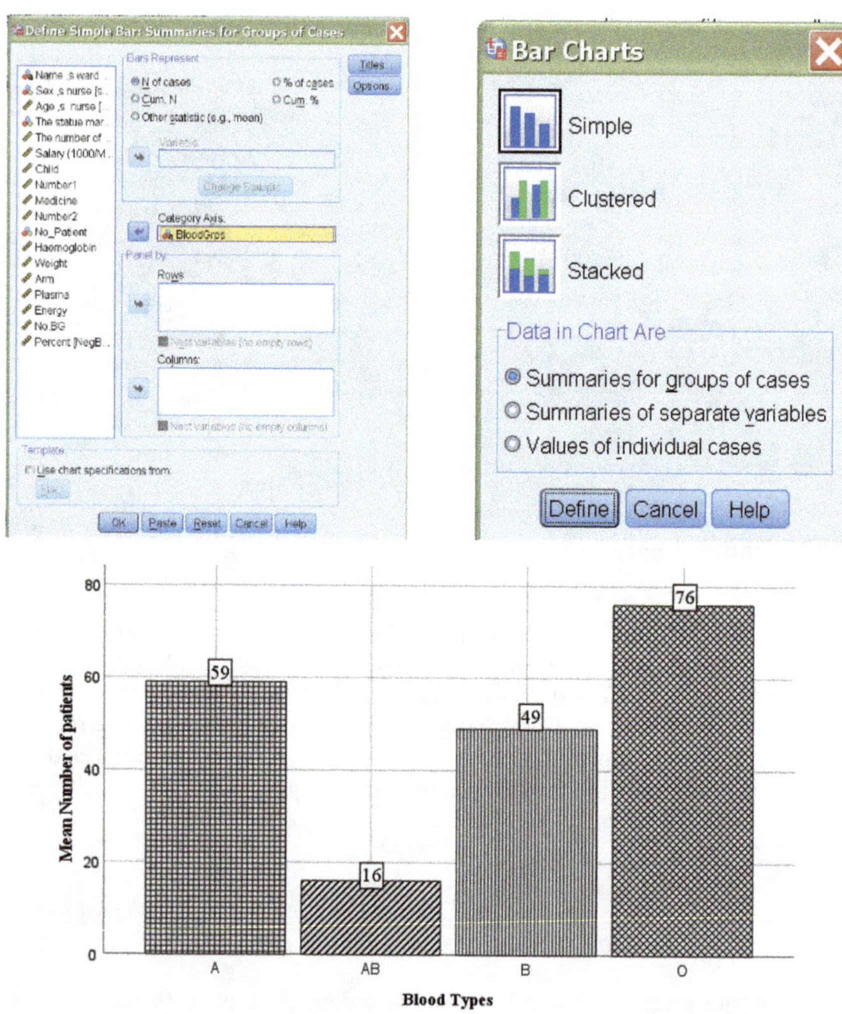

Fig. 2.2   Bar chart of blood types for a sample of 200

Stem-and-Leaf Plot/Stemplot

This type of chart is employed to showcase quantitative data, whether continuous or discrete. Stem-and-leaf displays serve to illuminate the relative density and shape of the data, providing readers with a swift overview of its distribution. In a stemplot (or stem-and-leaf plot), each quantitative value is separated into two components: the stem (typically the leftmost digit) and the leaf (usually the rightmost digit). The advantage of this plot lies in its ability to not only exhibit chart-like features but also allow readers to grasp the original data and interpret indicators such as the mode and median easily. Examples of stem-and-leaf diagrams can be seen in Figs. 3.2 and 4.2.

To construct a stem-and-leaf display, the observations must first be arranged in ascending order. Subsequently, the stems and leaves must be defined. Typically, the leaf comprises the last digit of each number (the rightmost digit), while the stem encompasses all other digits. For very large numbers or those with multiple digits, values may be rounded to a specific place value (e.g., hundreds), with the remaining digits on the left serving as the stem. It's worth noting that some of the remaining stems may be duplicates or like each other.

Then we draw a vertical bar and write the stems on the left side of this Bar. Then we read the sorted numbers in order and placed each leaf in front of its stem (to the right of the vertical line). If the numbers are repeated, the leaves should be written on the right side of the vertical line as many times as repeats. If the value of the leaf (the digit on the right) is zero, zero must also be written.

If the data includes two subgroups (such as sick and healthy or male and female), a back-to-back stemplot can be used to better compare these subgroups.

Example 8 In example 2, suppose the first 50 patients are female and the next 40 patients are male. Draw a back-to-back stemplot.

Solution: To draw this plot, we first sort 50 female patients and 40 male patients separately. Then in each group, we separate the right digit as "leaf". What remains is "stems", which include Figs. 1 to 9. We write them in a column in the middle of the page and, as in the previous exercise, we write down the values for women on the right and the values for men on the left side of the column (stem).

As in the previous exercise, we write the values for women on the right and the values for men on the left side of the stems' column.

Female	29, 31, 32, 35, 36, 36, 37, 38, 40, 41, 42, 45, 45, 47, 48, 48, 49, 51, 53, 55, 58, 59, 59, 60, 61, 61, 62, 63, 65, 66, 67, 68, 68, 69, 69, 72, 75, 79, 80, 80, 80, 81, 82, 83, 84, 84, 84, 85, 86, 88
Male	16, 21, 22, 26, 27, 27, 32, 34, 37, 38, 40, 42, 42, 45, 46, 47, 48, 50, 50, 52, 53, 54, 56, 57, 61, 61, 61, 62, 63, 64, 65, 66, 71, 72, 73, 76, 80, 81, 89, 90

Male age   **Female Age**

6	1	
7 7 6 2 1	2	9
8 7 4 2	3	1 2 5 6 6 7 8
8 7 6 5 2 2 0	4	0 1 2 5 5 7 8 8 9 9
7 6 4 3 2 0 0	5	1 3 5 8 9 9
6 5 4 3 2 1 1 1	6	0 1 1 2 3 5 6 7 8 8 9 9
6 3 2 1	7	2 5 9
9 1 0	8	0 0 0 1 2 3 4 4 4 5 6 8
0	9	

To draw this plot in SPSS, first choose the path **analyze\descriptive statistics\explore** to open a window like below. Then enter the desired quantitative variable in the **"dependent list"** box and the categorized quality variable in the **"factor list"** box. Click the **plots** button to open another window and select the **stem-and-leaf** option to draw a chart.

Note: To activate the plots option, both or plots options must be checked at the bottom of the left window.

Note: To activate the **plots** option, one of the **both** or **plots** options must be marked at the bottom left of this window (Fig. 2.3).

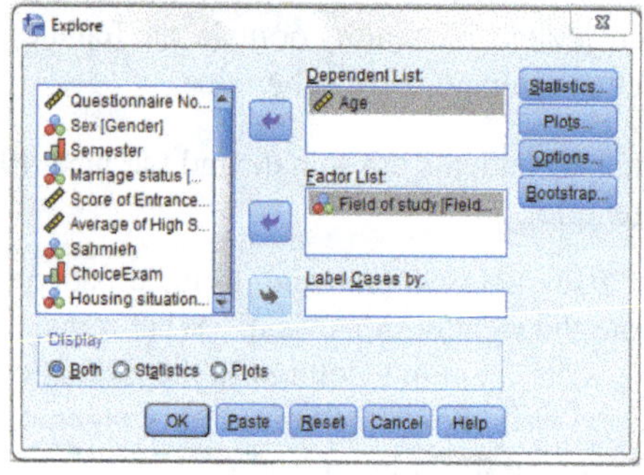
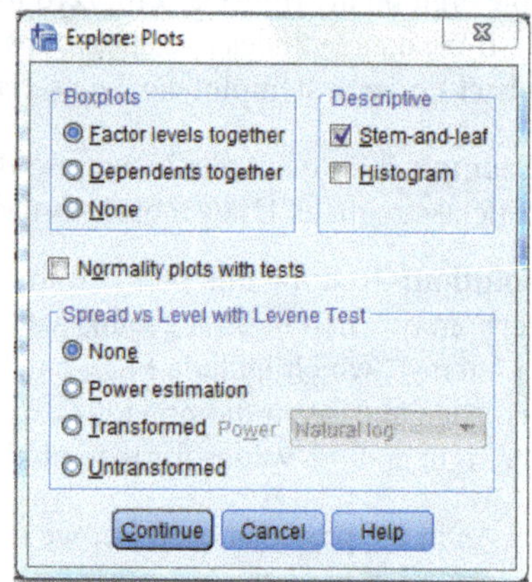

0	5 8
1	2 4 6 7
2	0 2 2 3 5 7
3	1 4 9
4	2 4 4
5	0 3
6	2

Fig. 2.3 Stem-and-leaf diagram of patients referred to a health center

Fig. 2.4 Stem-and-leaf plot
with frequencies

```
  2  |0| 5  8
  6  |1| 2  4  6  7
 (6) |2| 0  2  2  3  5  7
  9  |3| 1  4  9
  6  |4| 2  4  4
  3  |5| 0  3
  1  |6| 2
```

It is also recommended that on the left side of this diagram, a column is allocated to the cumulative frequencies of data up to the median. In this way, we add the frequencies of each row from the top with the frequencies of the previous row and write in the corresponding cell until we reach the median of the distribution. Then, from the bottom, we add the frequency of each row to the frequencies of the next row and write it in its corresponding cell to reach to the median from this side as well (the median is the value of an observation from which half of the observations are less and the other half are larger than it).

Example 9 For example, for patients referred to the health center, the eleventh value (25) is the median and is located in stem 2, and the absolute frequency of that stem (6) is written in brackets (Fig. 2.4).

Histogram

This chart is used for classifying quantitative data (such as age). In this diagram, the Y-axis is scaled in terms of frequency (absolute frequency or relative frequency or percentage of frequency) and the X-axis is scaled according to the boundaries of the classes. In this diagram, the rectangles are glued together, the width of each rectangle is proportional to the width of the classes, and their area is proportional to the observed frequency of the class. In this diagram, the height of the rectangles can only be proportional to the frequency of observations when the width of the classes is equal.

To avoid any ambiguity and to obtain the height and area of each rectangle relative to the frequency classes, it is recommended that the frequency of each class be divided by the width of its class.

To draw a histogram with SPSS, first choose the path graphs\legacy dialogs\histogram to open a window (Fig. 2.5).

Polygon and Ogive

These two similar charts are suitable for representing quantitative data. Firstly, let's discuss how to create a polygon diagram and then an Ogive.

> **Tip**
> The midpoint of a class is the average of its lower limit and the upper limit of the next class.

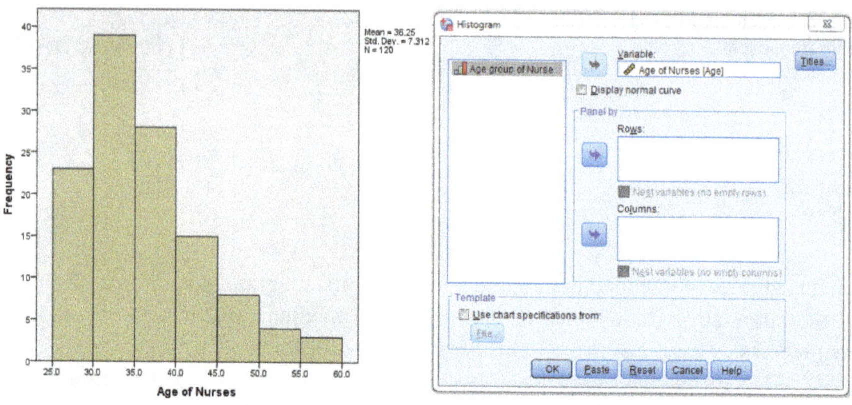

Fig. 2.5 Histogram of the age of 120 nurses working in the hospital

To draw a polygon diagram, we start by calculating the class midpoints, which are the averages of the lower and upper boundaries of each class. Next, we position each class along the X-axis based on its midpoint. The Y-axis is scaled according to the observed frequency (either absolute frequency, relative frequency, or frequency percentage). After determining the points where the X-coordinates represent the class midpoints and the Y-coordinates represent the frequencies, we connect these points sequentially with straight lines to form the frequency polygon.

$$\text{Class midpoint} = \frac{\text{Lower limit} + \text{Lower limit of next class}}{2}.$$

An **Ogive** plot, pronounced "oh jive", is another type of polygon that utilizes cumulative frequencies and cumulative relative frequencies instead of absolute frequency and relative frequency. To draw a polygon with SPSS, first choose the path graphs\legacy dialogs\line to open a window. After the new window was opened, if you choose one of the two options in the first row (% of cases or no. of cases), a polygon plot is drawn, and if you choose one of the two options in the next row (Cum N or Cum%), the Ogive plot is drawn (Fig. 2.6).

Scatterplot

A scatter plot, also known as a scatterplot, scatter graph, scatter chart, point graph, X–Y plot, or scattergram, is a type of plot used to depict the relationship between two quantitative variables (x, y). It offers a visual representation of how these variables are related and can suggest the presence and nature of a linear relationship between them.

To construct a scatterplot, the horizontal X-axis is dedicated to the first variable (x), while the vertical Y-axis represents the second variable (y). If both variables share the same measurement scale, it is advisable to maintain uniform scaling on both the X and Y axes.

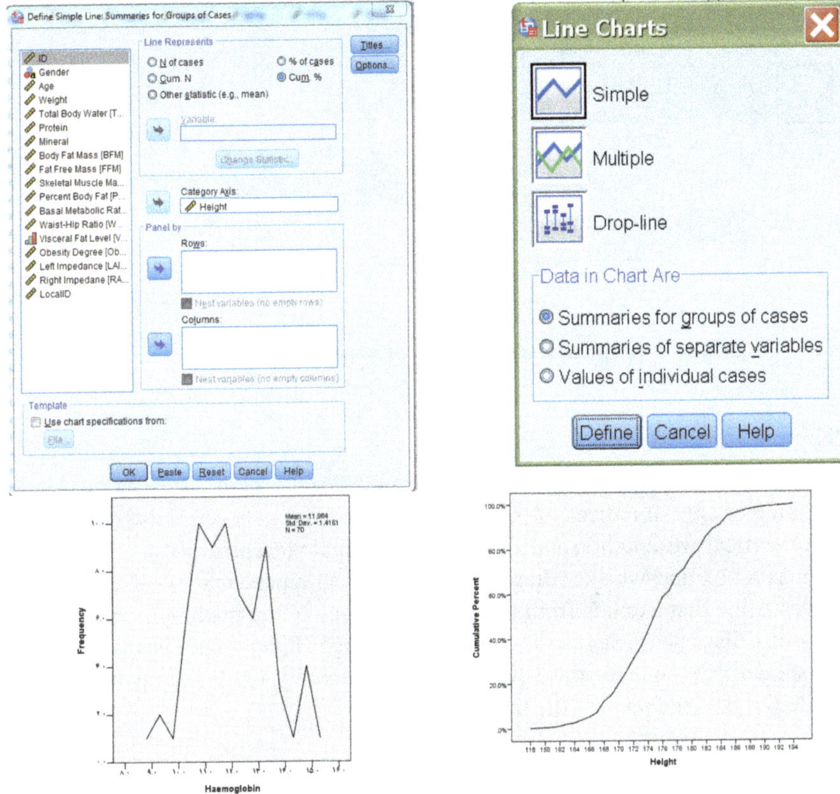

Fig. 2.6 Polygon and Ogive diagram for hemoglobin level of 70 females

Then we find the coordinates of all the observations in pairs (Y and X) on the coordinate system and mark it with a symbol. The set of specified points will give a general glance of the relationship between the two variables.

Obviously, if the values of two or more pairs of observations are the same and equal, there will be the same number of overlapping or very close points on the graph. For this reason, some points on the chart may appear more colorful than other non-duplicate points.

This scatterplot can be plotted for different subgroups of a dataset with different symbols to reveal if the relationship between two variables is different from each other in various subgroups. This plot can be drawn for a variable that has been measured at two different times (before and after an event).

To draw the scatterplot with SPSS, choose the path graphs\legacy dialogs\scatter/dot to open the following window (Fig. 2.7).

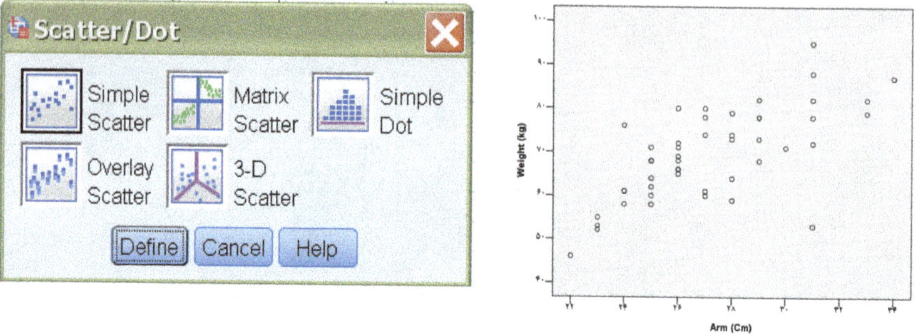

Fig. 2.7 Weight and arm circumference scatterplot of 50 persons

Boxplot

This plot is used for quantitative data (continuous or discrete). To draw the graph, calculating of central indices of the data is required and only one axis is scaled (Y-axis for vertical presentation and X-axis for horizontal presentation). A boxplot, also known as a box-and-whisker diagram, is a graphical representation of a dataset. It includes a line that extends from the minimum value to the maximum value, along with a box that encompasses the interquartile range (IQR)—the distance between the first quartile Q_1 and the third quartile Q_3. Additionally, lines are drawn at the first quartile Q_1, the median, and the third quartile Q_3.

> **Tip**
>
> An **outlier** is a value that is considerably larger or considerably smaller than most of the values in a dataset. **In other words,** observed values that lie very far away from the vast majority of the other sample values.

To draw the boxplot, we have to calculate first quartile (Q_1), second quartile or median ($Q_2 = \text{Med}$) and the third quartile (Q_3) of the dataset.

Then we mark the location of the first, second and third quartiles on the Y-axis (these indices are non-parametric and do not depend on the value of observations and their distribution, but only depend on the order in which they are placed. If the assumption of normality is met for the distribution of data, mean and standard deviation can be used instead). Then specify the maximum and minimum values of the data and draw a line from each one to the box.

Steps of Construction of a Boxplot

Step 1: Compute the first quartile, the median, and the third quartile.

Step 2: Draw X-axis and then vertical lines at the first quartile, the second and the third quartile.

Connect horizontal lines between the first and third quartiles to complete the box and create a rectangle.

Step 3: Compute the lower outlier boundaries (LOB) and upper outlier boundaries (UOB).

Step 4: Find the largest data value that is less than the UOB. Draw a horizontal line from the third quartile to this value. This horizontal line is called a **whisker**.

Step 5: Find the smallest data value that is greater than the LOB. Draw a horizontal line (whisker) from the first quartile to this value.

Step 6: Determine and plot any values, which are outliers, separately.

The width of the box in a boxplot is often proportional to the sample size, with a wider box indicating a larger sample size. It is common to relate the width of the box to the square root of the sample size, although this proportionality is not always strictly followed. Some statistical software packages offer modified boxplots that represent outliers as distinct points, such as asterisks or dots, and the solid horizontal line extends only to the minimum and maximum data values that aren't outliers.

The most commonly used method for identifying outliers is the interquartile range (IQR) method. The outlier boundaries are determined as follows:

- Lower outlier boundary $= Q_1 - 1.5 \times \text{IQR}$
- Upper outlier boundary $= Q3 + 1.5 \times \text{IQR}$.

Any observed data value falling below the lower outlier boundary or exceeding the upper outlier boundary is classified as an outlier.

When data can be partitioned into subsets according to another qualitative feature (such as gender or marital status), it is very useful to draw a boxplot for each subset and compare the subgroups.

In the boxplot, the height or length of the box contains 50% of the middle of the dataset and shows the interquartile range ($Q_3 - Q_1$).

To draw boxplot with SPSS, first choose the path graphs\legacy dialogs\boxplot to open a window (Fig. 2.8).

Although a boxplot is determined by the five indexes (minimum, first quartile, median, third quartile, and maximum value), it does not provide detailed information about the distribution of the data compared to histograms or stemplots. However, boxplots excel in comparing several datasets. To effectively compare two or more different datasets using boxplots, it is essential for them to share the same scale, facilitating straightforward comparisons.

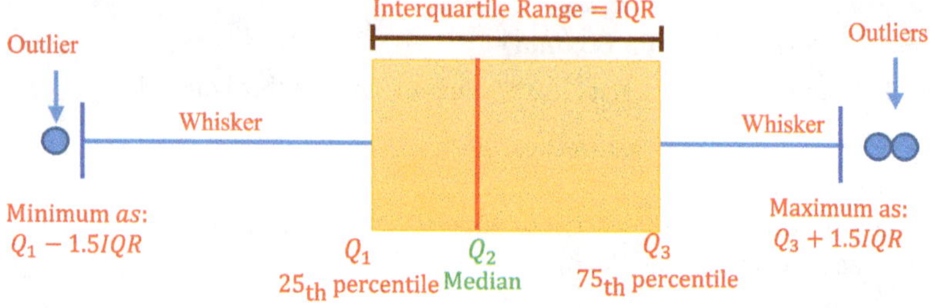

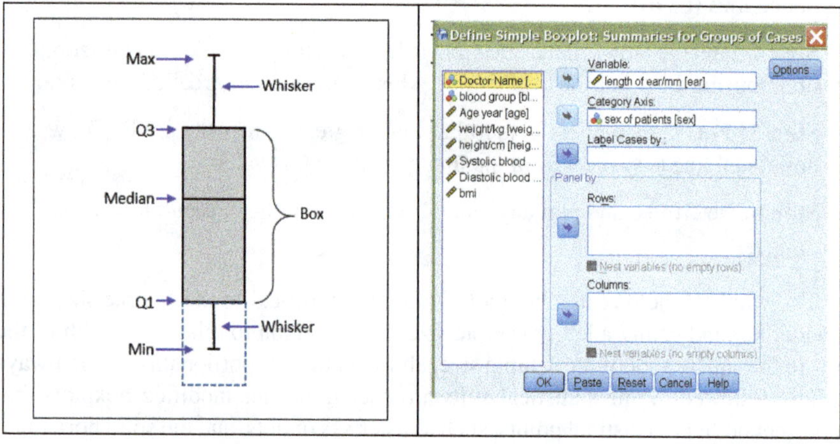

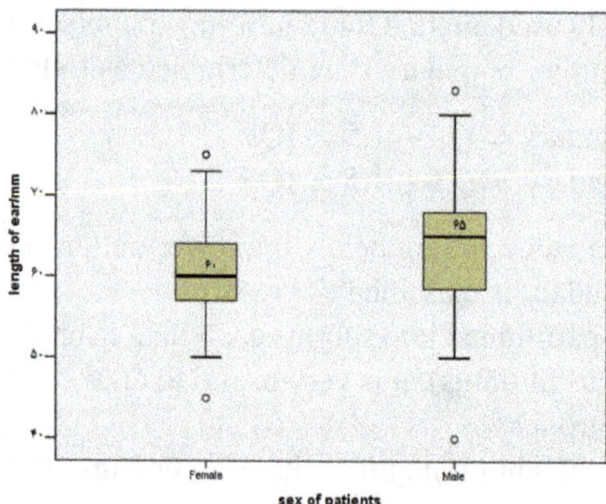

Fig. 2.8 Boxplot for the length of individuals' ears by gender

Interperate the Boxplot

1. When the median is near the center of the box (approximately halfway between the first and third quartiles), that is, the lines are about the same length and the two whiskers are approximately equal in length, the distribution is approximately symmetric.
2. If the median falls to the left of the center of the box (the median is closer to the first quartile than to the third quartile), that is, the right line is larger than the left line or the upper whisker is longer than the lower whisker, the distribution is positively skewed or skewed to the right.
3. When the median falls to the right of the center (the median is closer to the third quartile than to the first quartile), that is, the left line is larger than the right line or the lower whisker is longer than the upper whisker, the distribution is negatively skewed or skewed to the left.

A boxplot is often employed to discern the skewness of a dataset's distribution. Skewness refers to a lack of symmetry in the distribution, where one tail extends farther than the other. In a right-skewed distribution (also known as positively skewed), the boxplot displays a longer right whisker, indicating that relatively few data values have high values, with the majority located toward the left. Conversely, in a left-skewed distribution (or negatively skewed), the boxplot exhibits a longer left whisker, suggesting that most data values are situated toward the right. Therefore, by observing the lengths of the whiskers relative to each other, one can infer the skewness of the distribution.

2.6 Exercises

1. The following data is the blood glucose level (in millimoles per liter) of 40 first-year students.

4.1– 4.0– 3.6– 3.8– 4.6– 4.4– 3.3– 3.6– 3.8– 4.1– 3.9– 4.7– 3.4– 3.7– 2.2– 5.0– 4.7– 3.3–
4.2– 3.6– 4.1– 3.8– 4.8– 3.3– 4.7– 4.4– 4.9– 4.5– 3.7– 4.0– 4.9– 3.4– 2.9– 6.0– 4.0– 3.4–
5.1– 4.3– 4.8– 5.0

   A. Draw a stem-and-leaf diagram for this data.
   B. Tabulate this data and make its frequency distribution.
   C. Draw a histogram for this data.

2. The method of maternal delivery of 600 mothers in a hospital is as follows. Draw a suitable chart for how these children are born.

Type of delivery	No.
Cesarean	57
Normal	476
With aid	67
Total	600

3. Floating particles in the air at a sample of 49 cities in micrograms per cubic meter
   are as follows:

   A. Tabulate them using the Fisher suggestion.
   B. Draw a stem-and-leaf diagram for it.

   22– 42– 43– 49– 50– 27– 22– 23– 24– 25– 44– 63– 57– 27– 30– 27– 45– 49– 46– 12– 42–
   21– 32– 27– 16– 23– 74– 25– 32– 79– 51– 31– 33– 23– 19– 12– 38– 38– 24– 65– 51– 47–
   25– 36– 28– 42– 28– 30

4. In a sample of 50, the size of the arm circumference in centimeters and the weight
   in kilograms are as described in Table 2.9.

   A. Tabulate the above data in two-dimensional table (e.g., weights in rows and
      arm circumference in columns).
   B. Plot a scattergram for weights and arm circumference.

Table 2.9 Arm circumference (cm) and weight (kg) of 50 samples

Weight	Arm	Weight	Arm	Weight	Arm
78	29	66	26	71	30
68	25	68	26	66	26
72	26	73	29	78	31
61	24	66	26	58	24
79	33	60	25	76	24
88	31	78	29	58	25
69	26	82	31	59	28
71	25	66	26	60	27
68	29	64	25	55	23
73	28	64	28	79	28
53	31	74	27	65	26
53	23	62	25	68	25
46	22	82	33	61	27
87	34	72	31	80	27
52	23	95	31	71	26
82	29	74	28	80	26

   C.   Draw a stem and leaf for each variable separately.

   D.   Interpret the box-and-whiskers plot for these variables.

5. Draw a suitable chart for the number of medications prescribed in 1000 prescriptions of patients referred to a pharmacy shown in Table 2.10.

6. Draw a suitable chart for the reasons why people did not get the Covid-19 vaccine.

Reason	Freq
Religious believes	5
Fear of vaccine side effects	6
No time	18
Crowded queues for vaccination	24
Lack of approved vaccine by WHO	30
Propaganda of traditional anti-vaccine groups	29
Total	112

7. Draw a box plot for the number of painkiller pills consumed by rural households in the last month. Is the stem-and-leaf plot suitable for the data which are mostly one digit?

6– 11– 2– 3– 11– 8– 9– 11– 4– 3– 7– 12– 4– 3– 0– 1– 8– 2– 8– 4– 9– 4– 6– 9– 5– 4– 4– 5– 3–
2– 4– 8– 0– 6– 8– 9– 3– 6– 5– 8–

8. Table 2.11 shows the calcium level of two groups of 75 married women by age. The first group used contraceptive pills and the other group did not use the pill.

   A.   Classified the calcium levels in two-dimensional table (e.g., consuming pill in rows and calcium level in columns).

   B.   Classified by age and calcium level.

Table 2.10 Number of medications prescribed in 1000 prescriptions of patients

No. of medications	1	2	3	4	5	6	7	Sum
Frequency	37	137	218	340	192	56	20	1000

9. From the database of an insurance company, we have selected a sample of 40 patients who have undergone diagnostic imaging tests. The samples of patients from three types of CT scan, MRI, and radiology (R) are listed below. Categorize this data and draw a suitable chart for it.

> CT, R, MRI, R, R, R, R, R, CT, MRI, R, CT, MRI, MRI, R, CT, CT , R, MRI, CT, R, R, MRI, R, CT, MRI, CT, CT, CT, R, MRI, CT, MRI, R, CT, R, R, R, CT.

Table 2.11 Calcium levels in two groups of 75 women by age (group 1: taking contraceptive pill, group 2: not taking pills)

Age	Group 1	Group 2	Age	Group 1	Group 2	Age	Group 1	Group 2
25	96	104	27	94	101	30	96	99
19	105	99	27	96	97	30	99	104
20	93	95	27	103	98	31	91	98
21	106	101	28	105	106	31	99	111
21	106	98	28	93	100	32	108	103
21	95	108	28	101	97	32	97	99
21	99	102	29	98	99	32	90	94
21	105	96	29	99	94	54	104	108
22	97	98	30	98	107	33	99	100
22	98	98	30	99	99	33	95	106
22	105	102	30	96	104	33	104	101
22	95	100	30	99	95	34	95	93
22	106	104	31	91	98	35	100	91
23	100	92	31	99	111	35	105	107
23	101	103	31	95	96	36	103	90
24	103	106	32	99	103	36	98	104
24	104	108	32	97	99	37	99	105
24	102	96	32	90	94	37	91	100
25	93	101	54	106	86	38	94	102
25	103	101	26	99	103	39	90	95
54	106	86	27	98	95	39	94	108
27	101	106	28	105	106	39	99	91
26	99	103	29	98	99	40	99	97
26	95	96	29	99	94	40	100	100
27	98	98	30	98	107	40	96	91

Chapter 3
Summarizing Data

3.1 Introduction

A large amount of data without processing and summarizing is not understandable and sensible and even is ambiguous and confusing to be. In order to make sense and tangible the set of observations, it is necessary to summarize them with statistical indicators in order to provide a clear picture of the entire dataset. These measures also condense the dataset down to a few representative indexes and are usually brought into two general headings, "central tendency" and "dispersion indicators", to represent the entire data distribution and form the basis of descriptive statistics.

One reason we explain central tendency before dispersion indices is that many measures of dispersion involve measures of central tendency and they cannot be computed and understood without central tendency. Another reason, in descriptive statistics, measures of central tendency come up early on before measures of dispersion.

3.2 Measures of Central Tendency

Central tendencies are the most common statistical indicators that measure and determine the location central, midpoint, and center of gravity of a dataset. Mean, median, mood, and quartiles are the most important central indices.

> **Tip**
>
> When the data includes the whole population, the mean is denoted by the symbol μ and N is used in the denominator of its fraction. But when data is taken from a sample, the mean is represented by $\bar{x}$ and n is used instead of N in the denominator of its fraction.

Mean

The most common indicator for summarizing data is the mean, also known as the average, center of gravity or midpoint. There are different types of mean such as arithmetic mean, geometric mean and harmonic mean with various uses. This book deals only with arithmetic mean weight (Fig. 3.1).

The mean of a population is denoted by μ and is obtained from the sum of observations divided by a total number of data and is equal to:

$$\mu = \frac{x_1 + x_2 + x_3 + \cdots + x_N}{N} = \frac{\sum_{i=1}^{N} x_i}{N},$$

where the $\sum_{i=1}^{N} x_i$ is the sum of the observations from 1 to N. We will round the mean to one more decimal place than the raw data.

The formula for the mean in the frequency tables, where the observations of each class are given with frequency, is as follows:

$$\mu = \frac{\sum_{i=1}^{N} f_i x_i}{\sum_{i=1}^{N} f_i},$$

where f_i and x_i are the absolute frequency and the midpoint of each class. The first column of each table shows the class width. Calculate the average of lower and upper limits of each class to get the "midpoint of the class" and replace it with the above formula.

To facilitate the calculation of the mean, assign a column for $f_i x_i$ (product of the frequency and the midpoint of the each class). Then divide the sum of this column by the total number of observations to get the average.

Fig. 3.1 Center of gravity of a dataset

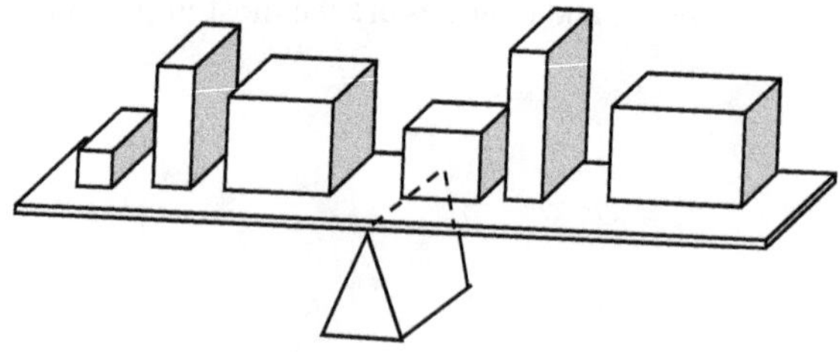

Table 3.1 Calculation for hemoglobin levels in 70 women

Hemoglobin level	x_i	f_i	$f_i.x_i$
8.5–9.4	9.0	3	27.0
9.5–10.4	10.0	6	66.0
10.5–11.4	11.0	19	209.0
11.5–12.4	12.0	16	192.0
12.5–13.4	13.0	15	195.0
13.5–14.4	14.0	6	84.0
14.5–15.4	15.0	5	75.0
Total	–	70	842.0

Tips

Rounding Rule for the Central Tendency:

Central tendency indexes should be rounded to one more decimal place than occurs in the raw data. For example, if the raw data are given in whole numbers (such as family size of 3, 4, 2, … the mean should be rounded to the nearest tenth ($\mu = 3.6$). If the data are given in tenths (such as exam weight of children in Kg 17.5, 13.0, 16.5, 18.6), the mean should be rounded to the nearest hundredth ($\mu = 14.65$), and so on.

The mean for a sample with size of n is shown with $\bar{x}$ and equals to:

$$\bar{x} = \frac{x_1 + x_2 + x_3 + \cdots + x_N}{n} = \frac{\sum_{i=1}^{n} x_i}{n}.$$

Example 1 For the information in Table 2.5, we assign a column for x_i and a column for $f_i.x_i$. The mean of hemoglobin level of 70 women will be equal to $\frac{842}{70} = 12.03$. These calculations are shown in Table 3.1.

When we merge two populations with N_1 and N_2 and mean of μ_1 and μ_2, the mean of the whole merged population is $\mu = \frac{N_1\mu_1 + N_2\mu_2}{N_1 + N_2}$. It can easily be generalized into the K populations with $N_1, N_2, \cdots$ and N_k and means of $\mu_1, \mu_2, \cdots$ and μ_k.

The means of pooled K populations is

$$\mu = \frac{N_1\mu_1 + N_2\mu_2 + \cdots + N_K\mu_K}{N_1 + N_2 + \cdots + N_K}.$$

Example 2 The mean of birth weight of 14 infants whose mothers smoked during pregnancy is 3.200 and the mean of birth weight of 15 infants whose mothers did not smoke is 3.590. The grand mean of birth weight of these 29 babies is:

$$\mu = \frac{N_1\mu_1 + N_2\mu_2}{N_1 + N_2} = \frac{14 \times 3.200 + 15 \times 3.590}{14 + 15} = 3.402.$$

Median

The median, denoted by Med, is widely used in survival studies and splits the number of observations into two equal parts. Thus, half of the observations are smaller and the other half are larger than median. The median is the midpoint of the sorted and does not deal directly with the size of the observations, but only with their order and ranking. Therefore, first all observations should be sorted in ascending order, and the middle number should be considered as the median. The procedure for calculating the median depends on whether the number of observations in the dataset is even or odd.

When the number of dataset is odd, the median is equal to the value of the middle number (number in row $\frac{N+1}{2}$), but if there are an even number of values in the dataset, the median will fall between two given values ($\frac{N}{2}$ and $\frac{N}{2} + 1$), that is the average values of the middle two numbers.

Example 3 Birth weights of 9 babies are as follows:
   3.05, 3.50, 2.90, 3.20, 3.75, 3.80, 4.10, 3.90, 3.30

After sorting them in ascending order, we have:
   2.70, 2.90, 3.05, 3.20, 3.30, **3.50**, 3.75, 3.80 3.90, 4.10

In this example, the number of subject $N = 9$ is odd. Therefore, the value of the fifth observation ($\frac{N+1}{2} = \frac{9+1}{2} = 5$) is the median (3.50) in this dataset.

Example 4 Birth weights of 10 babies are as follows:
   3.05, 3.50, 2.90, 3.30, 3.75, 3.80, 4.10, 3.90, 3.20, 2.70

After sorting them in ascending order, we have:
   2.70, 2.90, 3.05, 3.20, **3.30, 3.50**, 3.75, 3.80 3.90, 4.10

Because the number of observations ($N = 10$) is even, the average value of the fifth observation ($\frac{N}{2} = \frac{10}{2} = 5$) and the sixth observation ($\frac{N}{2} + 1 = \frac{10}{2} + 1 = 6$), that is, $\frac{3.30+3.50}{2} = 3.40$ will be the median of this data.
   $\text{Med} = \frac{3.30+3.50}{2} = 3.40$.

Calculating the median in frequency tables is a little more difficult. Since the nature of these tables is sorted, you don't need to sort them. Then we form the cumulative frequency column to identify the class that contains the median. The class containing median is that one includes the $\frac{N+1}{2}$th observation. That is, the class whose cumulative frequency is immediately equal to or greater than $\frac{N+1}{2}$. We then assume that the values in that class is 0 uniform distributed and we interpolate. We specify the following components in this table.

> **Tip**
> The median class is a class whose cumulative frequency is immediately equal
> or greater than $\frac{N+1}{2}$.

$$\text{Med} = L_i + \frac{\left(\frac{N+1}{2} - F_{i-1}\right)}{f_i} \times w_i,$$

where:

L_i is the lower class boundary of the class containing the median.

N is the total number of data.

F_{i-1} is the cumulative frequency of precede class of the median class that should
be excluded.

f_i is the frequency of the median class.

w_i is the class width of the median class.

Example 5 Table 3.2 determines the albumin level of 185 women based on the
contraceptive pill. Calculate the median albumin level of these 185 women.

Solution: According to the number of samples and cumulative frequency, the median
class is a group whose cumulative frequency is equal to or greater than $\frac{N+1}{2} = \frac{185+1}{2} = 93$ (the gray row). Then we substitute the values $L_i = 39.5, f_i = 43, w_i = 3$ and $F_{i-1} = 69$ in the above formula:

$$\text{Med} = L_i + \frac{\left(\frac{N+1}{2} - F_{i-1}\right)}{f_i} \times w_i$$

$$= 39.5 + \frac{\left(\frac{185+1}{2} - 69\right)}{43} \times 3 = 41.17.$$

Therefore, the exact value of the median is 41.17.

Table 3.2 Frequency distribution of albumin level of 185 females in terms of pill consuming

Albumin level	Consumed	Not consumed	Total (f_i)	F_i
31–34	3	1	4	4
35–36	8	5	13	17
37–39	34	18	52	69
40–42	20	23	43	112
43–45	25	23	48	160
46–48	4	19	23	183
49–51	0	2	2	185
Sum	94	91	195	–

In this example, the albumin level is a quantitative continuous value, and since the limits of the classes are given by an integer, we subtract 0.5 units from the lower limit of the median class (40) to calculate the actual boundary for the median calculation.

The main difference between the mean and the median

Indeed, the calculation of the mean involves the actual values of the data, whereas the median relies solely on their ranks, disregarding the actual values. Consequently, extremely large or small values exert a lesser influence on the median compared to the mean. The mean's formula incorporates every value in the dataset, making it sensitive to outliers, whereas the median's formula hinges only on the middle value or values, thereby offering more resilience against the impact of outliers. This aspect is especially crucial when analyzing datasets containing outliers.

In skewed distributions where the mean is affected by very large or very small values (such as income), the median is a more desirable index.

Since the median divides data into two equal parts, it is also called the second quartile (Q_2) or the fifth decile (D_5) or the fiftieth percentile (C_{50}). Therefore, by generalizing the above formula, percentiles, deciles, and quartiles can be calculated.

Quartiles

Quartiles are points that divide the dataset into four equal parts. There are 3 quartiles, which are called the first quartile (Q_1), the second quartile (Q_2) and the third quartile (Q_3) and determine the points of 25, 50 and 75% of the data distribution, respectively. Their computation formula is like the median formula.

$$Q_1 = L_i + \frac{\left(1 \times \frac{N+1}{4} - F_{i-1}\right)}{f_i} \times w_i$$

$$Q_2 = L_i + \frac{\left(2 \times \frac{N+1}{4} - F_{i-1}\right)}{f_i} \times w_i$$

$$Q_3 = L_i + \frac{\left(3 \times \frac{N+1}{4} - F_{i-1}\right)}{f_i} \times w_i,$$

where:

L_i is the lower class boundary of the class containing the related quartile.

N is the total number of data.

F_{i-1} is the cumulative frequency of precede class of the related quartile class that should be excluded.

f_i is the frequency of the related quartile class.

w_i is the class width of the related quartile class.

Example 6 Find the first and third quarters of the female albumin level data in Table 3.2.

Solution: To find the third quarter ($j = 3$) and according to the number of samples, the location of the third quarter is in the rank $\frac{3 \times (N+1)}{4} = \frac{3 \times (185+1)}{4} = 139.5$. The cumulative frequency of the class (43–45) is immediately greater than that. Therefore, $L_i = 42.5, f_i = 48, w_i = 3, F_{i-1} = 112$. Then by substitution the above values in the formula, we have:

$$Q_3 = L_i + \frac{\left(3 \times \frac{N+1}{4} - F_{i-1}\right)}{f_i} \times w_i$$

$$= 42.5 + \frac{\left(3 \times \frac{185+1}{4} - 112\right)}{48} \times 3 = 44.22.$$

The exact value of the third quarter is equal to 44.22, which means that 75% of the observed values are less than 44.22 and only 25% of them will be more than 44.22.

Similarly, to calculate the first quarter, $L_i = 36.5, f_i = 52, w_i = 3, F_{i-1} = 17$ and by substitution the above values in the formula, we have:

$$Q_3 = L_i + \frac{\left(1 \times \frac{N+1}{4} - F_{i-1}\right)}{f_i} \times w_i$$

$$= 36.5 + \frac{\left(1 \times \frac{185+1}{4} - 17\right)}{52} \times 3 = 38.20.$$

The exact value of the first quarter is equal to 38.2, which means that 75% of the observed values are less than 38.2 and only 25% of them will be more than 38.20.

Deciles

Deciles are points that divide the dataset into 10 equal parts. There are 9 deciles, which are called the first decile (D_1), the second decile (D_2) ... and the ninth decile (D_9) and determine the points of 10, 20, ... and 90% of the data distribution, respectively. Their computation formula is similar to the quartile's formula.

$$D_1 = L_i + \frac{\left(1 \times \frac{N+1}{10} - F_{i-1}\right)}{f_i} \times w_i$$

$$D_2 = L_i + \frac{\left(2 \times \frac{N+1}{10} - F_{i-1}\right)}{f_i} \times w_i$$

$$\vdots$$

$$\vdots$$

$$D_9 = L_i + \frac{\left(9 \times \frac{N+1}{10} - F_{i-1}\right)}{f_i} \times w_i,$$

where:

L_i is the lower class boundary of the class containing the related decile.

N is the total number of data.

F_{i-1} is the cumulative frequency of precede class of the related decile class that should be excluded.

f_i is the frequency of the related decile class.

w_i is the class width of the related decile class.

Example 7 Find the ninth decile of the female albumin level data in Table 3.2.

Solution: To find the third quarter ($j = 9$) and according to the number of samples, the location of the third quarter is in the rank $\frac{9 \times (N+1)}{10} = \frac{9 \times (185+1)}{10} = 167.4$. The cumulative frequency of the class (46–48) is immediately greater than that. Therefore, $L_i = 45.5, f_i = 23, w_i = 3, F_{i-1} = 160$. Then by substitution the above values in the formula, we have:

$$D_9 = L_i + \frac{\left(9 \times \frac{N+1}{10} - F_{i-1}\right)}{f_i} \times w_i$$

$$= 45.5 + \frac{\left(9 \times \frac{185+1}{10} - 160\right)}{23} \times 3 = 46.46.$$

The exact value of the first quarter is equal to 46.46, which means that 75% of the observed values are less than 46.46 and only 25% of them will be more than 46.46.

Centiles

Centiles are points that divide the dataset into 100 equal parts. There are 99 deciles, which are called the first centile (C_1), the second centile (C_2) ... and the ninety ninth centile (C_{99}) and determine the points of 1, 2 .... and 99% of the data distribution, respectively. Their computation formula is similar to the quartiles and deciles formula.

$$C_1 = L_i + \frac{\left(1 \times \frac{N+1}{100} - F_{i-1}\right)}{f_i} \times w_i$$

$$C_2 = L_i + \frac{\left(2 \times \frac{N+1}{100} - F_{i-1}\right)}{f_i} \times w_i$$

$$\vdots$$

$$\vdots$$

$$C_{99} = L_i + \frac{\left(99 \times \frac{N+1}{100} - F_{i-1}\right)}{f_i} \times w_i,$$

where:

L_i is the lower class boundary of the class containing the related centile.

N is the total number of data.

F_{i-1} is the cumulative frequency of precede class of the related centile class that should be excluded.

f_i is the frequency of the related centile class.

w_i is the class width of the related centile class.

Example 8 Table 3.3 determines the level of urea acid in 186 women based on the contraceptive pill. Get the 20th percentile.

Solution: To find the third quarter ($j = 20$) and according to the number of samples, the location of the third quarter is in the rank $\frac{20 \times (N+1)}{100} = \frac{20 \times (185+1)}{100} = 37.4$. The cumulative frequency of the class (30–39) is immediately greater than that. Therefore,

Table 3.3 Frequency distribution of urea acid in 186 women in terms of contraceptive pill consumption

Urea acid	Consumed pill	Not consumed	Total	Cumulative freq
20–29	3	4	7	7
30–39	20	22	42	49
40–49	42	43	85	134
50–59	19	13	32	166
60–69	5	9	14	180
70–79	1	1	2	182
80–89	3	0	3	185
90–99	1	0	1	186
Total	94	92	186	–

$L_i = 29.5, f_i = 42, w_i = 10, F_{i-1} = 7$. Then by substitution the above values in the formula, we have:

$$C_{20} = L_i + \frac{\left(20 \times \frac{N+1}{100} - F_{i-1}\right)}{f_i} \times w_i$$

$$= 29.5 + \frac{\left(20 \times \frac{186+1}{100} - 7\right)}{42} \times 3 = 36.74.$$

The exact value of the twentieth centile is equal to 36.74, which means that 20% of the observed values are less than 36.74 and only 80% of them will be more than 36.74.

The corresponding points of median, quartiles, deciles, and percentiles can be specified by Ogive plot (cumulative frequency curve) was determined. This can be done by the following steps:

Specify the desired quantile on the Y-axis.

From that point, draw a line parallel to the X-axis to intersect the Ogive plot.

From the intersection point, draw a vertical line to the X-axis.

The point of collision of this line (M) with the X-axis is the value of the desired index (Fig. 3.2).

Midpoint Range (MPR)

The midpoint range is another measure of central tendency that has less usage than the median and mean. This index is denoted by MPR and is the average of the minimum and maximum values, i.e.:

$$MPR = \frac{X_{Max} + X_{Min}}{2},$$

where the X_{Max} and X_{Min} are the maximum and minimum values, respectively. It is named sometimes mid-range (MR).

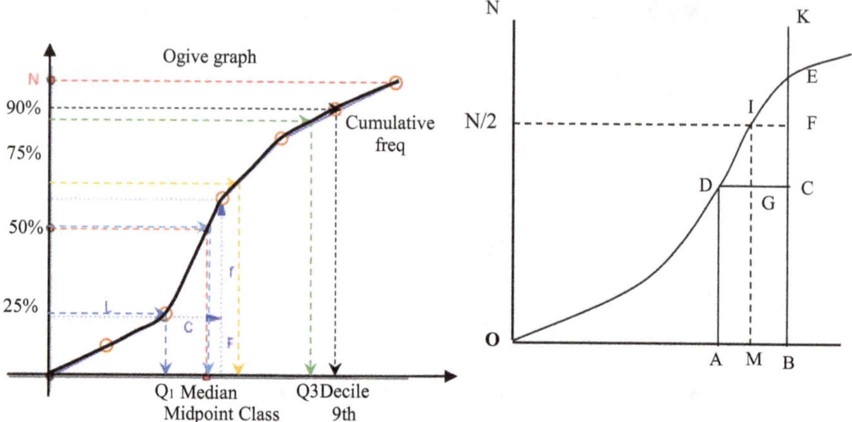

Fig. 3.2 Geometrical interpretation of the median. AB = Class width of median class, AD = BC = Cumulative frequency of precede class of median class, N = Total frequency, CE = Absolute frequency of median class, BE = AD + CE = Lower boundary of median class, OA = L = Difference between median value and lower boundary in median class, = FC = $\frac{N}{2}$ − BC

Example 9 The number of pain relief pills used by 40 patients in one day is as follows. Get the midpoint range of pain relief for these 40 patients (Table 3.4).

Solution: $X_{\text{Min}} = 0$, $X_{\text{Max}} = 12$ and the midpoint range for these 40 patients is:

$$\text{MPR} = \frac{X_{\text{Max}} + X_{\text{Min}}}{2} = \frac{1211 + 0}{2} = 6.$$

Mode

Mode is the value of an observation which has the highest frequency (peak) and is denoted by Mod. This index is not unique and may have two or more modes in a dataset (distribution). If a distribution has a single mode, it is called a "unimodal" distribution, if it has two modes, it is called a "bimodal" distribution, and if it has several views, it is called a "multimodal" distribution (Fig. 3.3).

Special attention should be paid to the distribution with more than one mode, because it may be possible to better understand and interpret its data by separating the subgroups (Fig. 3.4).

Determining and calculating the mode in raw data does not require a special formula. Because after sorting the data, the value that has most frequency is selected as the mode. But the value of the mode in the frequency tables and categorized data is

Table 3.4 Number of pain relief pills taken daily in 40 patients

6, 11, 2, 3, 11, 8, 9, 11, 4, 3, 7, 12, 4, 3, 0, 1, 8, 2, 8, 4, 9, 4, 6, 9, 5, 4, 4, 5, 3, 2, 4, 8, 1, 6, 8, 9, 3, 6, 5, 1

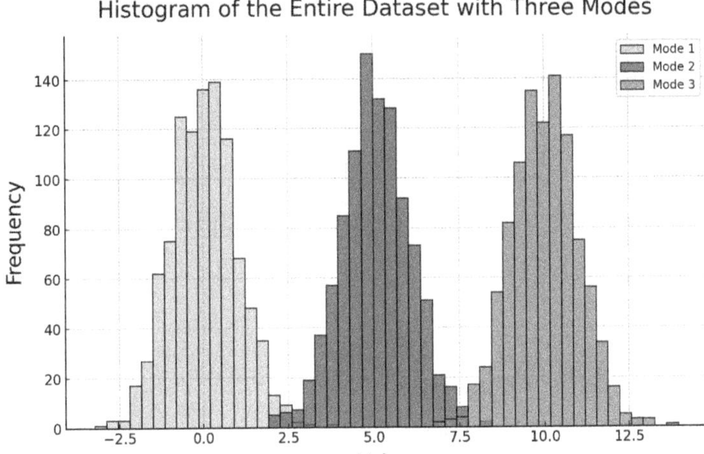

Fig. 3.3 Example of multimodal distribution

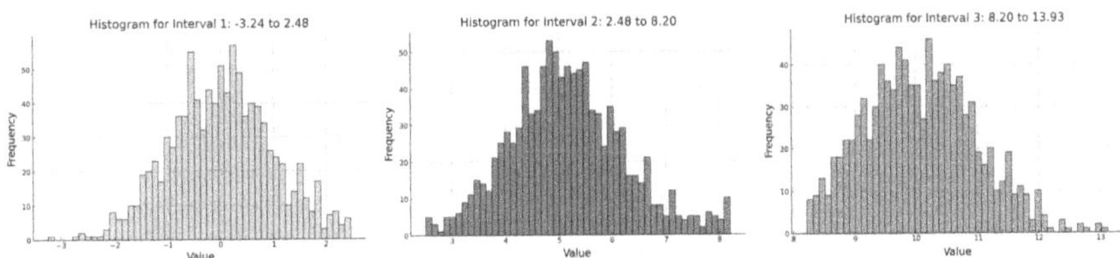

Fig. 3.4 Separated subgroups of Fig. 3.3

obtained using proportionality and interpolation, according to the following formula:

$$\text{Mod} = L_i + \frac{d_b}{d_b + d_a} \times w_i,$$

where
    L_i is the lower class boundary of the modal class.
    d_b is the frequency difference between the modal class and the previous one.
    d_a is frequency difference between the modal class and the next one.
    w_i is the class width.

Tip
The modal class is the class that has the largest ratio of "absolute frequency" to "class width".

Fig. 3.5 Geometrical
interpretation of the mode

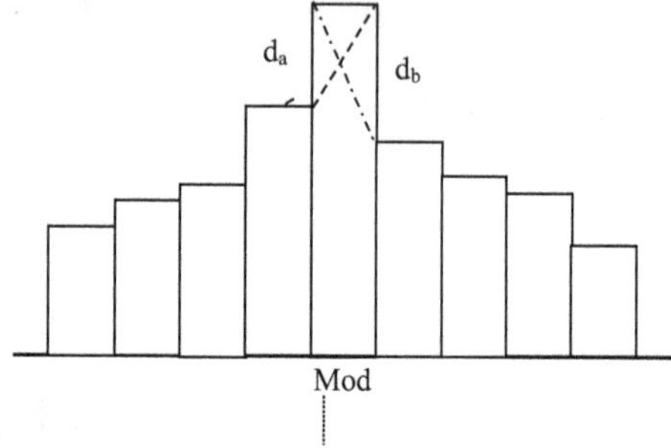

Table 3.5 Age distribution
of population in a
hypothetical village

Age groups	Freq
Less than 1 Y	36
1–5	155
6–10	185
11–14	110
15–24	215
25–64	340
65+	42
Sum	1083

In tables with equal class width, the modal class is the class that has the highest absolute frequency. But in tables with unequal class width, the modal class is the class whose has the largest ratio of "absolute frequency" to "class width". This geometric interpretation of the mode is shown in the figure below (Fig. 3.5).

Example 10 Table 3.5 shows the age distribution of the population in a village. Find out the mode of this population.

Solution: To calculate the mode, you must first find the modal class. When the class width is equal, the modal class is the same class with the highest frequency. However, if the class width is not equal, the modal class is a class with the highest ratio of absolute frequency to the class width. Therefore, we add another column to the table above to show the ratio of absolute frequency to class width.

According to the fourth column of Table 3.6, the third class 6–10 years is the modal class (the gray class). Since the studied variable of age is a discrete variable, the lower boundary of this class is equal to the lower class limit. Therefore, we substitute the values of $w_i = 5, L_i = 6, d_a = 185 - 155 = 30, d_b = 185 - 110 = 75$ in the mode formula.

$$Mod = L_i + \frac{d_b}{d_b + d_a} \times w_i$$

$$= 6 + \frac{30}{30 + 75} \times 5 = 7.43.$$

This means that in this village, the highest frequency is related to the age of 7.43 (7–8) years old.

In most distributions, the following empirical relationship is settled between the mean, median, and mode. By having the median and mean, the value of the mode can be obtained.

$$\mu - Mo = 3(\mu - Med) \quad \text{or} \quad Mo = 3Med - 2\mu.$$

To calculate the central tendencies (mean, median, mode, and centiles) with SPSS, choose the path of **analyze\descriptive statistics\frequencies** to open a new window.

Now select the desired variables from the left box and move to the right box. Then click the statistics option to open another window. At the top right of this window, mark the mean, median, and mode options.

In the percentile values box, the first option (quartiles) to calculate the quartiles, the second option (cut points for …. Equal groups) is for data cut points for equal groups. So, in this example, it specifies 5 groups with equal numbers and the percentile(s) option is assigned to the specific percentiles that the user writes the desired percentile and moves to the below box by pressing the add button.

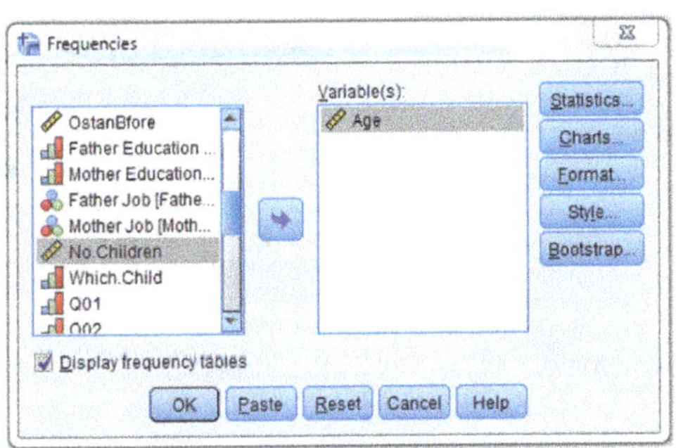
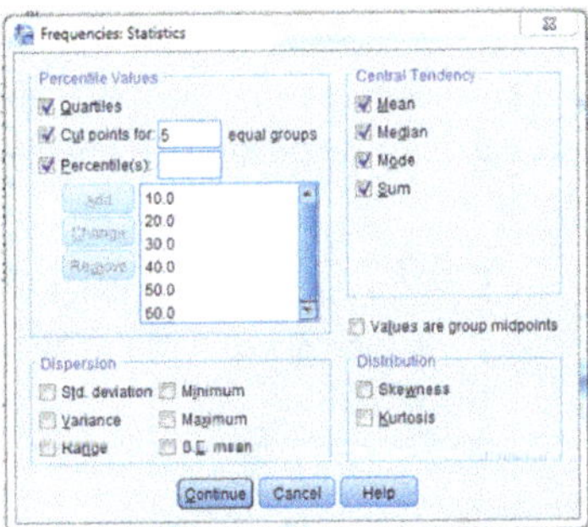

Table 3.6 Necessary calculations for Table 3.5 to find the mode

Age groups	Freq	Class width	Ratio of frequency to the class width
Less than 1Y	36	1	$36/1 = 36$
1–5	155	5	$155/5 = 31$
6–10	185	5	$185/5 = 37$
11–14	110	4	$110/4 = 27.5$
15–24	215	10	$215/10 = 21.5$
25–64	340	40	$340/140 = 8.5$
65+	42	35	$42/35 = 1.2$
Sum	1083	–	–

3.3 Measures of Dispersion

Sometimes two or more datasets which have very close measures of central tendency may differ from each other. The similarity of these measures, without paying attention to their differences raises interpretive problems. For example, consider the birth weight data of two samples of 5 newborn babies as follows:

$$A: 2.70,\ 2.90,\ 3.50,\ 3.90,\ 4.10.$$
$$B: 3.00,\ 3.20,\ 3.45,\ 3.65,\ 3.80.$$

Despite their means and medians are very close together ($\text{Med}_A = 3.50$, $\text{Med}_B = 3.45$, $\mu = 3.42$), these two datasets are different and not similar. Also look at the weight of the following two samples of 10.

$$C: 65,\ 66,\ 67,\ 68,\ 71,\ 73,\ 74,\ 77,\ 77,\ 77.$$
$$D: 100,\ 93,\ 85,\ 77,\ 77,\ 67,\ 62,\ 58,\ 54,\ 52.$$

Although the two datasets appear to be different, the measures of central tendency are exactly equal to $\mu = 71.5$, $\text{MPR} = 71$, $\text{Mod} = 77$ and $\text{Med} = 72$. Therefore, the measures of central tendency alone are not able to provide a clear and comprehensive glance of the data and other indicators should be used to reveal these differences. The following figure further illustrates this necessity (Fig. 3.6).

The measures that describe the spread and variation of the dataset are called "measures of dispersion", which we described below.

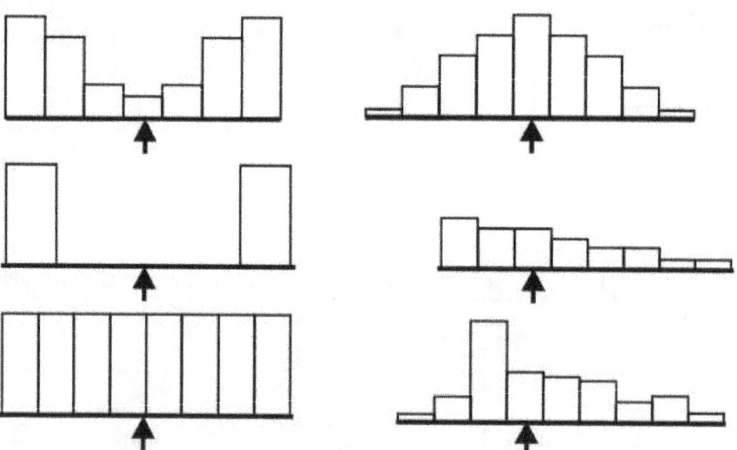

Fig. 3.6 Different frequency distributions with identical means

Range

The range (R) represents the most straightforward measure of dispersion in a dataset, determined by subtracting the smallest observation value from the largest.

$$R = X_{\text{Max}} - X_{\text{Min}}.$$

The value of R is always non-negative, because the smaller number must be subtracted from the larger number. It is very simple to calculate because it only involves and is affected by two values. It is also very sensitive to extreme values. So, the index is very weak because its value varies only by changing the two values of X_{Max} and X_{Min} observations. If the rest of the data change, its value will not change.

Interquartile range

The interquartile range (IQR) serves as a measure of spread and is invariably non-negative, calculated as the difference between the first and third quartiles.

$$\text{IQR} = Q_3 - Q_1,$$

where Q_1 and Q_3 are the first and third quartiles, respectively. IQR is a useful indicator to display the middle 50% of the data distribution.

Example 11 Find the interquartile range for data on female albumin levels in Table 3.2.

Solution: In example 6, the first and third quartiles of that table were 38.20 and 44.22. So the interquartile range is:

$$\text{IQR} = Q_3 - Q_1 = 44.22 - 38.20 = 6.02$$

This means that the distance between Q_1 and Q_3 (middle 50% of the data distribution) is 6.02 units (Fig. 3.7).

Employing the interquartile range (IQR) to detect outlier values in a dataset is a widely used and effective method. To achieve this, subtract 1.5 times the IQR from the first quartile and add 1.5 times the IQR to the third quartile to determine the upper

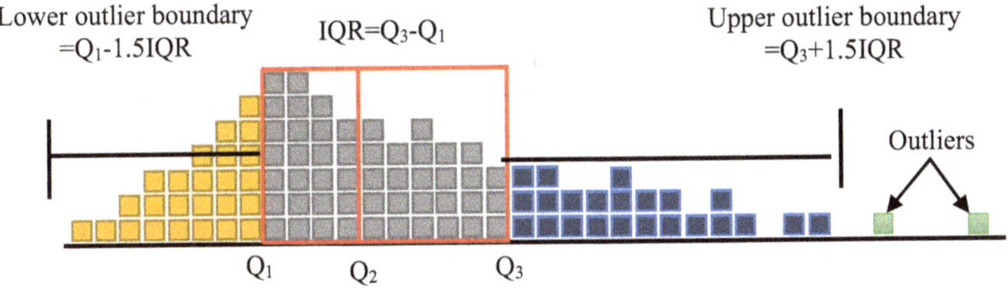

Fig. 3.7 Location of quartiles, median, and IQR

and lower outlier boundaries. Any value falling outside these boundaries (below the lower boundary or above the upper boundary) should be classified as an outlier.

$$IQR = Q_3 - Q_1.$$

Lower outlier boundary $= Q_1 - 1.5 \times IQR$.
Upper outlier boundary $= Q_3 + 1.5 \times IQR$.
Any value outside these boundaries (less than the lower boundary or greater than the upper boundary) should be considered to be an outlier.

Example 12 The data in Table 3.7 shows floating particles in micrograms per cubic meter (air pollution) in 51 cities. Find outlier observations.

Solution: First, we sort the data and obtain the first and third quartiles and interquartile range as in previous examples.

9, 12, 12, 16, 19, 21, 22, 22, 23, 23, 23, 24, 24, 25, 25, 25, 27, 27, 27, 27, 28, 28, 30, 30, 31, 32, 32, 33, 36, 38, 38, 42, 42, 42, 43, 44, 45, 46, 47, 49, 49, 50, 51, 51, 57, 63, 65, 74, 79, 83, 85

$$Q_3 = 47, Q_1 = 24$$
$$IQR = Q_3 - Q_1 = 47 - 24 = 23.$$

The outliers lower limit and upper limit are:

$$Q_1 - 1.5 \times IQR = 24 - 1.5 \times 23 = -10.5$$
$$Q_3 + 1.5 \times IQR = 47 + 1.5 \times 23 = 82.5.$$

Thus, values 83 and 85, which fall outside the limits of outliers, are considered as outliers.

Mean absolute deviation

When quantifying the dispersion of a dataset, it is logical to start by considering the individual differences between each value and the mean. Then mean deviations may be the first idea for measuring dispersion and spread of a dataset. That is, subtract each value from the mean $(x_i - \bar{x})$ and then obtain the average of residuals. But it proves mathematically that the sum of these deviations in any dataset equals always zero $(\sum(x_i - \bar{x}) = 0)$. Because the negative and positive numbers cancel each other

Table 3.7 Floating particles in the air in 51 cities in kilograms per cubic meter

---

27, 22, 23, 24, 25, 44, 63, 57, 27, 30, 9, 83, 27, 23, 74, 25, 32, 79, 51, 31, 22, 85, 42, 43, 49, 50, 28, 42, 28, 30, 45, 49, 46, 12, 42, 21, 32, 27, 16, 33, 23, 19, 12, 38, 38, 24, 65, 51, 47, 25, 36

---

out. To avoid this problem, we can take the average value of the absolute value of residuals ($\sum |x_i - \bar{x}|$) or the square of the residuals ($\sum (x_i - \bar{x})^2$). If the absolute value of these residuals (deviations) is used and we divide it by the number of data, "mean deviations" are obtained and are denoted by MD. In some books, it is also represented by mean absolute deviation (MAD). The following two formulas are used for raw data and for tables and classified data:

$$MD = \frac{\sum_{i=0}^{N} |x_i - \bar{x}|}{N} \quad \text{or} \quad MD = \frac{\sum_{i=0}^{N} f_i |x_i - \bar{x}|}{N}$$

Example 13 The birth weight of 5 newborns was 4.10, 3.90, 3.50, 2.90 and 2.70 kg. Find the mean deviation for these observations. $\bar{x} = 3.42$

Solution: The mean of this dataset is equal to $\bar{x} = 3.42$, and the mean deviation is calculated as follows:

$$MD = \frac{|2.70 - 3.42| + |2.90 - 3.42| + \cdots + |4.10 - 3.42| +}{5} = 0.496.$$

In the above formula, when the median is replaced to the mean, the "median deviations" is obtained and denoted by MdD.

$$MdD = \frac{\sum_{i=0}^{N} |x_i - \text{med}|}{N} \quad \text{or} \quad MdD = \frac{\sum_{i=0}^{N} f_i |x_i - \text{med}|}{N}.$$

It is proved that MdD is always minimal in any dataset and less than any other average deviation. This feature is one of the most important and prominent features of the median, and it is used in planning the establishment of public benefit centers with which the public is involved, such as locating clinics, health centers, hospitals, gas stations, bus stations, trains, and subways.

Variance and standard deviation

As mentioned in the preceding section, when we use the square of residuals (deviation of the mean) and divide the sum of the squares by the number of data, what is obtained is called variance and is denoted by σ^2 or $Var(x)$. The variance is influenced by all the data, and if only the value of one of the observations changes, the value of the variance changes. The following two formulas are used to calculate the variance for raw data and for tables and classified data:

$$Var(x) = \sigma^2 = \frac{\sum (x_i - \bar{x})^2}{N} \quad \text{and} \quad \sigma^2 = \frac{\sum f_i (x_i - \bar{x})^2}{N}.$$

The square of deviations is used to calculate the variance. In the squaring process, the residuals greater than one become larger and the numbers less than one become smaller. Therefore, the farther the data is from the mean, the greater the effect on variance. Conversely, smaller deviations that are closer to mean have less of an effect on variance.

The primary issue with variance as a measure of dispersion lies in its computation using squared deviations, resulting in units that are the squared units of the data. Consequently, the interpretation of the dataset becomes unrealistic as the units differ from the original data units. In many cases, it is preferable to utilize a measure of dispersion that shares the same units as the data. This can be achieved by simply taking the square root of the variance. The resulting measure is known as the standard deviation, denoted by Sd or σ for a population.

$$\sigma = \sqrt{\sigma^2} = \sqrt{\frac{\sum(x_i - \bar{x})^2}{N}} \quad \text{and} \quad \sigma = \sqrt{\frac{\sum f_i(x_i - \bar{x})^2}{N}}.$$

Example 14 Obtain the variance and standard deviation of data in Example 13.

Solution: The mean of this dataset is equal to $\bar{x} = 3.42$, and the variance and standard deviation are calculated as follows:

$$\sigma^2 = \frac{(2.70 - 3.42)^2 + (2.90 - 3.42)^2 + \cdots + (4.10 - 3.42)^2 +}{5} = 0.298$$

$$\sigma = \sqrt{0.298} = 0.546.$$

The above formulas are theoretical and conceptual formulas of variance and standard deviation and do not have a practical aspect for manual calculations. When the distance between the numbers and the mean is large, or the mean is decimal, especially the alternating decimal that never reaches zero (e.g., $\frac{16}{3} = 5.\bar{3}$), the calculation of variance in addition to being difficult is not accurate. For this reason, the more practical formulas for variance and standard deviation, which make calculations easier, are as follows.

$$\text{Var}(x) = \sigma^2 = \frac{1}{N}\left[\sum f_i x_i^2 - \frac{(\sum f_i x_i)^2}{N}\right] = \frac{\sum f_i x_i^2}{N} - (\mu)^2.$$

To calculate variance and standard deviation in tables and classified data, it is recommended to assign columns x_i^2, $f_i x_i$ and $f_i x_i^2$ to and and replace the sum of the two columns ($\sum f_i x_i$ and $\sum f_i x_i^2$) directly from the table in the above formula. This method is the most practical, simplest, and most accurate method of calculating variance and standard deviation.

> **Tip**
>
> When the data includes the whole population, the standard deviation is denoted by the symbol σ and N is used in the denominator of its fraction. But when the data is taken from the sample, the standard deviation is shown with Sd and $(n-1)$ is used instead of N for the denominator of its fraction (also for variance).

Example 15 Calculate the variance and standard deviation of hemoglobin level in Table 3.1.

Solution: We denote the midpoint class by x_i and add two more columns ($f_i x_i$ and $f_i x_i^2$) to the table for ease of calculating variance. Then we substitute the sum of these two columns in the variance formula (Table 3.8).

We now replace the values $N = 70$, $\sum f_i x_i = 842$ and $\sum f_i x_i^2 = 10342$ in the variance formula:

$$\text{var}(x) = \sigma_x^2 = \frac{1}{N}\left[\sum f_i x_i^2 - \frac{\left(\sum f_i x_i\right)^2}{N}\right]$$

$$= \frac{1}{70}\left[10342 - \frac{(842)^2}{70}\right] = 3.056$$

$$\sigma = \sqrt{\sigma_x^2} = \sqrt{3.056} = 1.75.$$

When we have two sets of data with size of N_1 and N_2 from two different populations (e.g., sick and healthy) with means μ_1 and μ_2 and variances σ_1^2 and σ_2^2 and combine them, the total variance of the pooled data is equal to:

$$\sigma^2 = \frac{N_1\left(\sigma_1^2 + d_1^2\right) + N_2\left(\sigma_2^2 + d_2^2\right)}{N_1 + N_2},$$

where d_1 and d_2 are the difference between the mean of the first population from the pooled mean ($d_1 = \mu_1 - \mu$) and the difference between the mean of the second population and the pooled mean ($d_2 = \mu_2 - \mu$). But if the mean of these two datasets are the same ($\mu_1 = \mu_2 = \mu$), then d_1 and d_2 are omitted and we have:

$$\sigma^2 = \frac{N_1 \times \sigma_1^2 + N_2 \times \sigma_2^2}{N_1 + N_2}.$$

Table 3.8 Calculation for variance of hemoglobin levels

Hemoglobin level	x_i	f_i	$f_i x_i$	$f_i x_i^2$
8.5–9.4	9.0	3	27.0	243.0
9.5–10.4	10.0	6	66.0	660.0
10.5–11.4	11.0	19	209.0	2299.0
11.5–12.4	12.0	16	192.0	2304.0
12.5–13.4	13.0	15	195.0	2535.0
13.5–14.4	14.0	6	84.0	1176.0
14.5–15.4	15.0	5	75.0	1125.0
Total	–	70	842.0	10342.0

If the number of these two datasets is also equal ($N_1 = N_2$), N_1 and N_2 also omitted and we have:

$$\sigma^2 = \frac{\sigma_1^2 + \sigma_2^2}{2}.$$

Example 16 Assuming that the mean and variance of albumin levels in a sample of 94 subjects were 40.17 and 12.13 and in another sample of 91 were 42.19 and 14.81, respectively, get the mean and variance of the sum of these two samples (185 people).

Solution: First, we get the mean of two merged samples:

$$\mu = \frac{N_1 \mu_1 + N_2 \mu_2}{N_1 + N_2} = \frac{94 \times 40.17 + 91 \times 42.19}{94 + 91}$$

$$= 41.16.$$

The difference between the mean of each sample and the grand mean is equal to:

$$d_1 = 40.17 - 41.16 = -0.99$$
$$d_2 = 42.19 - 41.15 = 1.03.$$

Then, by replacing the above values, the total variance is obtained:

$$\sigma^2 = \frac{N_1 \left(\sigma_1^2 + d_1^2\right) + N_2 \left(\sigma_2^2 + d_2^2\right)}{N_1 + N_2}$$

$$= \frac{94\left(12.13^2 \pm 0.99^2\right) + 91\left(14.81^2 + 1.03^2\right)}{94 + 91}$$

$$= 14.47.$$

NOTE: When the data is obtained from the sample, the variance and standard deviation of the sample (S^2, S or Sd) is an estimate of the variance and standard deviation of the population. Because the variance and standard deviation of the population is a given and specified value, so $n - 1$ subjects of n sample subjects can take any value, but the last value (for the nth subject) will then be resulted from the rest. The $n - 1$ is called the degree of freedom (df). Therefore, $n - 1$ should be substituted for n in the denominator of variance and standard deviation.

$$S^2 = \hat{\sigma}^2 = \frac{\sum f_i (x_i - \bar{x})^2}{n - 1}$$

$$= \frac{1}{n - 1}\left[\sum f_i x_i^2 - \frac{\left(\sum f_i x_i\right)^2}{n}\right]$$

$$Sd = \sqrt{\frac{\sum f_i(x_i - \bar{x})^2}{n-1}},$$

where $\bar{x}$ is the sample mean:

$$\bar{x} = \frac{\sum f_i x_i}{n}.$$

Dividing the standard deviation by the square root of the sample size, "standard error" is obtained and denoted by Se and is one of the essential components for estimating the confidence interval.

$$Se(x) = \frac{Sd}{\sqrt{n}}.$$

The standard error of estimated proportion for the population ($p\frac{r}{n}$) is:

$$Se(p) = \sqrt{\frac{p(1-p)}{n}}.$$

To comprehend the conceptual meaning of standard deviation and interpret it effectively, we can utilize two approaches: the "empirical rule" and "Chebyshev's theorem".

The empirical rule posits that for datasets exhibiting an approximately bell-shaped distribution, the following characteristics hold true:

- Approximately 68% of all values fall within 1 standard deviation of the mean.
- Roughly 95% of all values fall within 2 standard deviations of the mean.
- About 99.7% of all values fall within 3 standard deviations of the mean.

Chebyshev proved that the proportion of any set of data (not only bell-shaped distributions) lying within K standard deviations of the mean is always at least $1 - \frac{1}{K^2}$ (K is a positive number greater than 1). This is the advantage of Chebyshev's theorem to the empirical rule. But, results from Chebyshev's theorem are only approximate. For example, consider $k = 2$, then At least $1 - \frac{1}{2^2} = 0.75$ (or 75%) of all values lie within 2 standard deviations of the mean.

For example, if in a study height was measured for a large sample of adults. The mean was $\bar{x} = 178$ and $Sd = 7.6$. From Chebyshev's inequality, we can conclude that at least 75% $(1 - \frac{1}{2^2})$ of adults are between 162.8 to 193.2 cm ($178 - 2 \times 7.6$ to $178 + 2 \times 7.6$) and at least 88.9% of adults are between 155.2 to 200.8 cm ($178 - 3 \times 7.6$ to $178 + 3 \times 7.6$).

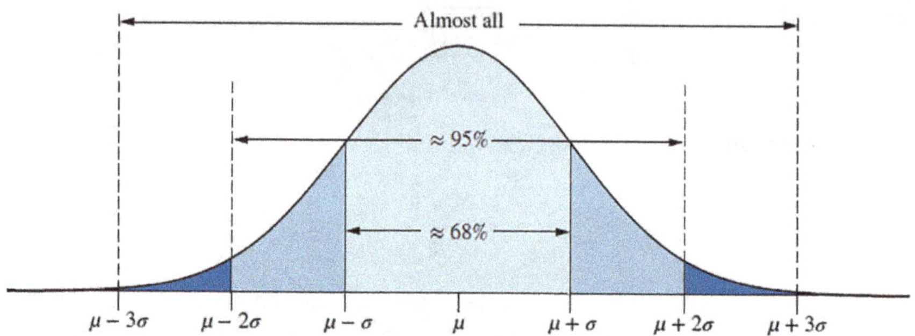

Coefficient of Variation

Standard deviation is not useful when comparing dispersion in two different population or at two different times or two different variables in a population, because standard deviation is affected by the mean value of those variables. Therefore, by dividing the standard deviation of each variable or population by its mean, the "coefficient of variation" index is obtained, and it is represented by CV. For ease of expression of the cv, it is sometimes multiplied by 100 and displayed as a percentage.

$$cv = \frac{\sigma}{\mu} \times 100 \quad \text{or} \quad cv = \frac{Sd}{\overline{x}} \times 100$$

$$cv = \frac{\sigma}{\mu} \times 100 \quad \text{or} \quad cv = \frac{Sd}{\overline{x}} \times 100.$$

Example 17 The mean and standard deviation of height in a population are 155.5 and 17, respectively, and the mean and standard deviation of weight in the same population are 56 and 12, respectively. Compare the height and weight dispersion in this population.

Solution: Weight and height measurement scales are different and cannot be compared. In order to be comparable to each other, their unit of measurement must be removed. For this purpose, we obtain the coefficient of variation in height and weight in this population.

$$CV_{Weight} = \frac{\sigma_w}{\mu_w} \times 100 = \frac{17}{155.5} \times 100 = 10.93$$

$$CV_{Height} = \frac{\sigma_H}{\mu_H} \times 100 = \frac{12}{56} \times 100 = 21.43.$$

Now, despite the different measurement units of height and weight, the dispersion of these two traits can be compared in the population. In this population, weight dispersion around the mean is more than height dispersion around its mean.

Example 18 Estimation of mean and standard deviation of weight of infants with smoking mothers at birth is 3.200 and 0.493, respectively, and infants of non-smoking mothers are 3.590 and 0.371, respectively. Compare the weight dispersion of these two groups.

Solution: Although the measurement unit is the same in both groups, but because they are from two different populations with different means, their dispersion cannot be easily compared. It is recommended to calculate the coefficient of variation in these two groups first and then compare.

$$CV_{Smoker} = \frac{\sigma_S}{\mu_S} \times 100 = \frac{0.493}{3.200} \times 100 = 15.41$$

$$CV_{Nonsm} = \frac{\sigma_{Non}}{\mu_{Non}} \times 100 = \frac{0.371}{3.590} \times 100 = 10.33.$$

Thus, the weight dispersion of newborns whose mothers smoke is more than the weight dispersion of infants with non-smoker mothers.

To calculate measures of dispersion by SPSS, select the path of **analyze\descriptive statistics\frequencies** until a window like the one below opens. In this window, select the desired quantitative variables from the left box and enter to the box in the right. Then select the "statistics" option at the top right corner to open another window. At the top right of this window are the std. deviation, variance, range, minimum, maximum, and SE mean.

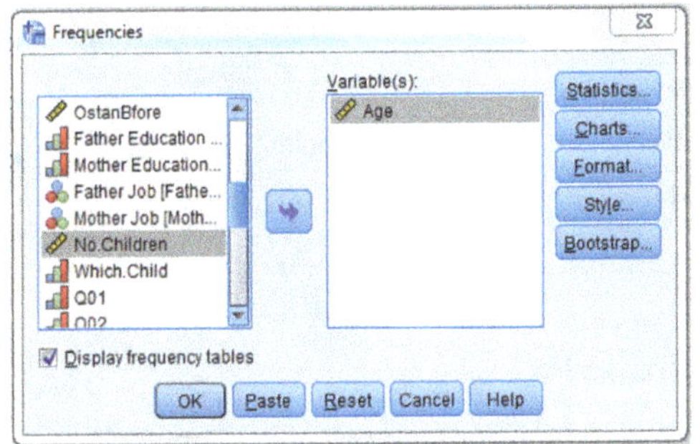

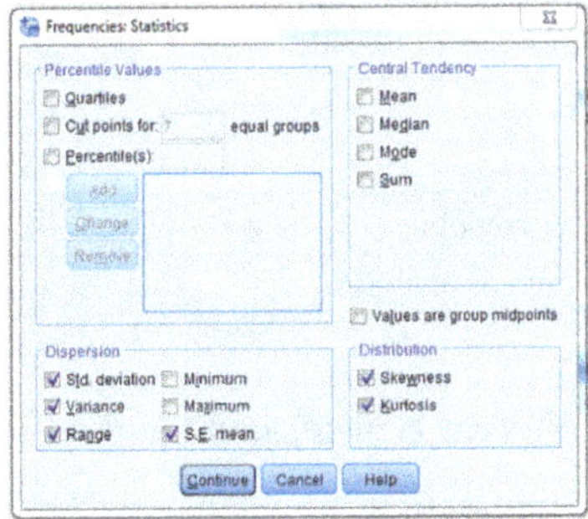

You can also use the path of analyze\descriptive statistics\descriptive to calculate measure of dispersion and be to open the following window. Then select the "options" button to open another window. In this window, choose the requested options.

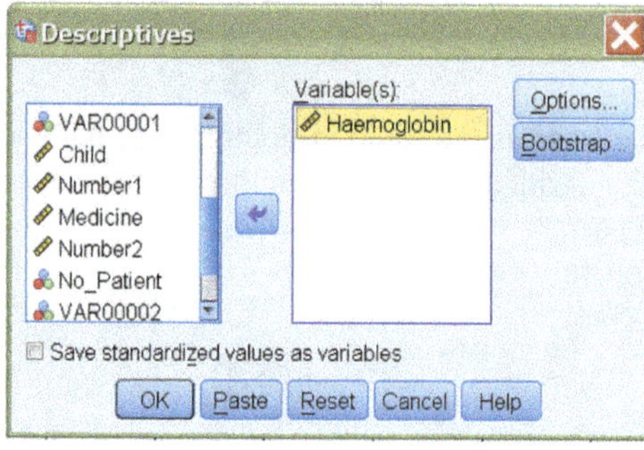

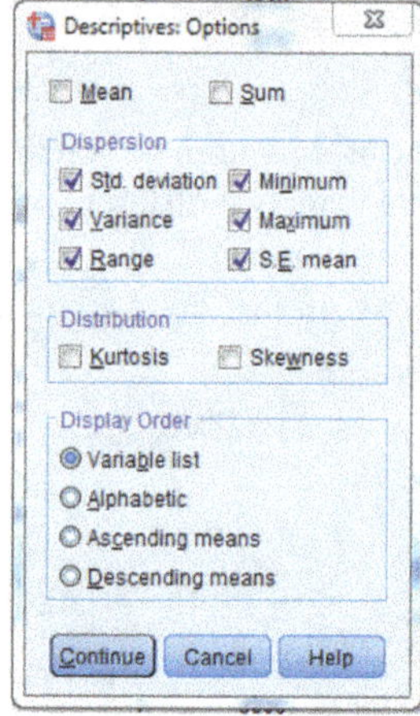

Descriptive statistics

	N	Range	Minimum	Maximum	Mean	
	Statistic	Statistic	Statistic	Statistic	Statistic	Std. Error
Hemoglobin	70	6.3	8.8	15.1	11.984	0.1693
Valid N (listwise)	70					

Descriptive statistics

	Std. deviation	Variance
	Statistic	Statistic
Hemoglobin	1.4161	2.005
Valid N (listwise)		

Different type of scales and descriptive index

In this section, we summarize the application of different types of descriptive indexes in Table 3.8 according to the type of data.

Scales statistics	Nominal	Ordinal	Interval	Ratio
Mode	✓	✓	✓	✓
Median	–	✓	✓	✓
Mean	–	–	✓	✓
Quartiles	–	✓	✓	✓
IQR	–	–	✓	✓
Range	–	–	✓	✓
Standard deviation	–	–	✓	✓
Coefficient of variation	–	–	–	✓

This table shows that:

For nominal data, only the mode is defined and used.

For ordinal data, the mode, median, and quartiles are defined and applicable.

For interval data, all statistics are defined and conceptual except the coefficient of variations, which shows the variations around the mean. Because in this scale, starting point of zero is contractual and the mean is also a function of it.

For ratio data, all statistics are defined and conceptual.

3.4 The Effect of Constant Changes on Measures of Central Tendency and Dispersion

Sometimes statistical calculations are heavy and difficult. To simplify the calculations, you can first transform them to new data by performing some simple mathematical operations (addition, subtraction, multiplication, and division) on the original data. It then calculated the statistical indices on the newly transformed data and then retransformed the original changes. For example, in the study of people's height, subtract the constant value of 150 from all observations. The calculation of the mean and variance for the newly transformed data will then be very short and simple and then applied the effect of the reduced value of 150 on the mean and variance.

Effect of addition and subtraction

If the constant value is subtracted from all observations and the mean and standard deviation for the new variable $(x_i - C)$ are calculated, the variance and standard deviation of the original data and the transformed data do not change, but the new mean $\bar{x}_{\text{New}}$ decreases as much as the constant value of C, which C value should be added to it, namely:

$$\bar{x} = C + \bar{x}_{\text{New}}.$$

The effect of addition is inverse of the effect of subtraction.

$$\bar{x} = C + \bar{x}_{\text{New}}.$$

Example 19 The height of 11 people is equal to 189–180–177–153–154–156–148–168–174–171–167, by subtracting the fixed number 160 from all observations, find the relevant indicators.

Solution: Mean, mean, variance, and standard deviation of the initial data are:

$$\text{Med} = 168, \quad \mu = 167, \quad \sigma^2 = 151.45, \quad \sigma = 12.307$$

If the constant value $C = 160$ is subtracted from all the above observations, the new values $(x_i - C)$ are equal to:

$$7, \ 11, \ 14, \ 8, \ -12, \ -4, \ -6, \ -7, \ 17, \ 20, \ 29.$$

Mean, mean, variance, and standard deviation of new data are equal to:

$$\text{Med} = 168, \quad \mu = 167, \quad \sigma^2_{\text{New}} = 151.45, \quad \sigma_{\text{New}} = 12.307.$$

It can be seen that:

$$\text{Med}_{\text{New}} + C = \text{Med} \quad \text{i.e.} \quad 8 + 160 = 168$$
$$\mu_{\text{New}} + C = \mu \quad \text{i.e.} \quad 7 + 160 = 167.$$

That is, the new mean and median change by the same constant value.

$$\sigma^2_{\text{New}} = \sigma^2 = 151.45, \quad \sigma_{\text{New}} = \sigma = 12.307.$$

That is, the new variance and standard deviation do not change.

Adding and subtracting a constant value from all observations means changing the origin of the coordinate system, while multiplying and dividing the observations by a constant value means changing the scale of the measurement.

The effect of multiplication and division

If we multiply all the observations by a constant value of C and calculate the measures of central tendency and dispersion for the new variable (Cx_i), the variance of the newly transformed data (σ^2_{New}) is increased by the square of that fixed number (C^2) and the standard deviation The mean and the median become C times larger. Since the multiplication and division are inverse of each other, to achieve the variance of the initial data, we must divide the resulting variance by (C^2) and the standard deviation, the mean and the median by the constant number C, i.e.:

$$\sigma^2_{\text{New}} = \sigma^2 \times C^2$$
$$\text{Sd}_{\text{New}} = \text{Sd} \times C$$
$$\mu_{\text{New}} = \mu \times C$$
$$\text{Md}_{\text{New}} = \text{Md} \times C.$$

Example 20 The body temperature of 9 people is: 37.7, 37.3, 36.4, 36.5, 37.2, 38.3, 36.2, 39.3, and 38.6. First subtract a constant value of 36 from all observations and multiply the remainder by a constant value of 10, and then calculate the measures of central tendency and dispersion.

Solution: Mean, mean, variance, and standard deviation of the initial data are:

$$\text{Med} = 37.3, \quad \mu = 37.5, \quad \sigma^2 = 1.018, \quad \sigma = 1.009.$$

If the constant value $C = 36$ is subtracted from all the above observations and the result is multiplied by the constant value ($a = 10$), the new values $(x_i - C) \times a$ are equal to:

$$26, \ 33, \ 2, \ 23, \ 12, \ 5, \ 4, \ 17, \ 13, \ 29.$$

Mean, mean, variance, and standard deviation of new data are equal to:

$$\text{Med}_{\text{New}} = 13, \quad \mu_{\text{New}} = 15, \quad \sigma^2_{\text{New}} = 101.8, \quad \sigma_{\text{New}} = 10.09.$$

It can be seen that:

$$(\text{Med}_{\text{New}} \div a) + C = \text{Med} \quad \text{i.e.} \quad (13 \div 10) + 36 = 37.3$$
$$(\mu_{\text{New}} \div a) + C = \mu \quad \text{i.e.} \quad (15 \div 10) + 36 = 37.5.$$

That is, the new mean and median should be magnified to the opposite value, and then added to the opposite value as they have been reduced.

$$\left(\sigma^2_{\text{New}} \div a^2\right) = \frac{101.8}{100} = 1.018 = \sigma^2,$$
$$(\sigma_{\text{New}} \div a) == \frac{101.8}{10} = 10.18 = \sigma.$$

That is, we need to shrink the new variance to the square of the value as we have enlarged, but shrink the new standard deviation by the same value we have enlarged.

Conclusion: Subtracting the constant value of $C = 36$ from any data did not affect the standard deviation and variance, but multiplying the constant value by 10 affected them and made the standard deviation as much as its (10 times) and the variance as large as its square (100 times) has done. Therefore, in order to achieve the standard deviation and variance of the initial data, they should be divided into 10 and 100, respectively.

$$\sigma = \frac{\sigma_{\text{New}}}{a^2} = \frac{101.8}{100} = 1.018$$
$$\text{Sd} = \frac{\text{Sd}_{\text{New}}}{a} = \frac{10.09}{10} = 1.009.$$

On the other hand, both the values of $a = 10$ and $C = 36$ have affected the mean and median. Therefore, reverse operations should be done with priority for them. That is, first divide the new mean by $a = 10$ and then add the constant number $C = 36$.

3.5 Exercises

1. We have two series of observations from two populations (urban and rural) with equal numbers of $N = 40$ with means of μ_1 and μ_2 and variances of σ_1^2 and σ_2^2. We merge these populations. Obtain the total variance of this whole merged population. Calculate the same calculation with a sample of 16.

2. The number of previous deliveries of 120 women aged 30–40 years is given in the table below:

No. of deliveries	0	1	2	3	4	Total
Frequency	20	32	36	25	7	120

Find out its mean, median, third quartile, and standard deviation.

3. Consider a series of hypothetical observations and investigate $\sum_{i=1}^{N}(x_i - \mu)$ is equal to zero.

4. The table below shows the value of systolic blood pressure (SBP) and diastolic blood pressure (DBP) of 200 people in the population. Calculate the standard deviation of systolic and diastolic blood pressure. Perform systolic blood pressure calculations by subtracting the most appropriate constant from all measures and diastolic blood pressure calculations by dividing all data by the most appropriate constant.

SBP	90–	100–	110–	120–	130–	140–	150–	160–	180–	Total
N	3	20	64	57	31	10	5	7	3	200
DBP	60–	65–	70–	75–	80–	85–	90–	95–	100–	Total
N	5	7	30	22	66	14	42	2	12	200

5. If we want to compare the dispersion of two types of systolic and diastolic blood pressure, which measure is appropriate?

6. The table below shows the height and weight of men in population. Compare the dispersion of height and weight in this population using the coefficient of variation. In this comparison, make sure to use formulas related to uniform effects.

Height (Cm)	155–	160–	165–	170–	175–	180–	185–	190–	Total
N	1	7	32	61	53	31	9	6	200
Weight (Kg)	60–	65–	70–	75–	80–	85–	90–	95–	Total
N	27	44	46	41	30	6	2	4	200

7. According to the formula of the mode in the tables and categorized data, interpret the value of the mode in the following situations and explain where the mode is located:

   A. $d_1 = 0$
   B. $d_2 = 0$
   C. $d_1 = d_2$

8. The table below shows the level of blood cholesterol in the population. Calculate the first and third quartiles and the seventeenth and sixtieth percentiles. Draw a box–whiskers diagram.

Blood cholesterol	150–	200–	250–	275–	300–	325–	350–	400–	450–	Sum
No	19	35	45	25	23	21	23	6	3	200

9. If in the categorized data, the cumulative frequency of one category is exactly equal to $\frac{N+1}{2}$, investigate and show that the median of that dataset will be equal to upper limit of that category or the lower limit of the next category.

10. Find the mean, variance, standard deviation, and median of blood glucose level in the following data and draw a box-and-whisker diagram.

4.1– 4.0– 3.6– 3.8– 4.6– 4.4– 3.3– 3.6– 3.8– 4.1– 3.9– 4.7– 3.4– 3.7– 2.2– 5.0– 4.7– 3.3– 4.2– 3.6–
4.1– 3.8– 4.8– 3.3– 4.7– 4.4– 4.9– 4.5– 3.7– 4.0– 4.9– 3.4– 2.9– 6.0– 4.0– 3.4– 5.1– 4.3– 4.8– 5.0

11. Investigate with a numerical example that the coefficient of variation (CV) does not depend on the unit of measurement. That is, if the data is multiplied by a fixed number, the CV does not change. To do this, calculate the CV in height of 10 people on both the meter and inch scales (equal to 2.54 cm).

Chapter 4
Probability

4.1 Introduction

Achieving a complete separation among scientific disciplines proves practically challenging, as many fields draw upon established principles from one another for validation and support. Statistics, in particular, shares an intrinsic connection with both mathematics and probability. A profound grasp of various statistical concepts often hinges on a comprehensive understanding of probability. This interdependence is so significant that statistics and probability are frequently intertwined. In certain contexts, they are even viewed as a unified scientific discipline, underscoring their crucial role in comprehending, applying, advancing, and expanding the realm of knowledge in this field.

To acquaint students of statistics with the intricacies of probability, this book features a dedicated chapter. Within this chapter, foundational definitions and principles of probability are presented independently, offering learners a broader perspective to comprehend and conceptualize various statistical issues.

4.2 Definitions

Random Event

Random events refer to occurrences or non-occurrences that are inherently unpredictable, varying in each experiment. Examples include the outcome of flipping a coin, a patient's recovery from a treatment, the outcome of twins in a delivery, the gender of a newborn, or the vaccination status of a child within a community. These events embody the essence of unpredictability, introducing an element of chance into the outcomes of diverse experiments.

S. H. Saneii and H. Doosti, *Practical Biostatistics for Medical and Health Sciences*,
https://doi.org/10.1007/978-981-97-3083-4_4

Sample Space (Possible Outcomes)

In any given experiment, a range of events may unfold, each carrying varying probabilities. The comprehensive set encompassing all conceivable outcomes, leaving no room for any other potential result, is termed the "possible outcomes" or "sample space", denoted by Ω. It is assured that the experiment will yield a result within this set. For instance, in the context of the birth of a baby, the possible outcomes are {girl, boy}. When rolling a dice, the sample space consists of {6, 5, 4, 3, 2, 1}, and in drawing a card from a deck, the sample space includes {face, 10, 9, 8, 7, 6, 5, 4, 3, 2, A}. This delineation of possible outcomes is integral to understanding and analyzing the inherent variability within diverse experiments.

Favorable Outcomes

Favorable outcomes refer to a subset of the sample space that holds particular interest for the experimenter. Designated as "favorable outcome" and denoted by M, these outcomes are those the experimenter wishes to observe. It is important to note that the set of favorable outcomes (M) is a subset of the overall "possible outcomes" (Ω), and as such, the cardinality of M will never exceed that of Ω.

For instance, when rolling a dice and focusing on the appearance of an even number, the favorable outcomes are {6, 4, 2}. In the context of drawing a card and desiring a numeric card, the favorable outcomes encompass {10, 9, 8, 7, 6, 5, 4, 3, 2}. These favorable outcomes are pivotal in guiding the experimenter's attention toward specific results within the broader spectrum of possible outcomes.

Mutually Exclusive Events and Non-exclusive Events

Events that can occur simultaneously are termed "non-exclusive". For instance, in rolling a dice, the events "even number" and "multiple of 3" can logically and practically co-occur when the number 6 is rolled, as it is both even and a multiple of 3.

Conversely, events that cannot occur simultaneously are labeled "mutually exclusive". Taking the example of rolling a dice again, the events "even number" and "multiple of 5" are mutually exclusive since it is logically and practically impossible for a single outcome to be both even and a multiple of 5. Simply put, two events are not mutually exclusive (or disjoint) if the occurrence of one event does not preclude the occurrence of the other. Graphically, the presence of an intersection signifies non-mutually exclusive events, while the absence of an intersection indicates mutually exclusive or disjoint events.

Equally Likely Events

Events are considered "equally likely" when there is no inherent favoritism toward the occurrence of one event over another, and their chances of happening are the same. For instance, in rolling a dice, each of the outcomes {6, 5, 4, 3, 2, 1} has an equal probability of occurring, with none appearing to take precedence.

Conversely, in a deck of cards, the events "a face" and "numeric" when drawing a card are not equally likely. The probability of drawing a face card is different from that of drawing a numeric card, indicating that these events do not share equal likelihood. The concept of equally likely events is fundamental in probability theory, providing a basis for understanding and calculating probabilities in various scenarios.

Independent Events

Events are considered independent when the outcome or non-outcome of one event has no influence on the occurrence or non-occurrence of another. For instance, the sex of a baby is often assumed to be independent of the sex of subsequent or previous infants. This assumption is grounded in the idea that the sex of each baby is not influenced by the sex of previous babies, and similarly, it does not affect the sex of future babies. The independence of events is a crucial concept in probability theory, enabling the analysis and calculation of probabilities in situations where events do not impact each other.

Impossible and Certain (Sure) Events

An event is deemed "impossible" when its occurrence is practically inconceivable, meaning that it cannot happen, and the probability of its occurrence is consistently zero. For instance, in throwing a dice, obtaining a number of 7 is considered an "impossible" event.

On the contrary, an event that is practically certain to occur is termed a "certain" or "sure" event. In other words, the probability of its occurrence is always one. A sure event encompasses the entire sample space. For example, in rolling a dice, getting a number less than 6 is a "sure" event, as it covers all possible outcomes and is guaranteed to happen. These concepts of impossible and sure events help delineate the extremities of probability scenarios.

Definition and Concepts of Probability

The term "probability" is commonly employed in everyday language to express varying degrees of certainty about events, as seen in phrases like "perhaps today is raining", "he may be sick", and others. In a formal context, "probability is a measure of the chance (proportion of times) of an event occurring in the long run", capturing the certainty associated with the likelihood of an event.

Mathematically, the classical definition of probability, denoted as $P(A)$, is applicable when the number of outcomes is finite and each outcome is equally likely. In this scenario, it is expressed as the ratio of the number of favorable cases ($n(A)$) to the total number of possible cases ($N = n(S)$):

$$P(A) = \frac{n(A)}{N},$$

where S is a super set that contains all possible elements.

Obviously, the maximum value of $n(A)$ can be equal to N. That is, all possible cases are favorable. In this situation, the probability will be equal to 1, which is the probability of the certain (sure) event. The minimum value of $n(A)$ is also zero, that is, when none of the possible cases is favorable. In this situation, the probability is equal to the probability of impossible event. Therefore, the probability varies between zero and 1 ($P(A) \leq 1$). That is, it can never be negative or greater than one.

It's important to note that this classical definition has limitations, especially when dealing with situations where outcomes are not equally likely or the sample space is not finite. In such cases, alternative probability measures, such as the ratio of the length of event A to the length of the sample space or other probability density functions, become more appropriate.

In more advanced probability theory, especially when handling continuous probability distributions, a broader set of tools and measures, including probability density functions and cumulative distribution functions, are employed to provide a more nuanced and comprehensive understanding of probability in diverse scenarios.

To better and easier understand the principles and laws of probability, it is necessary to express concepts and rules about sets:

The empty set is a set which has no member and written { } or $\emptyset$.

The empty set is a proper subset of any sets except itself.

If set A has one or more elements of set B, then A is a subset of B and is denoted by $A \subset B$. B is then a **superset** of A.

Any set is its own subset $A \subset A$.

Two sets are equal when all their elements are the same.

The union of sets A and B is a set that contains all the elements of set A and set B or both. The union of two sets A and B is denoted by $A \cup B$.

The **intersection** of two sets A and B is the set containing all elements of A that also belong to B (or equivalently, all elements of B that also belong to A) and denoted by $A \cap B$.

If A is a subset of superset S, the **complement** of a set A, often denoted by $\overline{A}$ (A^c or A'), is a set containing the elements not in A.

4.3 Principles of Probability

The probability of any event is non-negative and varied between 0 and 1 ($0 \leq p(A) \leq 1$).

The probability of superset is always equals to 1 ($p(S) = 1$).

The probability of empty set is always equals to 0 ($p(\emptyset) = 0$).

For any mutually exclusive events, we have $(A + B) = p(A) + p(B)$ (Fig. 4.1).

In general, it is always true for two or three not exclusive events:

$$p(A \cup B) = p(A) + p(B) - p(A \cap B),$$
$$p(A \cup B \cup C) = p(A) + p(B) + p(C)$$
$$- p(A \cap B) - p(A \cap C)$$
$$- p(B \cap C) + p(A \cap B \cap C).$$

Fig. 4.1 Graphical
presentation of set and subset

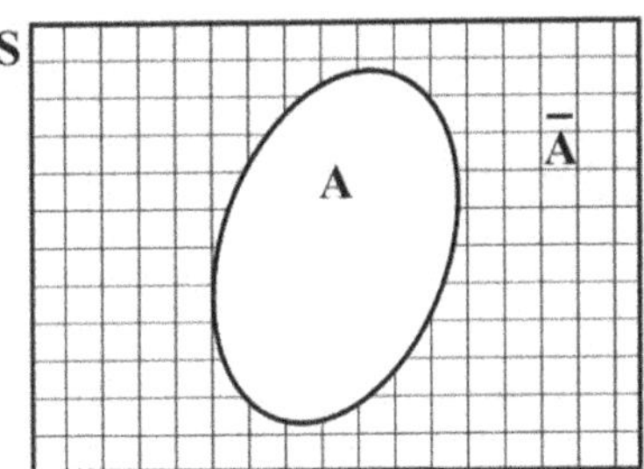

Fig. 4.2 Graphical
presentation of joint events

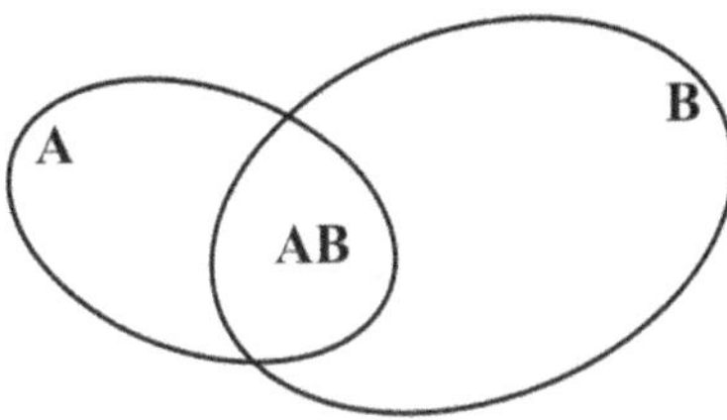

The following figure easily shows the concept of joint and disjoint of the events
A and B (Fig. 4.2).

Example 1 In a rehabilitation clinic, the probability of having a patient with muscle
pain is 0.5 and the probability of having high blood pressure is 0.6. What is the
probability that someone has at least one disease (muscle pain or hypertension or
both)?

Solution: Obviously, the probability of having at least one disease (muscle pain or
hypertension or both) in that clinic cannot be equal to $0.6 + 0.5 = 0.1$, because
the probability always varies between 0 and 1 and cannot be greater than 1. The
problem arises from the fact that patients who have both diseases "together" have
been counted twice, once with patients with muscle pain and once with patients with
hypertension. Assuming that these two diseases are independent, the probability of
both diseases jointly should be subtracted from the sum.

$$p(\text{both diseases}) = 0.5 \times 0.6 = 0.30.$$

Therefore, the probability of having at least one type of disease in this clinic is
equal to:

$$p(\text{Muscle} \cup \text{Hypertension}) = p(\text{Muscle})$$
$$+ p(\text{Hypertension}) - p(\text{both diseases})$$
$$= 0.5 + 0.6 - 0.3 = 0.8.$$

The probability of two complementary events is A and is always equal to 1.

$$p(A) + p(\overline{A}) = 1.$$

For two independent events, the following relationship is always established:

$$p(A \cap B) = p(A) \times p(B).$$

Example 2 In a family with 5 children, what is the probability that the eye color of the first 3 children will be hazel and the eye color of the other children will be blue? Assuming that the probability that a newborn's eye is hazel or blue is 1.6 and 5.6.

Solution: We note $A =$ hazel eye color of first 3 children and $B =$ blue eye color of the next 2 children. So:

$$P(A) = \frac{5}{6} \times \frac{5}{6} \times \frac{5}{6} = \left(\frac{5}{6}\right)^3,$$

$$P(B) = \frac{1}{6} \times \frac{1}{6} = \left(\frac{1}{6}\right)^2.$$

Since 2 events are independent, therefore:

$$P(A \cap B) = P(A) \times P(B) = \left(\frac{5}{6}\right)^3 \times \left(\frac{1}{6}\right)^2.$$

Generally, for some independent events the following relation is established:

$$P(A \cap B \cap C \cdots \cap Z) = P(A) \times P(B) \times P(B) \cdots \times P(Z).$$

Example 3 In a family of 5 children, what is the probability that the first child is a girl and the other children are a boy one every other?

Solution: We denote $A =$ first child be girl and $B =$ the rest are a boy one every other. The event of be may occur in 2 situations $B_1 = (B, G, B, G)$ or $B_2 = (G, B, G, B)$. Then:

$$p(A) = \frac{1}{2}, \quad p(B_1) = \left(\frac{1}{2}\right)^4, \quad p(B_2) = \left(\frac{1}{2}\right)^4$$

$$P(B) = p(B_1 \cup B_2) = p(B_1) + p(B_2)$$

$$= \left(\frac{1}{2}\right)^4 + \left(\frac{1}{2}\right)^4 = \left(\frac{1}{2}\right)^3 = \frac{1}{8}.$$

Since the A and B events are independent, then we have:

$$p(A \cup B) = p(A) \times p(B) = \frac{1}{8} \times \frac{1}{2} = \frac{1}{16}.$$

4.4 Conditional Probability

Conditional probability is a measure of the probability of an event occurring, given that another event (by evidence) has already occurred. It is a tool that lets you comment about some related uncertain events are. It also lets you explain about how the probability of an event can vary under different conditions. In probability theory, conditional probability is very important and a is one of the fundamental concepts.

If the interested event is A and the you know that event B has occurred, "the conditional probability of A Given B", or "the probability of A under the condition B", is usually written as $P(A|B)$.

$$P(A|B) = \frac{\text{Joint probability of } A \text{ and } B}{\text{marginal probability of } B} = \frac{P(A \cap B)}{P(B)}.$$

The following relationships can be easily obtained:

$$P(A \cap B) = P(A|B) \times P(B),$$
$$\frac{P(A|B)}{P(A)} = \frac{P(B|A)}{P(B)},$$
$$P(A \cap B) = P(A|B) \times P(B) = P(B|A) \times P(A).$$

Note that conditional probability does not state that there is always a causal relationship between the two events, as well as it does not indicate that both events occur simultaneously.

Example 4 Table 1.4 shows the number of people in a population of 100,000 in an area and the mortality rate per thousand populations by age. What is the probability that a person from this area will die at the age of 51?

Solution: If A = "probability of survival up to 50 years" and B = "probability of death between 50 and 51 years", we have:

Age	No. of alive	Mortality rate
50	90,718	6.43
51	90,135	7.00
52	89,504	7.62
53	88,822	8.30
54	88,085	9.03
55	87,290	9.77
56	86,437	10.60
57	85,521	11.56
58	84,532	12.71
59	83,448	13.98
60	82,291	15.42

$$P(A) = \frac{90718}{10000} = 0.90718.$$

The probability of dying before the age of 50 is equal to:

$$P(\overline{A}) = 1 - P(A) = 1 - 0.90718 = 0.09282.$$

The probability of dying between the ages of 50 and 51 is equal to:

$$P(B) = \frac{90718 - 90135}{10000} = 0.00583.$$

In this case $P(A \cap B) = P(B)$, then:

$$P(A \cap B) = P(B) = 0.00583.$$

Therefore, the probability of death at any particular age is obtained using conditional probability, because it means that the person in question must be alive before that year. In other words, the probability of a person dying at the age of 51 as long as we know he was alive before that age (50 years).

$$P(B|A) = \frac{P(A \cap B)}{p(A)} = \frac{0.00583}{0.90718} = 0.00643.$$

This means that the probability that a person who has passed the age of 50 will die between the ages of 50 and 51 is 0.00643.

Example 5 Sixty % of the students in a university are female and a 30% of them are local students. We know that 10% of the students are local females. What is the probability that a local student be female?

Solution: We note "F" for Female, "L" for local students, $F \cap L =$ for local female students, and "$F|L$" for being a female given that the student is local. Then $P(F) = 0.60$, $P(L) = 0.30$, and $P(F \cap L) = 0.10$.

Using conditional probability, the probability of being a female given that the student is local is:

$$P(F|L) = \frac{P(F \cap L)}{P(L)} = \frac{0.10}{0.30} = 0.33$$

4.5 Bayes' Theorem

Bayes' theorem represents an extension of conditional probability, serving as a method for categorizing phenomena based on supplementary information regarding the event's occurrence. This additional information typically comprises prior knowledge concerning conditions that may be associated with the event, including probabilities of other phenomena occurring or not. Essentially, Bayes' theorem facilitates the determination of a probability (posterior) when certain other probabilities (prior) are known. It operates as a mathematical principle governing the adjustment of beliefs considering evidence presented.

The logic of equation is:

$$\text{Posterior} = \text{Prior} \times \frac{\text{Likelihood}}{\text{Marginal probability}},$$

$$P(A|B) = P(A) \times \frac{P(B|A)}{P(B)},$$

where:

The **prior** probability, also known as the previous probability, refers to the probability of an event before new data is gathered. It represents the most rational assessment of the probability of an outcome based on existing knowledge before conducting an experiment.

On the other hand, the **posterior** probability denotes the revised or updated probability of an event occurring after incorporating new information. It's essential to note that posterior probability is distinct from likelihood and reflects the updated probability following the consideration of fresh data.

Likelihood is another conditional probability. It is the probability of the evidence being present, given that the hypothesis is true. Remember, this is not the same as the posterior!

Marginal probability is the probability of the evidence, under any circumstance. This means the probability of the evidence being present, whether the hypothesis is true or false.

Bayes' theorem is important and widely used in probability theory. The Bayesian method is used when the values of $P(A)$ and $P(A \cap B)$ are not directly available to obtain $P(B|A)$. Indeed, the probability of an event can be computed by conditioning it on the occurrence or non-occurrence of another event. Often, calculating the probability of an event directly can be challenging. However, by leveraging Bayes' theorem and conditioning one event on another, it becomes possible to derive the probability. This approach enables a systematic and effective method for calculating probabilities, especially in scenarios where direct computation is impractical.

The relation $p(A) = p(A \cap B) + p(A \cap \overline{B})$, which its accuracy is easily proved, is used to prove Bayesian rule. The following rule, known as Bayesian rule, is derived from the conditional probability formula.

$$p(B|A) = \frac{p(B \cap A)}{p(A)} = \frac{p(A \cap B)}{p(A \cap B) + p(A \cap \overline{B})}$$

$$= \frac{p(A|B)p(B)}{p(A|B)p(B) + p(A|\overline{B})p(\overline{B})} .$$

Example 6 The physician's diagnosis is usually based on the patient's laboratory results. Both the lab results and the doctor's diagnosis may be erroneous (that is, a person may be really healthy, but the doctor may diagnose him/her ill, or s/he is really ill, but the doctor diagnoses him/her as healthy). Assume that the probability of being actually ill is 0.3 and the probability of error in the laboratory tests is 0.06. Find the probability of being actually ill for a person whom was diagnosed as ill by the doctor has.

Solution: Suppose A is defined as "the person is diagnosed ill", B is "the person is actually ill", and $\overline{B}$ is "the person is healthy".

According to Bayes' rule:

The probability that the person is actually ill $= p(B) = 0.3$.

The probability that the person is healthy $p(\overline{B}) = 1 - 0.3 = 0.7$.

Probability that the relevant tests are correct $p(A|B) = 0.9$.

Probability of error in tests $p(A|\overline{B}) = 0.1$.

Therefore, the probability of correctly diagnosing a patient, i.e., the actually ill person diagnosed by the doctor sick is equal to:

$$p(B|A) = \frac{p(A|B)p(B)}{p(A|B)p(B) + p(A|\overline{B})p(\overline{B})}$$

$$= \frac{0.3 \times 0.9}{0.3 \times 0.9 + 0.7 \times 0.06} = 0.865.$$

Example 7 Medical tests can occasionally yield inaccurate results, prompting us to introduce uncertainty by employing probabilities. Let's denote A as the event signifying a positive test outcome for a virus in a person, while B represents the event indicating the individual's actual infection with the virus. Assuming that the virus affects 2% of the population and given a conditional probability of 0.995 for a positive test result given that the person has the virus (denoted $p(A|B)$), we establish these conditions to model uncertainty in the testing process. Also, suppose the conditional probability that a person not having the virus tests positive (a false positive) is $p(A|\overline{B}) = 0.01$. Find out the probability that the test result of someone who has actually the virus is positive $p(B|A)$. In other words, when a person tests positive, how likely is it that person is actually infected with the virus?

Solution: In this example, the following probabilities are given:

The probability that a person is infected is equal to $(B) = 0.02$.

Probability of not having the virus is: $p(\overline{B}) = 1 - p(B) = 1 - 0.02 = 0.98$.

Conditional probability that the test is positive, given that the person has the virus, is $p(A|B) = 0.995$.

Conditional probability that the test is positive, given that the person has not the virus, is $p(A|\overline{B}) = 0.01$.

$$p(A) = p(A|B).p(B) + p(A|\overline{B})p(\overline{B})$$
$$= (0.995 \times 0.02) + (0.01 \times 0.98) = 0.0297.$$

In this example, A is the event that the test of a person is positive and B is the event that the person actually has the virus. Supposing a person test is positive and using Bayes' theorem, we are able to update the probability that the person has the virus, the posterior probability $p(B|A)$:

$$p(B|A) = \frac{p(A|B)p(B)}{p(A|B)p(B) + p(A|\overline{B})p(\overline{B})}$$
$$= \frac{0.995 \times 0.02}{0.995 \times 0.02 + 0.01 \times 0.98} = 0.67.$$

The probability of having the virus has varied from 0.02 to 0.67.

$$p(\text{Disease}|\text{test}+) = \frac{p(\text{test} + |\text{disease})p(\text{disease})}{p(\text{test}+)}$$
$$= \frac{p(\text{test} + |\text{disease})p(\text{disease})}{p(\text{test} + |\text{disease})p(\text{disease}) + p(\text{test} + |\text{disease})p(\text{no disease})}.$$

Example 8 A test to diagnose and screen arthritis in people over the age of 50 has been suggested. According to national studies, 10% of this age group suffers from this type of arthritis. The test result is 85% positive for people with confirmed arthritis and 4% for healthy people without the disease (false positive). Obviously, for this test to be useful in screening, the positive results of that test must indicate a high percentage of disease, i.e., the $P(D|T+)$ must be high.

Solution: We noted the event of arthritis $= D$, being healthy (no arthritis) $= \overline{D}$, positive test result $= T+$, and negative test result $= T-$. The given information in this example is: $(D) = 0.90$, $p(\overline{D}) = 0.90$, $p(T - |D) = 0.15$, $p(T - |\overline{D}) = 0.96$, and $p(T + |D) = 0.85$, $p(T - |\overline{D}) = 0.04$.

In this example, $P(D \cap T+)$ and $P(T+)$ are not given and cannot be obtained directly from the conditional probability formula. They must be obtained in another way.

This information can be seen in the "probability path" diagram.

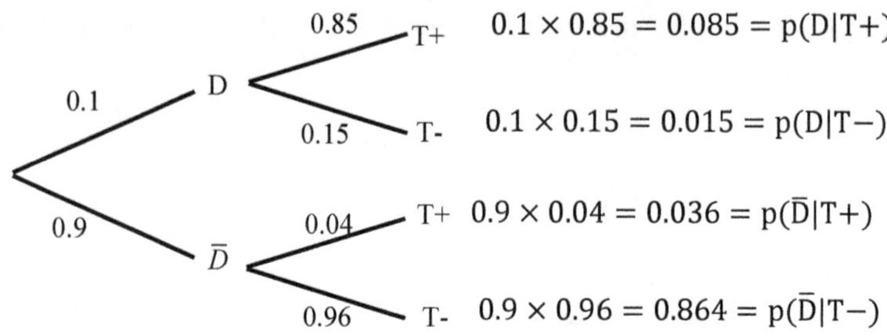

There are four paths in this chart. The sum of paths 1 and 3 gives the probability of positive test result, $p\ (T+)$, and the sum of paths 2 and 4 will the probability of negative test result $p\ (T-)$.

$$p(T+) = 0.085 \times 0.036 = 0.121,$$
$$p(T-) = 0.015 \times 0.864 = 0.879.$$

We now draw a tree path in the opposite direction.

According to this tree diagram and the conditional probability formula, the probability of being ill a person with positive test result is obtained by dividing the number 0.085 by 0.121:

$$X = p(D|T+) = \frac{0.085}{0.121} = 0.70.$$

Thus, if a person's test is positive, there is a 70% chance that the person is actually ill. The same value can be obtained by placing the above values in the Bayesian formula.

$$p(D|T+) = \frac{p(T+|D) \times p(D)}{p(T+|D) \times p(D) + p(T+|\overline{D}) \times p(\overline{D})}$$
$$= \frac{0.85 \times 0.10}{0.85 \times 0.10 + 0.04 \times 0.9} = 0.70.$$

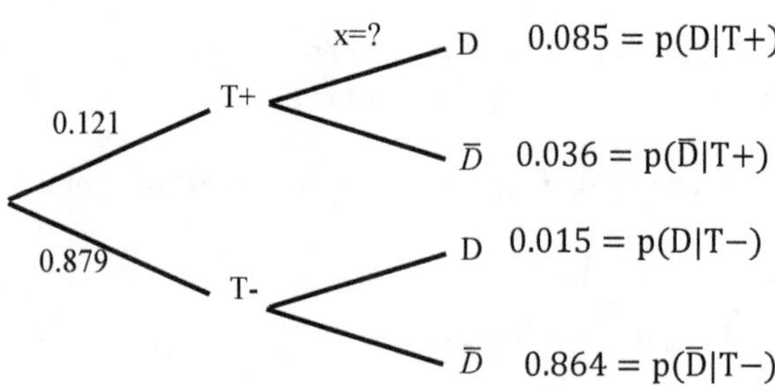

Example 9 There are 3 red and 2 green marbles in the first urn and 2 red and 8 green marbles in the second urn. Someone throws the dice. If the dice turns up multiple of 3 (3 and 6) sides, he chooses a marble from the urn I, otherwise a marble is chosen from the urn II (but he does not reveal whether it has turned up multiple 3 or no, that means we don't know which urn has chosen). If the selected marble is red, what is the probability that urn I was chosen?

Solution: If we denote R for event "red", I for "urn I", and II for "urn II", the following probability is given:

$$p(R|I) = \frac{3}{3+2} = 0.6, \quad p(R|II) = \frac{2}{8+2} = 0.2.$$
$$p(I) = \frac{2}{6} = \frac{1}{6} \quad \text{and} \quad p(II) = \frac{4}{6} = \frac{2}{3}.$$

Calculating $p(I|R)$ is the answer for the question, which is obtained as below.

$$
\begin{aligned}
p(I|R) &= \frac{p(I) \cdot p(R|I)}{p(I) \cdot p(R|I) + p(II) \cdot p(R|II)} \\
&= \frac{\left(\frac{1}{3}\right)(0.6)}{\left(\frac{1}{3}\right)(0.6) + \left(\frac{2}{3}\right)(0.2)} = 0.6.
\end{aligned}
$$

Example 10 An ambulance takes patients after inquiry, with a probability of 40% to hospital A, a probability of 25% to hospital B, and a probability of 35% to hospital C. The probability of infection of the patient in hospitals A, B, and C is equal to 0.6, 0.4, and 0.3, respectively. If your patient became infected after discharge from the hospital, what is the probability that the ambulance took him or her to hospital C?

Solution: We denote "I" for event of "infection" and A, B, and C for events of different hospitals A, B, and C:

$$p(A) = 0.40, \ p(B) = 0.25, \ p(C) = 0.35,$$
$$p(I|A) = 0.60, \ p(I|B) = 0.40, \ p(I|C) = 0.30.$$

Calculating $p(I|C)$ is the answer for the question.

$$
\begin{aligned}
p(I|C) &= \frac{p(I) \cdot p(C|I)}{p(I) \times p(A|I) + p(B) \times p(R|B) + p(I) \times p(C|I)} \\
&= \frac{0.35 \times 0.3}{(0.4 \times 0.6) + (0.25 \times 0.4) + (0.35 \times 0.3)} \\
&= \frac{0.105}{0.24 + 0.10 + 0.105} = \frac{0.105}{0.445} \\
&= 0.236.
\end{aligned}
$$

4.6 Bayes' Theorem and Epidemiology

Bayesian theorem has many applications in epidemiology and public health. For example, diagnostic and screening tests that vary along with the prevalence of diseases can be obtained using the Bayes' theorem. Suppose the following contingency table in which the columns represent the disease status and rows show the result of the diagnostic test.

		Disease status		
		Positive	*Negative*	*Total*
Test	*Positive*	$a = true\ positive$	$b = false\ positive$	$a + b$
	Negative	$c = false\ negative$	$d = true\ negative$	$c + d$
	Total	$a + c$	$b + d$	n

Tips

Sensitivity (true positive rate) is the proportion of diseased people who have been also diagnosed correctly positive by a diagnostic test.

Specificity (true negative rate) is the proportion of disease-free people who have been also diagnosed correctly negative by a diagnostic test.

Positive predictive value (PPV) is the ratio of patients whose diagnostic test results are also correctly diagnosed as positive.

Negative predictive value (NPV) is the ratio of patients whose diagnostic test results are also correctly diagnosed as negative.

By dividing true values of positive or true negative (a or d) to marginal total gives useful indices of sensitivity and specificity in epidemiology:

$$\text{Sensitivity} = \frac{a}{a + c} \text{ and } \text{Specificity} = \frac{d}{b + d}.$$

And dividing false values of positive or true negative (b or c) to marginal total gives useful indices in epidemiology:

$$\text{PPV (positive predictive value)} = \frac{a}{a+b},$$
$$\text{NPV (negative predictive value)} = \frac{d}{c+d}.$$

The proportion of the observed total disease ($a + c$) to the sample size of the study is named prevalence:

$$\text{Prevalence} = \frac{a+c}{n}.$$

The PPV and NPV give a direct assessment of the usefulness of the test in practice.

By employing Bayes' theorem, we can assess the predictive capability of the test across various population groups with different disease prevalences. This approach allows us to calculate the probability of disease given a positive test result, enabling us to understand how the test performs in different demographic subsets, such as distinct age groups or the overall population.

$$
\begin{aligned}
p(D+|T+) &= \frac{p(T+|D+) \times p(D+)}{p(T+)} \\
&= \frac{p(T+|D+) \times p(D+)}{p(T+|D+) \times p(D+) + p(T+|D-) \times p(D-)}.
\end{aligned}
$$

Conceptually we know that:

$$
\begin{aligned}
p(D+) &= \text{prevalence of disease,} \\
p(D+|T+) &= \text{PPV,} \\
p(D-|T-) &= \text{NPV,} \\
p(T+|D+) &= \text{sensitivity,} \\
p(T+|D-) &= 1 - \text{specificity.}
\end{aligned}
$$

During the application stage of a screening test in a target population, individual disease statuses are typically unknown, as the purpose of screening is to identify undetected cases. However, disease prevalence data can often be sourced from national agencies and health surveys. Predictive values are then computed based on this prevalence information.

$$
\text{PPV} = \frac{(\text{sensitivity}) \times (\text{prevalence})}{(\text{sensitivity}) \times (\text{prevalence}) + (1 - \text{specificity}) \times (1 - \text{prevalence})},
$$

$$
\text{NPV} = \frac{(\text{specificity}) \times (1 - \text{prevalence})}{(\text{specificity}) \times (1 - \text{prevalence}) + (1 - \text{sensitivity}) \times (\text{prevalence})}.
$$

Bayes' theorem enables the calculation of predictive values without requiring data from the application stage. Instead, all that's necessary are the disease prevalence (accessible from health ministries or other sources) and the sensitivity and specificity of the screening test.

Example 11 Suppose the prevalence of Covid-19 in population A is 1% and in population B is 0.5%, a diagnostic test (assume PCR) has 95% sensitivity and 99% specificity. Calculate and interpret PPV and NPV for the disease.

Solution: We know that specificity $= 99\%$, sensitivity $= 95\%$, $p(A) = 1\%$, and $p(B) = 0.5\%$.

For population A with 1% prevalence, we have:

$$\text{PPV} = \frac{(\text{sensitivity}) \times (\text{prevalence})}{(\text{sensitivity}) \times (\text{prevalence}) + (1 - \text{specificity}) \times (1 - \text{prevalence})}$$

$$= \frac{0.95 \times 0.01}{0.95 \times 0.01 + (1 - 0.99) \times (1 - 0.01)} = \frac{0.95 \times 0.01}{0.95 \times 0.01 + 0.01 \times 0.99}$$

$$= \frac{0.0095}{0.0095 + 0.0099} = 0.489.$$

Comment: In this population, about 48.9% of patients with positive PCR are correctly classified as diseased.

$$\text{NPV} = \frac{(\text{specificity}) \times (1 - \text{prevalence})}{(\text{specificity}) \times (1 - \text{prevalence}) + (1 - \text{sensitivity}) \times (\text{prevalence})}$$

$$= \frac{0.99 \times (1 - 0.01)}{0.99 \times (1 - 0.01) + (1 - 0.95) \times 0.01} = \frac{0.0099}{0.0099 + 0.0005}$$

$$= 0.952.$$

Comment: In this population, about 95.2% of patients with negative PCR are correctly classified as disease free.

For population B with 0.5% prevalence, we have:

$$\text{PPV} = \frac{(\text{sensitivity}) \times (\text{prevalence})}{(\text{sensitivity}) \times (\text{prevalence}) + (1 - \text{specificity}) \times (1 - \text{prevalence})}$$

$$= \frac{0.95 \times 0.005}{0.95 \times 0.005 + (1 - 0.99) \times (1 - 0.005)} = \frac{0.95 \times 0.005}{0.95 \times 0.005 + 0.01 \times 0.995}$$

$$= \frac{0.00475}{0.00475 + 0.00995} = 0.323.$$

Comment: In this population, about 32.3% of patients with positive PCR are correctly classified as diseased.

$$\text{NPV} = \frac{(\text{specificity}) \times (1 - \text{prevalence})}{(\text{specificity}) \times (1 - \text{prevalence}) + (1 - \text{sensitivity}) \times (\text{prevalence})}$$

$$= \frac{0.99 \times (1 - 0.005)}{0.99 \times (1 - 0.005) + (1 - 0.95) \times 0.005} = \frac{0.98505}{0.98505 + 0.00025}$$

$$= 0.999.$$

Comment: In this population, about 99.9% of patients with negative PCR are correctly classified as disease free.

Tips

The PPV and NPV describe the performance of a diagnostic test or other statistical measure.

They depend also on the prevalence of a disease.

A high result can be interpreted as indicating the accuracy of such a statistic.

4.7 Odds and Probability

Although the words odds and probability are closely related, they are not equivalent. These two terms are widely used in biostatistics and epidemiology and this section gives a brief description of the two terms. The term of odds is defined as the probability that the event will occur divided by the probability that the event will not occur. In other words, odds is the ratio of chance of "winning" a bet by the chance of "losing" that bet. We know that the probability of something not happening (or not winning) is equal to the "one minus the probability of happening or winning" of that thing. Therefore:

$$\text{Odds} = \frac{p(A)}{1 - p(A)}.$$

By simple algebraic operations in this equation, the probability of occurrence of the event is equal to:

$$p(A) = \frac{\text{Odds}}{1 + \text{Odds}}.$$

The relationship between odds and probability is summarized in the table below.

Probability	Odds	Interpretation
1.00	∞	The event will definitely happen
0.90	9.0	An event happens in 90% of cases (that is, it happens nine times versus not happening once)
0.75	3.0	The event occurs in 75% of cases (i.e., three times against one time)
0.667	2.0	An event occurs in two-thirds of cases (i.e., twice versus once)
0.50	1.0	The event happens in half of the cases (that is, it happens once versus not happening once)
0.333	0.5	The event happens in a third of cases (that is, it happens once versus not happening twice, it is half)
0.25	0.333	The event happens in 25% of cases (i.e., it happens once versus not happening three times, one third)
0.10	0.111	The event happens in 10% of cases (that is, it happens once versus nine times not happening, one ninth)
0.00	0	An event never happens

As can be seen from the above table, the probability varies between zero and one, while "odds" can take any non-negative value. For very low probabilities, the value of odds is very close to the probability value. But as the probability increases, so does the odds. So that when the probability reaches 0.5, the odds is equal to one. When the probability reaches 1, the value of odds tends to infinity. This definition of odds is also consistent with the dialog language. For example, if the hope of losing a bet is three to one (3:1), it means that your hope of "losing" in the bet is three times your hope of "winning". So you lose the bet with a probability of 0.75.

4.8 Counting Principle

The counting principle is a rule and very helpful to figure out the total number of possible outcomes (sample space) in a situation. For example, if there are m possible ways for an event to occur and n possible ways for another event to occur, there are m x n possible ways for both events to occur.

In other words, the counting principle is used to determine how many different ways one can choose/do certain events and is very useful in calculating the probability of an event. It is especially necessary to calculate the number of possible events and favorable events. These methods include factorial, permutation, and combination and are briefly described below.

Factorial

A factorial, noted by $n!$ with integer number of n (between 1 and infinite), is a product of consecutive counting numbers starting at 1 and ending at n.

$$n! = 1 \times 2 \times 3 \times \cdots \times (n - 2) \times (n - 1) \times n.$$

The number 0 is an exception. $0! = 1$ and $1! = 1$.
It can be easily said:

$$n! = n \times (n - 1)!.$$

It is very useful for when we're trying to count how many different orders there are for things or how many different ways we can combine things.

Example 12 How many different ways can we arrange 5 books in the shelf?

Solution: We have 5 choices for the first book. For second book, we are left with 5 $- 1 = 4$, so that we have 5×4 choices for the first two books, in order. Now, for the third book, we have $5-2 = 3$ choices giving us $5 \times 4 \times 3$ in order. And so on, until we get down to just one book remains, and then just one thing remaining. Altogether, we have $5 \times 4 \times 3 \times 2 \times 1 = 120$.

Permutation

The number of ways to select and arrange objects from among distinct objects is permutation and denoted by P_n^r. In permutation, order of objects matters. This means that ABC, ACB, BAC, BCA, CBA, and CAB are different.

$$P_n^r = \frac{n!}{(n-r)!}.$$

Example 13 In how many ways, we can pick 3 letters from 5 letters (A, B, C, D and E)?

Solution: We have 5 choices for the first letter. For the second letter, we are left with $5 - 1 = 4$, so that we have 5×4 choices for the first two letters, in order. For the third letter, we have $5 - 2 = 3$ choices giving us $5 \times 4 \times 3$ in order. Then, we have $5 \times 4 \times 3 = 60$.

$$P_n^r = \frac{n!}{(n-r)!} = \frac{5!}{(5-3)!} = 60.$$

Example 14 A coded lock is made from digits 0–9 (without repetition). We choose a three-digit code by chance. What is the probability that the lock will be open?

Solution: The number of possible outcomes (sample space) is equal to $n(S) = P_{10}^3 = 10 \times 9 \times 8 = 720$, and the number of favorable outcomes is equal to 1. So the probability of unlocking is $\frac{1}{720}$ in the first time.

Combination

Combination is the total number of ways of selecting r distinct combinations of n objects, irrespective of order (i.e., order NOT important). That is, AB is the same as BA and is not different. The number of combinations of r object and n object is obtained from the following formula:

$$C_n^r = \binom{n}{r} = \frac{n!}{r!(n-r)!}.$$

It can be easily shown that the following relations are always established.

$$C_n^0 = C_n^n = 1 \quad C_n^1 = n$$
$$C_n^r = C_n^{n-r} \quad C_{n-1}^r + C_{n-1}^{r-1} = C_n^{r+1}.$$

Example 15 A health center has 5 employees. With these 5 staffs, in how many ways we can create:

A. A queue of 5 including all 5 staffs?
B. A queue of 3 staffs?
C. A committee of 3 staffs.

Solution:

A. Forming a queue means that when people swap places, a new queue is formed. Therefore, factorial should be used.

$$5! = 5 \times 4 \times 3 \times 2 \times 1 = 120.$$

B. To form a queue of 3 out of 5 people, you must use the permutation of 5 members 3 to 3. Since in the queue, the order of people is matter:

$$P_5^3 = \frac{5!}{(5-3)!} = \frac{5 \times 4 \times 3 \times 2 \times 1}{2 \times 1} = 60.$$

C. In order to form a three-member committee, the arrangement of individuals is not important, i.e., the ABC is no different from the BCA and CAB. So we use combination and have:

$$\binom{5}{3} = \frac{5!}{3!(5-3)!} = \frac{5!}{2!3!} = 10.$$

Example 16 We have 4 books with different cover designs. With these 4 books, in how many ways we can:

Arrange them on the library shelf?

Gave two of them as a gift to someone?

Solution:

A factorial should be used to count the number of cases in which all four books can be arranged on the shelf. $4! = 4 \times 3 \times 2 \times 1 = 24$.

To donate two of the four available books, we need to use the permutations of 4 books 2 by 2, because the order of their placement creates a different donation. $P_4^2 = \frac{4!}{(4-2)!} = \frac{4 \times 3 \times 2 \times 1}{2 \times 1} = .$

Example 17 There are two populations, the first one with 7 patients and 18 healthy individuals and the second with 11 patients and 28 healthy individuals. One person randomly migrates from the first population to the second population and settles in that population. After that, we select another person from the second population. What is the probability that the selected person is ill?

Solution: We denote event A = immigrant person and event B = the illness of the selected person. The probability of being ill the immigrant person is $p(A) = \frac{7}{25}$ and being healthy is $(\overline{A}) = \frac{18}{25}$.

If the immigrant person is ill, the second population will have 12 patients and 28 healthy ones. As a result, the probability that the second person be ill is equal $p(B|A) = \frac{12}{40}$.

If the immigrant person is healthy, the second population will have 11 patients and 29 healthy. As a result, the probability that the second person be ill is equal $p(B|\bar{A}) = \frac{11}{40}$.

Therefore, the probability that the second person be ill is equal to:

$$p(B) = p(B|A) \times p(A) + p(B|\bar{A}) \times p(p(B|\bar{A}))$$

$$= \frac{12}{40} \times \frac{7}{25} + \frac{11}{40} \times \frac{18}{25} = 0.282.$$

Example 18 From a city where 60% of the population are literate, we randomly select 10 people. Find the following probabilities:

i.   Every 10 people be literate.
ii. The first 3 selected people only be literate.
iii. Only the first two and the last two be literate.

Solution: If we denote event $A =$ be literate, $P(A) = 0.6$ and:

i.   In this case, $x = 10$ and $p = 0.6$, then $p(x = 10) = (p)^x = 0.6^{10} = 0.006$.
     Since we want only the first three selected persons to be literate, which means that the other seven are illiterate.
ii. In this case, $p = 0.6$ and $x = 3$, then $p(x_1, x_2, x_3) = (p)^x \cdot (1 - p)^{n-x} = 0.6^3 \times (1 - 0.6)^7 = 0.00035$, because only the first three people are important to us, and the other 7 people are not important to be literacy or illiteracy.
iii. Because we want the first two and the last two to be literate, which means that having literacy and not having the rest of the six don't matter, and the probability of that six is equal to one. $p(x_1, x_2, x_9, x_{10}) = 0.6^2 \times 1^6 \times 0.6^2 = 0.6^4 = 0.1296$.

4.9 Expected Value

Expected value or mathematical expectation of a discrete random variable of X characterized by $E(x)$ is a generalization of the weighted average. When the variable X can take the values of $x_1 \cdot x_2 \cdot \cdots \cdot x_n$, respectively, with the corresponding probabilities of $p_1 \cdot p_2 \cdot \cdots \cdot p_n$ (sum of probabilities is 1), means that the probability of having x_1 is p_1, the probability of having x_2 is p_2 and .... the probability of having x_n is p_n. The following table can be drawn.

X-values	x_1	x_2	x_3	$\cdots$	x_n
Probability	p_1	p_2	p_3	$\cdots$	p_n

Then, the expected value of X is:

$$E(X) = p_1 x_1 + p_2 x_2 + p_3 x_3 + \cdots + p_n x_n = \sum_{1}^{n} p_i x_i.$$

There are two important properties for expected value:

The expected value is always between the smallest ($x_{\min}$) and the largest values ($x_{\max}$), i.e., $X_{\min} \le E(X) \le X_{\max}$.

For any constant values a and b, we have: $E(a + bX) = a + bE(X)$.

Insurance companies operate with expected value to estimate the cost of life insurance policy.

An insurance firm offers a 1-year term life insurance plan to a 50-year-old male customer. The customer pays a premium of \$300. In the event of his death within the year, the company will provide a payout of \$60,000 to his beneficiary. According to the U.S. Centers for Disease Control and Prevention, the likelihood of a 70-year-old man being alive 1 year later is 0.9715. Let's denote X as the profit earned by the insurance company. Determine the probability distribution and the expected value of the profit.

Example 19 An insurance company wants to figure out the break-even point without any expenses (premium money) for 50 years old for the next year. If he dies within one year, the company will pay \$60,000 to his beneficiary. Assuming that the probability of dying 50 years old is 0.005 and the probability of living for the next year is 0.995, figure out the break-even point.

Solution: A break-even point means that the total sales equal the total expenses. This means that the insurance company is bringing in the same amount of money that they need to cover all of their expenses. To figure out it, the mathematical expectation must be zero.

$$E(X) = 0,$$
$$p(\text{living}) \times \text{paid amount} + p(\text{death}) \times \text{premium money} = 0,$$
$$0.005 \times 60000 + 0.995 \times \text{premium money} = 0,$$
$$300 + 0.995 \times \text{premium money} = 0,$$
$$\text{premium money} = \tfrac{300}{0.995} = 301.507.$$

Then, the minimum expected premium without any extra expenses is 301.507.

$$p(\text{living}) \times \text{paid amount} + p(\text{death}) \times \text{premium money} = 0,$$
$$0.005 \times 60000 + 0.995 \times \text{premium money} = 0,$$
$$300 + 0.995 \times \text{premium money} = 0,$$
$$\text{premium money} = \tfrac{300}{0.995} = 301.507.$$

Example 20 Variable X is the cost of treatment for each patient in a hospital. In this hospital, six types of treatments A, B, C, D, E, and F are performed which cost 1.5, 4, 2.5, 7, 5.2, and 3.5 currencies, respectively. If patients refer to treatments A, B, C, D, E, and F, respectively, obtain the mathematical expectation of this variable (the cost per patient in this hospital).

Solution: We create the table below and write down in another row the multiplication of cost of any type of treatment by the probability of it. The sum of this row is equal to the mathematical expectation of the above variable $E(x) = 3.495$.

Treatment	A	B	C	D	E	F	Sum
X	1.5	4	2.5	7	2.5	3.5	–
P	0.25	0.18	0.20	0.10	0.15	0.12	1
x.p	0.375	0.72	0.50	0.70	0.78	0.42	3.495

4.10 Exercises

1. According to Table 4.1 of this chapter, find the probability that a 55 years old will die by the age of 60.
2. If a family has five children, find the probability that have:

   A. Two consecutive sons and 3 consecutive daughters.
   B. Four consecutive sons.

3. The class has 18 female students and 12 male students; 6 girls and 4 boys wear glasses. We randomly select 4 students from the attendance list. Find the probability that:

   A. All 4 students are wearing glasses.
   B. None of them wears glasses.

4. In the above class, 2 students are selected from the list, and find the conditional probability of:

   A. Both students are wearing glasses, if both of them are boys.
   B. Neither of them wears glasses, if both of them are girls.

Table 4.1 Distribution of blood type by gender and blood type

Blood type		Female	Male	Total
Known	A	15	2	17
	B	16	1	17
	O	1	2	3
	AB	27	0	27
	Total	59	5	64
Don't know		96	40	136
Grand total		155	45	200

5. With digits 0–9 how many figures can be formed:

   A. With 3 digits.
   B. With 3 digits without zero.
   C. With 3 digits and multiplicative of 5.
   D. With 3 digits but even.

6. In a survey of several GPs, 64 out of 200 people identified their blood type. If the number of visitors to a GP on a particular day is 7, find the probability that:

   A. Patients know their blood type one every other.
   B. Only the first three patients know their blood type.
   C. Only three patients know their blood type.

7. If in the above study the distribution of blood type is as shown in Table 4.1, calculate the probabilities below:

   A. A visitor knows his/her blood type.
   B. A female visitor knows her blood type.
   C. A male visitor has blood type of O.

8. Half of the population is over 50 years old. If the probability of developing hypertension in people over 50 is 35% and in people under 50 is 15% and a person from this population is chosen randomly, calculate the following probabilities:

   A. That person is over 50 year's old and developed hypertension.
   B. The person is less than 50 years old and does not have blood pressure.
   C. That person developed blood pressure.

9. The probability of recovery after a surgery operation is 60%. We randomly select 8 patients who have undergone surgery under the same condition. Find out the probability that:

   A. The first 3 patients have recovered and this method is ineffective for the other 5 patients.
   B. Only 3 patients recover.

10. There are 10 yellow apple seedlings and 5 red apple seedlings in a laboratory field. An unknowing person accidentally picked and planted three saplings. Find the probability that:

    A. All three seedlings are red apples.
    B. Maximum of two seedlings is yellow apples.
    C. At least one apple is yellow.

11. In a laboratory greenhouse, 640 of the 800 seeds sown germinated. We choose six seeds by accident. We select six seeds randomly. Find out the probability that:

    A. None of them has sprouted.
    B. All have sprouted.

12. A livestock vaccine with a probability of 0.80 has a positive effect. If we select 6 livestock that have been reacted in the same conditions, find out:

    A. The probability that this vaccine has had a positive effect on the first four livestock and is unaffected for the last two livestocks.
    B. Only affect 4 livestocks.

13. Assuming that the two events A and B are independent of each other, which of the following is independent of each other?

    A. A and $\overline{B}$.
    B. $\overline{A}$ and B.
    C. Both.
    D. A prosthetic factory produces 170 hand prostheses and 330 foot prostheses daily. We choose 30 prostheses randomly. Find out the probability that both prostheses are of the same type?

14. Two sports teams A and B play in a three-set match. The final winner of the match is the team that has won two sets. The history of the two teams shows that the two teams have faced each other 10 times, and team A has won 6 times and team B 4 times. Find out the following possibilities:

    A. The match will be finished in three sets.
    B. The match ends in two sets.

   To solve this exercise, a tree diagram with probability of 0.6 and 0.4 can be used.

15. Sixty % of patients with dental lesions prefer TENS to anesthetic. Among the 8 patients being treated, consider the following probabilities:

    A. The first three patients preferred the use of TENS and the rest preferred anesthetic.
    B. Three of these patients use TENS and the rest use anesthetic.

Notice: The rest of this chapter's exercises are suitable for medical equipment engineering students.

16. The probability of performing pneumatic, mercury, and gear rotations in a cobalt CO^{60} device is 0.40, 0.10, and 0.50, respectively. In the studies of this device, it has been found that 3, 4, and 2 percent of these rotations are faced with problems, respectively. Find out the probability that:

    A. A rotation that encounters difficulty, is it mercury type?
    B. Pneumatic rotation but without problems?

17. In a sample of 30 angiography devices, 9 devices have film changer. Four angiographic devices are selected randomly, find out the probability that:

    A. All 4 devices have film changer.
    B. None of them has a film changer.
    C. A maximum of 2 devices have a film changer.
    D. Have at least one film changer.

18. In one study, 70 signals out of 100 real and even signals in one book and 50 signals out of 120 signals in another book were detected without Fourier series coefficients. We now randomly select 15 signals from both books. What is the probability that these 15 signals have Fourier coefficients (if we know that the signals were selected from the second book)?

19. The probability of using Step Down model and Step Up model autotransformers by an engineer in a certain circuit is equal to 0.05 and 0.03. What is the probability that the engineer will use at least one of these models in his circuit structure?

20. Half of the diagnostic radiology devices available in a medical center have been used for more than 20 years. The probability of failure of devices with a lifespan of more than 20 years is equal to 0.35 and the probability of failure of devices with a lifespan of less than 20 years is equal to 0.15. If a device is selected at random, find the following possibilities:

    A. The device has been working for more than 20 years and is broken.
    B. Has worked for less than 20 years and is worked properly.
    C. Is broken.

21. Eighty % of medical devices produced in a factory have a certificate. We select 7 devices randomly. Find out the probability that the first 5 devices have certificates and the last 2 devices do not have certificates.

22. In an electronics lab, there are 2 analog multimeters and 3 digital multimeters. We choose a multimeter at random. Then add a multimeter opposite to what came out of the container. Then take another multimeter out of the container. What is the probability that the second multimeter is analog? If the second multimeter is analog, what is the probability that the first multimeter is digital?

23. In a specialized hospital, 640 out of every 800 patients suspected of cancer are imaging with PET system. Six patients will be selected to evaluate the test results. Find the probability that:

    A. No PET system has been used for any patient.
    B. All 6 patients used PET system.
    C. Only two patients used pet system.

24. The probability of a sustained response in orbit is equal to $\frac{1}{7}$. What is the probability that the number of sustained responses in the study of 5 of these circuits is more than 3?

25. Examining a number of continuous time intermittent signals, it was found that about 20% of these signals have dirikle conditions. Ten signals are selected randomly, what is the probability that exactly two signals have dirikle conditions?

26. There are 4 inductors and 11 capacitors in the circuit that we want to place these elements in parallel in the circuit. What is the probability that all four inductors will fall behind each other?

27. In the circuit, five inductors with values of 1, 2, 3, 4, and 5 H are connected in series. We choose two inductors at random and obtain their equivalents. What is the probability that the equivalent inductor size is greater than 5?

Chapter 5
Discrete Distributions

5.1 Introductions

The primary objective of probability models in any random experiment involves two fundamental components: the sample space and the assignment of probabilities to each elementary outcome. The method presented in the previous chapter for calculating the probability of a random outcome is not universally applicable to all "sample space" scenarios. Wider sample spaces necessitate more comprehensive approaches, and probability distributions serve as effective tools for such cases. These distributions are typically represented through tables, graphs, mathematical formulas, etc., encapsulating all possible outcomes of the random variable along with their assigned probabilities.

A random variable is a variable whose value is contingent upon chance, randomness, and the outcome of a random experiment. When a real number is assigned to each situation in a "sample space", the outcome is referred to as a random variable. Random variables are denoted by uppercase letters, such as X and Y, while observed values are indicated by lowercase letters, such as x and y.

Events unfold in specific patterns, and probability rules model the occurrence of these events, enabling the determination of probabilities across the entire sample space. These models, including binomial, geometric, hypergeometric, multinomial, Poisson distributions, as well as normal, t-student, Chi-squared, and others, provide a framework for understanding and predicting outcomes. This chapter focuses on the most essential discrete probability distributions, such as binomial, geometric, hypergeometric, multinomial, and Poisson distributions, presenting their key properties and applications.

© The Author(s), under exclusive license to Springer Nature Singapore Pte Ltd. 2024 95
S. H. Saneii and H. Doosti, *Practical Biostatistics for Medical and Health Sciences*,
https://doi.org/10.1007/978-981-97-3083-4_5

5.2   Binomial Distribution

To facilitate a better understanding of the binomial distribution, we begin with the concept of a Bernoulli trial, followed by an exploration of binomials raised to different powers.

Bernoulli trial

A Bernoulli trial, also known as a binomial trial, is a random experiment featuring precisely two possible outcomes—often framed as "success-failure" or "yes–no". In this trial, the probability of success remains constant with each iteration. Named after the seventeenth-century Swiss mathematician Jacob Bernoulli, these trials should be conducted independently. Let p represents the probability of success in a Bernoulli trial, and q denotes the probability of failure. Notably, the sum of the probability of success (p) and the probability of failure (q) is always equal to one, as success and failure are complementary and mutually exclusive. This relationship is expressed by the equation

$$p + q = 1.$$

Now, let's extend the concept to multiple trials. If we independently repeat the Bernoulli trial n times and denote the number of successes as X, where X is a discrete random variable ranging from 0 to n, we can calculate the probabilities associated with different outcomes. For instance, X can take values such as 0 (no success), 1 (one success), 2 (two successes), and so forth, up to n (success in all trials). While computing the probabilities for a small number of trials is straightforward, it becomes challenging and time consuming for a large number of experiments. In the following sections, we will delve into a general method for calculating these probabilities efficiently (Table 5.1).

In this table, the first column shows the number of trials from 1 to n and the first row shows the number of successes from 0 to the number of trials (n). For example, if the Bernoulli trial is repeated twice ($n = 2$), the variable X can have no success

Table 5.1   Corresponding probability for Bernoulli trial

No. of trials	No. of success							General form
	$X = 0$	$X = 1$	$X = 2$	$X = 3$	$X = 4$	$\cdots$	$X = n$	
$n = 1$	q	p	–	–	–	$\cdots$	–	$(p+q)^1$
$n = 2$	q^2	$p \cdot q$	p^2	–	–	$\cdots$	–	$(p+q)^2$
$n = 3$	q^3	$3pq^2$	$3p^2q$	p^3	–	$\cdots$	–	$(p+q)^3$
$n = 4$	q^4	$4pq^3$	$6p^2q^2$	$4p^3q$	p^4	$\cdots$	–	$(p+q)^4$
$\vdots$	$\vdots$	$\vdots$	$\vdots$	$\vdots$	$\vdots$	$\cdots$		$\vdots$
$n = n$	q^n	npq^{n-1}	$\cdots$	$\cdots$	$\cdots$	$\cdots$	p^n	$(p+q)^n$

Table 5.2 Pascal triangle and binomial expansion $(p+q)^n$

n	The coefficient of C												
1							1						
2						1		1					
3					1		2		1				
4				1		3		3		1			
5			1		4		6		4		1		
6		1		5		10		10		5		1	
$\vdots$	$\vdots$	$\vdots$	$\vdots$	$\vdots$	$\vdots$	$\vdots$	$\vdots$	$\vdots$	$\vdots$	$\vdots$	$\vdots$	$\vdots$	$\vdots$

$(X = 0)$ or one success $(X = 1)$ or two successes $(X = 2)$. Because Bernoulli's trials are independent, the probability of no success or two failures equals $qq = q^2$, the probability of one success and one failure equals $qp + pq = 2pq$, and the probability of two consecutive success equals $pp = p^2$.

Algebraically, the probability of the number of successes in n trials can be obtained by expanding the binomial coefficients of $(p+q)^n$.

$$(p+q)^1 = p+q,$$
$$(p+q)^2 = p^2 + 2pq + q^2,$$
$$(p+q)^3 = p^3 + 3p^2q + 3pq^2 + q^3,$$
$$(p+q)^4 = p^4 + 4p^3q + 6p^2q^2 + 4pq^3 + q^4.$$

The binomial expansion $(p+q)^n$ can be shown by the general form $Cp^x \cdot q^{n-x}$ that the coefficient C can be found from the "Pascal triangle" in Table 5.2. The initial combination of these coefficients was discovered by the Iranian mathematician Omar Khayyam.[1]

The general form of this expansion is as below:

$$(p+q)^n = \sum_{x=0}^{n} \binom{n}{x} p^x q^{(n-x)}.$$

That is the probability of x successes in n independent trials, denoted by $P_n(x)$ and called the binomial distribution, is equal to:

$$P_n(x) = C_n^x \times p^x q^{(n-x)} = \frac{n!}{x!(n-x)!} \times p^x \cdot q^{(n-x)}$$

The mean and variance of Binomial distribution are $\mu = np$ and $\sigma^2 = npq$ (Table 5.3).

---

[1] Kennedy, Evelyn (1966). "Omar Khayyam". The Mathematics Teacher. LIX(3):140–142. https://doi.org/10.5951/MT.59.2.0140. JSTOR27957296.

Table 5.3 Corresponding probability for Bernoulli experiment

$$P_n(x) = C_n^x \times p^x q^{(n-x)} = \underbrace{\frac{n!}{x!(n-x)!}}_{\substack{\text{The No. of events} \\ \text{with exactly } x \\ \text{successes in } n \text{ trials}}} \times \underbrace{p^x . q^{(n-x)}}_{\substack{\text{The probability} \\ \text{of } x \text{ successes} \\ \text{in } n \text{ trials}}}$$

Example 1 It is asserted that a new pain relief treatment effectively cures muscular pain in 70% of cases. When administered to 10 patients, determine the probability that: (a) No more than 3 patients will experience relief. (b) The number of patients cured falls between 6 and 8, inclusive. (c) At least 5 patients will be cured.

Solution: Suppose the patients are independent, the binomial distribution with $n = 10$ and $p = 0.7$ can applied for $X = $ number of patients who are cured. To compute the required probabilities, we consult the binomial table for $n = 10$ and $p = 0.70$.

(a)
$$
\begin{aligned}
p(X \leq 3) &= p(X = 0) + p(X = 1) \\
&\quad + p(X = 2) + p(X = 3) \\
&= C_{10}^0 p^0 q^{(10-0)} + C_{10}^1 p^1 q^{(10-1)} \\
&\quad + C_{10}^2 p^2 q^{(10-2)} + C_{10}^3 p^3 q^{(10-3)} \\
&= \frac{10!}{0!10!} 0.7^0 \times 0.3^{10} + \frac{10!}{1!9!} 0.7^1 \\
&\quad \times 0.3^9 + \frac{10!}{2!8!} 0.7^2 \times 0.3^8 \\
&\quad + \frac{10!}{3!7!} 0.7^3 \times 0.3^7 = 0.0000059 \\
&\quad + 0.000138 + 0.00116 + 0.0090 = 0.103,
\end{aligned}
$$

(b)
$$
\begin{aligned}
p(6 \leq X \leq 8) &= p(X = 6) + p(X = 7) + p(X = 8) \\
&= \frac{10!}{6!4!} 0.7^6 \times 0.3^4 + \frac{10!}{7!3!} 0.7^7 \times 0.3^3 \\
&\quad + \frac{10!}{8!2!} 0.7^8 \times 0.3^2 = 0.2001 + 0.2668 \\
&\quad + 0.2335 = 0.7004
\end{aligned}
$$

(c) $p(X \geq 4) = 1 - p(X \leq 3) = 1 - 0.103 = 0.897.$

Example 2 Thirty percent of university students have poor vision. We select 4 students randomly. Find the probability that:

(a) All 4 students have vision problem.
(b) No one has visual impairment.
(c) At least one person has visual impairment.

Solution: In this example $n = 4$, $p = 0.3$, and $q = 1 - p = 1 - 0.3 = 0.7$. So:

(a) $p(x = 3) = C_4^4 p^4 q^{(4-4)} = \frac{4!}{4!0!} 0.3^4 \times 0.7^0 = 0.0081.$
(b) $p(x = 0) = C_4^0 p^0 q^{(4-0)} = \frac{4!}{0!4!} 0.3^0 \times 0.7^4 = 0.2401.$
(c) $p(x \geq 1) = 1 - p(x = 0) = 1 - 0.2401 = 0.7599.$

Example 3 20% of digital blood pressure monitors do not have the required accuracy and are returned to the factory within 1-year warranty. If 6 units have been sold in a given week, what is the probability that in the 1-year warranty period?

(a) None of those devices is corrupted.
(b) Fail at least one of the devices.
(c) Four digital blood pressures' monitor breaks down and does not work properly.

Solution: Assuming the known values ($n = 0.6$, $p = 0.20$, and $q = 1 - p = 1 - 0.2 = 0.8$) in this example, we can apply the binomial distribution to calculate the probabilities.

(a) $p(x = 0) = p_6(0) = \binom{6}{0}(0.2)^0 (0.8)^6 = 0.262.$

$$p(x \geq 1) = p(x = 1) + p(x = 2) + p(x = 3)$$
$$+ p(x = 4) + p(x = 5) + p(x = 6)$$

(b)
$$= 1 - p(x = 0) = 1 - \binom{6}{0}(0.2)^0 (0.8)^6.$$

$$= 1 - 0.262 = 0.738.$$

(c) $p(x = 4) = p_6(4) = \binom{6}{4}(0.2)^4 (0.8)^2 = 0.001024.$

The binomial distribution is used for discrete and countable variables and has two parameters n and p. The shape of the binomial distribution depends on the values of n and p. The closer the values of q and p are to each other, the more symmetrical the distribution of the binomials becomes. So that if $q = p$, this distribution will be symmetric for all values of n. The farther the values q and p are from each other, the more skewed the distribution. As the value of n increases, the skewness of the distribution decreases, so much so that if n is large enough, the binomial distribution tends to return to the normal distribution.

It should be noted that the use of binomial distribution in infinite populations is desirable when the probability of an event can be assumed to be constant. But in finite populations, the replacement sampling method should be used. That is, each sample is returned to its place immediately and then the next sample is selected. But it often happens that this distribution is used in finite population without replacement which is problematic and disrupts the independence assumption of repeated experiments, unless the ratio of population size to sample size ($\frac{N}{n}$) is at least 10 (Fig. 5.1).

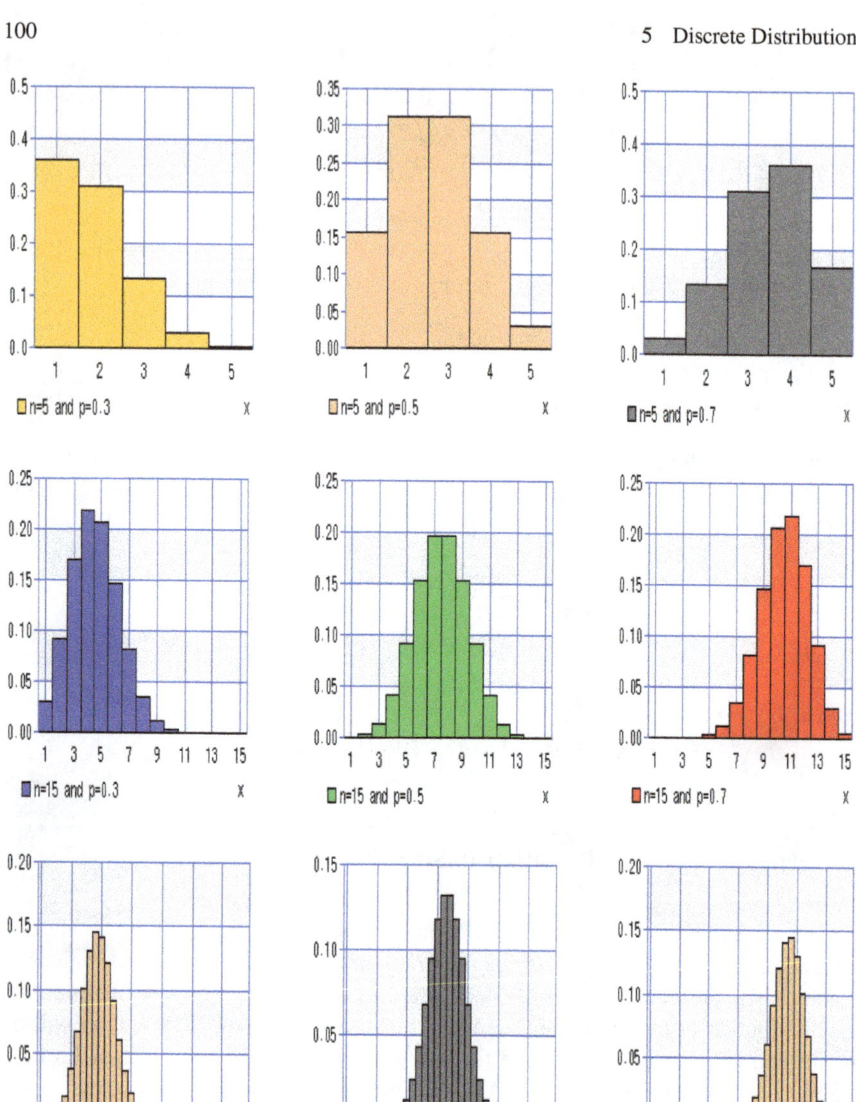

Fig. 5.1 Different binomial distributions by *p* and *n* values

5.3   Multinomial Distribution

As mentioned, the binomial distribution deals with variables that their outcome is only in two categories (yes and no or male and female), which is called binomial variable (Bernoulli trial). Now if the outcome of variable X involves 3 or more mutually exclusive types or categories or events $(E_1 \cdot E_2 \cdots \cdot E_k)$ with corresponding probability of occurring $(p_1 \cdot p_2 \cdots p_k)$, that $p_1 + p_2 + p_3 + \cdots + p_k = 1$, it named multinomial variable and we employ multinomial distribution instead of binomial distribution. For example, blood types can occur in types of A, B, AB, or O. The multinomial distribution is a generalization of the *binomial distribution*. If we repeat an experiment n times independently and each time one of the k disjoint events occurs, the probability that X will occur $(x_1 \cdot x_2 \cdots x_k)$ times is obtained from the following formula:

$$p(x_1 \cdot x_2 \cdots x_k; n \cdot p_1 \cdot p_2 \cdots p_k) = \frac{n!}{(x_1)!(x_2)! \ldots (x_k)!} p_1^{x_1} \times p_2^{x_2} \times \cdots \times p_k^{x_k},$$

where $x_1 + x_2 + \cdots + x_k = n$ and $p_1 + p_2 + \cdots + p_k = 1$.

Example 4 Consider drawing one card randomly from a standard deck of playing cards, replacing it in the deck, and repeating this process 5 times. Determine the probability of drawing 2 face cards, 1 ace card, and 2 numeric cards.

Solution: We apply the multinomial distribution formula to solve the example. The known information is below:

The experiment consists of 5 trials, so n $= 5$.

The 5 trials produce 2 face cards, 1 ace card, and 2 numeric cards; so $x_1 = 2$, $x_2 = 2$, and $x_3 = 1$.

On any particular trial, the probability of drawing a face, ace, and numeric cards is $p_1 = \frac{12}{52} = 0.231 \cdot p_2 = \frac{4}{52} = 0.077$ and $p_3 = \frac{36}{52} = 0.692$, respectively.

We substitute this information into the multinomial distribution:

$$p(x_1 \cdot x_2 \cdot x_3; n \cdot p_1 \cdot p_2 \cdot p_3) = \frac{n!}{x_1! \times x_2! \times x_3!} p_1^{x_1} \times p_2^{x_2} \times p_3^{x_3}$$

$$= \frac{5!}{2! \times 1! \times 2!} \times 0.231^2 \times 0.077^1$$

$$\times 0.692^2 = 0.059.$$

Example 5 In an urn, there are 4 white balls, 3 red balls, and 3 green balls. A ball is randomly selected, and its color is noted before being replaced each time. Determine the probability that, after selecting 5 balls, 2 are white, 2 are red, and 1 is green.

Solution: We apply the multinomial distribution to solve it. The known information is:

The experiment consists of 5 trials, so $n = 5$.

The 5 trials produce 2 white balls, 2 red balls, and 1 green ball, so $x_1 = 2$ $x_2 = 2$ and $x_3 = 1$.

On any particular trial, the probability of drawing a white, red, and green is $p_1 = \frac{4}{10} = 0.3$, $p_2 = \frac{3}{10} = 0.3$ and $p_3 = \frac{3}{10} = 0.3$, respectively.

We replace this information into the multinomial distribution:

$$p(x_1 \cdot x_2 \cdot x_3; n \cdot p_1 \cdot p_2 \cdot p_3) = \frac{n!}{x_1! \times x_2! \times x_3!} p_1^{x_1} \times p_2^{x_2}$$

$$\times p_3^{x_3} = \frac{5!}{2! \times 2! \times 1!} \times 0.4^2 \times 0.3^2$$

$$\times 0.3^1 = 0.1296.$$

Example 6 From a *standard deck of 52 playing cards*, we draw 9 cards. Find the probability that there are 2 ace, 3 face, and 4 numeric cards.

Solution: When the selection of cards from a playing deck card is with replacement, the probability of occurrence is equal each time. In this case: the probability of coming an ace equal to $p(A) = \frac{4}{52} = \frac{1}{13}$, the probability of coming a face card equal to $p(F) = \frac{12}{52} = \frac{4}{13}$, and the probability of coming a numeric card equal to $p(N) = \frac{36}{52} = \frac{9}{13}$. Also $x_1 + x_2 + x_3 = 2 + 3 + 4 = 9$ and $p(A) + p(F) + p(N) = \frac{1}{13} + \frac{4}{13} + \frac{9}{13} = 1$.

$$p(2, 3, 4) = \frac{n!}{(x_1)!(x_2)!(x_3)!} \times p_1^{x_1} \cdot p_2^{x_2} \cdot p_3^{x_3}$$

$$= \frac{9!}{2!3!4!} \times \left(\frac{1}{13}\right)^2 \times \left(\frac{4}{13}\right)^3 \times \left(\frac{9}{13}\right)^4$$

$$= 0.499.$$

Example 7 The distribution of blood types in population is assumed as the table below.

Blood type	Percentage	+RH	−RH
O	44	37.4	6.6
A	42	35.7	6.3
B	10	8.5	1.5
AB	4	3.4	0.6

At a blood donation center, 12 people donate blood daily. What is the probability that an equal number (3 individuals) of all four blood types will be donated on a given day?

Solution: In this example, $p(O) = 0.44$, $p(A) = 0.42$, $p(B) = 0.10$ and $p(AB) = 0.04$, $x_1 = x_2 = x_3 = x_4 = 3$, $(O) + p(A) + p(B) + p(AB) = 0.44 + 0.42 + 0.10 + 0.04 = 1$, $x_1 + x_2 + x_3 + x_4 = 3 + 3 + 3 + 3 = 12$.

$$p(3, 3, 3, 3) = \frac{n!}{(x_1)!(x_2)!(x_3)!(x_4)!} \times p_1^{x_1} \cdot p_2^{x_2} \cdot p_3^{x_3} \cdot p_4^{x_4}$$

$$= \frac{12!}{3!3!3!3!} \times (0.44)^3 \times (0.42)^3 \times (0.10)^3 \times (0.04)^3$$

$$= 0.00015.$$

5.4 The Limit of the Binomial Distribution

As mentioned, the binomial distribution has two parameters n and p. By changing (larger or smaller) these two parameters, it is sometimes difficult to calculate the probability. For example, by increasing n or decreasing the value of p too much, this calculation is difficult, impossible, time consuming, and approximate. To better understand this difficulty, pay attention to the following two examples.

Example 8 One of three patients with headache is treated with acupuncture. Find out the probability of recovery x out of 200 patients with acupuncture method.

Solution: In this example, $n = 200$, $p = \frac{1}{3}$ and $q = \frac{2}{3}$ and $x = x$. Using the binomial distribution, calculations $p_{200}(x) = \binom{200}{x}\left(\frac{1}{3}\right)^x\left(\frac{2}{3}\right)^{200-x}$ must be performed.

It is difficult to calculate this expression for any x-value. If $x = 0$, calculating the expression $\left(\frac{2}{3}\right)^{200}$ is difficult and impossible. If $x = 100$, it is also difficult to calculate the expression $\left(\frac{2}{3}\right)^{100}\left(\frac{1}{3}\right)^{100}$. Finally, if $x = 200$ is obtained, it is practically impossible to compute the expression $\left(\frac{1}{3}\right)^{200}$.

Example 9 According to a study in Kermanshah province, the prevalence of pulmonary tuberculosis in urban women is 0.002. We select a sample of 150 urban women in this province. Find the probability that x people have TB.

Solution: In this example, $n = 150$, $p = 0.002$, $q = 0.998$, and $x = x$.

$$p_{150}(x) = \binom{150}{x}(0.002)^x(0.998)^{150-x}.$$

The calculation of this probability is also difficult and very approximate.

It is mathematically proven that in some cases of binomial distribution, other distributions such as the normal distribution and the Poisson distribution can be used instead.

The calculation of these expressions (Examples 6 and 7) is very difficult due to the small p or large n and the answers are very approximate and other methods must be employed.

It is mathematically proved that in some cases, other distributions such as the normal distribution and the Poisson distribution can be used instead of the binomial distribution.

Also, when experiments are performed in a binomial distribution without replacement, i.e., the probability of success changes in each test, the Hypergeometric distribution is used.

In addition, if the number of experiments in the binomial distribution is not fixed and the experiments continue until the first success is achieved, the geometric distribution should be used.

> **Tip**
> Geometric distribution means that $(k - 1)$ unsuccessful (failure) trials must occur in order to achieve the first trial to be successful.

5.5   Geometric Distribution

Geometric distribution is used for discrete variables and is a process that meets all the conditions of binomial distribution; only the number of trials is not fixed. This means that we repeat the Bernoulli trial with probability of p on any one trial, until the first success occurs on the kth trial. Therefore, the random variable of x is the number of Bernoulli trial and must be the result of the first $(k - 1)$ trials be failure and only the kth trial be a success. But in a binomial distribution, the number of trials is predetermined and the random variable X is the number of successes. The probability of geometric distribution is obtained from the following formula.

$$p(x = k) = (1 - p)^{k-1} \times p.$$

The expected value (mean) and variance of the number of independent trials to achieve the first success are equal to:

$$E(x) = \frac{1}{p} \quad \text{and} \quad \text{Var}(x) = \frac{1-p}{p^2}.$$

The mean and variance of the number of failures before the first success are also equal to:

$$E(X) = \frac{1-p}{p} \quad \text{and} \quad \sigma^2 = V(X) = \frac{1-p}{p^2}.$$

The geometric distribution is also called the "waiting time" distribution. Because if we assume each Bernoulli trial as a unit of time, the waiting time for the first success is exactly equal to the random variable x, which has a geometric distribution. Geometric distribution is used in time management to reach the success before a certain time period. It is also useful for studying a rare phenomenon in population, such as the presence of a rare blood disorder.

Example 10 A pharmaceutical company makes syrups such that 98% of them are not defective. The produced syrups are selected at random and tested.

(a)   What is the probability that the second selected syrup is the first to be non-defective?

(b) What is the probability that the first non-defective syrup is randomly selected after the 10th selection?

Solution: Let "a non-defective syrup" be a "success" with $p = 98\% = 0.98$.

The second selected syrup is the first to be non-defective, which means that the first syrup is defective and the second syrup is non-defective. Therefore:

$$p(x = 2)(1 - 0.98)^{2-1} \times (0.98) = 0.0196.$$

The probability that the first non-defective syrup is randomly selected after the 10th selection means that the first to tenth selections must be defective. Therefore, after the 10th selection means: $P(x > 10)$. Substitute n by 10 and p by 0.98 in the formula $P(x > n) = (1 - p)^n$.

$$P(x > 10) = (1 - 0.98)^{10} = 10^{-20} \approx 0.$$

Tip

If we wait for predetermined number x of successes in geometric distribution rather than waiting only one success, the Negative Binomial distribution, which is generalized of geometric distribution, is used.

Example 11 A patient is waiting for a suitable matching kidney donor for a transplant. The probability that a randomly selected donor is a suitable match is $p = 0.125$. What is the expected number of donors who will be tested before a matching donor is found?

Solution: With $p = 0.125$, the mean number of failures before the first success is $E(Y) = \frac{1-p}{p} = \frac{1-0.125}{0.125} = 7.0$.

For the alternative formulation, where X is the number of trials up to and including the first success, the expected value is $E(X) = \frac{1}{p} = \frac{1}{0.125} = 8$.

Example 12 The probability of encountering a universal blood donor (with blood type O and Rh negative) is 0.05. Calculate the probability that the fifth person chosen will be the first to qualify as a universal blood donor.

Solution: In this example, $p = 0.05$. In order for the fifth blood donor to be the first person with universal donor, it is equal to

$$p(x = 5) = (1 - 0.05)^{5-1} \times 0.05 = 0.0407.$$

5.6 Hypergeometric Distribution

The hypergeometric distribution is one of the most widely used distributions and is very similar to binomial distribution. In the binomial distribution, the events were independent and its necessity was the infinity of population or sampling with replacement. However, in the hypergeometric distribution, the trials are not independent of each other.

If the sample is selected from a small finite population without replacement, the trials can no longer be considered independent of each other, the probability of selecting subjects varies from time to time, and the binomial distribution will not give exact probabilities. The smaller the population, the less accurate the calculations. Therefore, if the outcomes occur in one of two types and the selection is without replacement, the hypergeometric distribution should be used instead of the binomial distribution. Obviously, in the case of without-replacement selection, the larger the population size, the closer the accuracy of the binomial distribution will be to the hypergeometric distribution. Consider a population that has A objects of one type (such as female), while the remaining B objects are of the other type (such as male). If n samples are selected without replacement that x objects are from type A and $n - x$ other objects are type B, Table 5.4 can be considered:

In this case, the probability of getting x objects of type A and $n - x$ objects of type B is obtained from the following formula.

$$p(x) = \frac{\binom{A}{x}\binom{B}{n-x}}{\binom{A+B}{n}} = \frac{\frac{A!}{(A-x)!\times x!} \times \frac{B!}{(B-n+x)!\times(n-x)!}}{\frac{(A+B)!}{(A+B-n)!\times n!}}.$$

The easiest way to distinguish a hypergeometric distribution with a binomial distribution is how to select samples which are selected without replacement in a hypergeometric distribution. In the binomial distribution, trials must be independent of each other. That is, the probability of each selection is equal to the other selections.

Example 13 A medical diagnostic laboratory performs 10 tests a day, which 3 cases are suspected to Covid-19. If 4 samples are randomly selected from this laboratory, what is the probability that 2 cases are Covid-19?

Table 5.4 Objects of population and sample in terms of two types

	Type A	Type B	Total
Population	A	B	$A + B$
Sample	x	$n - x$	n

Solution: Since the sample selection is without replacement, the hypergeometric distribution is used. $N = 10$, $A = 3$, $B = 7$, $n = 4$, $x = 2$, and $n - x = 2$.

$$p(x = 2) = \frac{\frac{A!}{(A-x)!\times x!} \times \frac{B!}{(B-n+x)!\times(n-x)!}}{\frac{(A+B)!}{(A+B-n)!\times n!}}$$

$$= \frac{\frac{3!}{(3-2)!\times 2!} \times \frac{7!}{(7-4+2)!\times(4-2)!}}{\frac{(7+3)!}{(7+3-4)!\times 4!}}$$

$$= \frac{3 \times 21}{210} = 0.3.$$

Example 14 In a medical diagnostic laboratory, 12 tests are performed daily, 5 of which are not covered by any insurance. If we choose 3 of these tests randomly, what is the probability that 2 of them are not insured?

Solution: In such cases where population is finite and we want to identify people with a characteristic, the selection of people will be without replacement. Therefore, a hypergeometric distribution can be used. In this example $A = 5$, $B = 7$, $A + B = 12$, $n = 4$, and $x = 3$.

$$p(x) = \frac{\binom{A}{x} \cdot \binom{B}{n-x}}{\binom{A+B}{n}}$$

$$= \left(\frac{A!}{(A - x)!.x!} \times \frac{B!}{(B - n + x)!(n - x)!} \right) \div \frac{(A + B)!}{(A + B - n)!n!}$$

$$= \left(\frac{5!}{(5 - 3)!.3!} \times \frac{7!}{(7 - 4 + 3)!(7 - 3)!} \right) \div \frac{12!}{(12 - 4)!4!}$$

$$= \left(\frac{5!}{2!.3!} \times \frac{7!}{6!1!} \right) \div \frac{12!}{8!4!} = (10 \times 7) \div 495 = 0.1414.$$

Example 15 The classroom consists of 7 female students and 3 male students. Find the probability that the council will be consisted of two women and one man.

Solution: In the election of the council, individuals cannot be elected by replacement. Therefore, a hypergeometric distribution can be used. In this example, $A = 7$, $B = 3$, $A + B = 10$, $n = 3$, and $x = 2$.

$$p(x) = \frac{\binom{A}{x}\cdot\binom{B}{n-x}}{\binom{A+B}{n}}$$

$$= \left(\frac{A!}{(A-x)!.x!} \times \frac{B!}{(B-n+x)!(n-x)!}\right) \div \frac{(A+B)!}{(A+B-n)!n!}$$

$$= \left(\frac{7!}{(7-2)!.2!} \times \frac{3!}{(3-3+2)!(3-2)!}\right) \div \frac{10!}{(10-3)!3!}$$

$$= \left(\frac{7!}{5!.2!} \times \frac{3!}{2!1!}\right) \div \frac{10!}{7!3!} = (21 \times 3) \div 120 = 0.525.$$

5.7   Poisson Distribution

The Poisson distribution is a limiting form of **binomial distribution**. In the **binomial** variable x *counts* the number of successes from n independent **Bernoulli** trials. If n becomes extremely large and the probability of success becomes extremely small, applying the binomial distribution is not accurate and it is recommended to employ the Poisson distribution. Therefore, when rare events occur randomly in time or space, the number of events observed in a continuous interval is defined as the variable of interest. The interval may have the nature of time or refer to the length, or to an area or to volume and space.

If n is large and p is small, so that $np < 5$ (and cautiously less than 10), the probability of event occurrence is calculated by the Poisson distribution as follows:

$$p(x) = e^{-\lambda}\frac{\lambda^x}{x!},$$

where e is a constant value of Napierian approximately equals 2.71828 and $\lambda = n \cdot p$ and x is an integer figure of 0, 1, 2, ...n.

Matsunawa provides accuracy thresholds for the Poisson–binomial approximation. A general guideline indicates that the Poisson approximation to the binomial is reliable when n is at least 20 and p is less than or equal to 0.05. Moreover, it is highly accurate when n is at least 100 and p is less than or equal to 0.10.

Tip

A rule of thumb is that the Poisson approximation to the binomial is good, if $n \geq 20$ and $p \leq 0.05$, and very good if $n \geq 100$ and $p \leq 0.10$.

This distribution has only one non-negative parameter, because its two components, n and p, are both non-negative. The shape of the distribution depends on the quantity of λ. The mean and variance of this distribution are also equal. Because in Poisson distribution p is very small and therefore $q = 1 - p$ is very close to 1, then n.p.q approximately equals to n.p.

$$\mu = \sigma^2 = \lambda = n \cdot p.$$

The Poisson distribution is used in many areas of medical (particularly, in laboratory, toxicology, bacteriology, and epidemiology) research, which we deal with the expansion of a substance per unit of area, or particles per unit of volume and space, or with the occurrence of accidents per unit of time and generally with a rare event in a very large population.

Examples of rare events include the number of daily referrals from a fire to the emergency department of a hospital, the number of white blood cells in a drop of blood, the number of floating and suspended particles or the spread of bacteria in a liquid, the mortality rate of pregnant mothers, etc.

In this distribution, the probability of an event occurring at a given interval must be proportional to the length of that interval. This means that the distribution of events at different interval must be the similar. That is, they are accidentally spread everywhere. Therefore, if, for example, particles are precipitated in a liquid, the Poisson distribution cannot be assumed for them.

Example 16 The University of Medical Sciences has 1095 students. If the probability of birth is the same on different days of the year, find the probability of:

Four students were born today.

Less than 4 students were born today.

More than 4 students were born today.

Solution: Provided information is $n = 1095$ and $p = \frac{1}{365} = 0.0027$, $\lambda = np = 1095 \times \frac{1}{365} = 3$. $\lambda = 3$ means that we are expected to have 3 births a day in the medical university. Using Poisson distribution:

For $x = 4$: $p(x) = e^{-\lambda} \frac{\lambda^x}{x!} = 2.718^{-3} \frac{4^4}{4!} = 0.168$.

$X < 4$ means that x can take values of 0, 1, 2, or 3.

$$p(x < 4) = p(x = 0) + p(x = 1) + p(x = 2) + p(x = 3)$$

$$= 2.718^{-3} \frac{3^0}{0!} + 2.718^{-3} \frac{3^1}{1!} + 2.718^{-3} \frac{3^2}{2!} + 2.718^{-3} \frac{3^3}{3!}$$

$$= 0.0498 + 0.1494 + 0.2240 + 0.2240 = 0.6472.$$

$x > 4$ means that x can take the values 5, 6... and 1095. Calculating the probability for the above series is too long and time consuming, so using its complementary probability we have:

$$p(x > 4) = 1 - p(x \leq 4)$$
$$= 1 - \left[p(x < 4) + p(x = 4) \right]$$
$$= 1 - [0.6472 + 0.1680] = 0.1848.$$

Example 17 Two percent of MRIs taken from patients are not of good quality and the patient must be reimaged. If the center has 150 weekly visits, find the probability of all MRIs which are taken well.

Solution: Provided information is $n = 150$ and $p = 0.02$ and $\lambda = np = 150 \times 0.02 = 3$. $\lambda = 3$ means that we are expected to have 3 reimaging per week. When it is said that all MRIs are done correctly, it means that none of them are damaged ($x = 0$). Using the Poisson distribution, we have:

$$p(x) = e^{-\lambda} \frac{\lambda^x}{x!} = 2.718^{-3} \frac{3^0}{0!} = 0.0497.$$

Example 18 An average of 60 calls is made to an emergency center every hour. What is the probability that the center will not be called at all in 4 min?

Solution: Provided information is $n = 240$ and $p = \frac{1}{60}$, $\lambda = np = 240 \times \frac{1}{60} = 4$. $\lambda = 4$ means that we are expected to have 3 calls in 4 min (240 s). For ($x = 0$) using the Poisson distribution, we have:

$$p(x) = e^{-\lambda} \frac{\lambda^x}{x!} = 2.718^{-4} \frac{4^0}{0!} = 0.018.$$

Example 19 There are two maternity hospitals in a city. The daily capacity of the larger maternity hospital is 10 people, but at present it receives an average of 4 people per day. The city's health authorities plan to close the smaller maternity hospital, which has in average of two daily visits. Obviously, its clients will inevitably go to another maternity hospital. Rising demand for larger maternity hospital (more than a daily 10 women) has worried officials. Assuming that the Poisson distribution can be used for the number of daily maternity admissions with an average of 6, it is desirable to estimate the number of days of the year when patient admissions exceed the nominal capacity (10 people) of the maternity hospital (Table 5.5).

Solution: Officials should be worried if the daily maternity hospital exceeds 10 deliveries (i.e., $x = 11, 12, 13, \ldots$). Calculating the probability for all these x's is difficult, long and even inaccurate, because the value of x is unlimited and infinite. Therefore, we obtain the probabilities of 10 deliveries and less ($x = 0, 2, 3, \ldots 10$) and subtract the sum from 1 to obtain the probability of having more admissions.

Assuming $\lambda = 6.1$, these calculations are obtained, for example, for the probability of no daily delivery and the probability of daily admission of 3 patients as follows:

Table 5.5 Probabilities

Daily delivery	Probability
0	0.0022
1	0.0137
2	0.0417
3	0.0848
4	0.1294
5	0.1579
6	0.1605
7	0.1399
8	0.1066
9	0.0723
10	0.0440
Total	0.9530

$$p(x) = e^{-\lambda}\frac{\lambda^x}{x!} = 2.718^{-6.1}\frac{6.1^0}{0!} = 0.0022,$$

$$p(x) = e^{-\lambda}\frac{\lambda^x}{x!} = 2.718^{-6.1}\frac{6.1^3}{3!} = 0.0848.$$

The probabilities obtained for all x's are summarized in Table 6.5.

$$p(x \leq 10) = 0.953,$$
$$p(x > 10) = 1 - p(x \leq 10) = 1 - 0.953 = 0.047.$$

The probability that the maternity capacity will exceed 10 deliveries per day is equal to 0.047 or (4.7%) 17 days per year ($0.047 \times 365 = 17.16$) in excess of the capacity.

After ensuring that n is large and p is small, one way to investigate the use of Poisson is to calculate the mean and variance of the sample distribution. If these two measures are close together, the Poisson distribution can be used.

5.8 Exercises

1. From a radioactive object, an average of 15 electrons per minute (with Poisson distribution) is thrown. Calculate the probability of being thrown in a given second is 2 electrons?
2. Below a microscope field, the number of cells in 500 cells is as shown in Table 5.6:

Table 5.6 Number of cells in a microscope field cells in 500 cells

No. of cells	0	1	2	3	4	5	6
Freq.	134	182	110	58	9	4	3

    A. Can the distribution of this data be considered a Poisson distribution?

    B. Assuming that the data distribution is Poisson, find the theoretical distribution in the table above.

3. The probability of death after a certain surgery in a given period of time is equal to 0.2. If we have 6 patients who have had surgery, find the following probability:

    A. None of these patients die.

    B. A maximum of 2 patients die.

    C. Maximum of 2 patients survive.

    D. Exactly one patient dies.

4. Among angiography devices that need repair, 7 devices have cardiac monitoring systems. We selected a sample of 20 of these devices. Find the probability that 5 devices have a cardiac monitoring system.

5. The number of white blood cells in healthy people is 6000 per cubic millimeter of blood. We take a drop of blood with a volume of 0.001 cubic millimeter of sample blood. How many white blood cells are expected to be present? Find the probability that there are a maximum of 2 white blood cells in this drop of blood?

6. Each series of ointment production in the pharmaceutical factory contains 25,000 tubes. According to previous inspections, for every 9000 tubes of product, 2 tubes are filled with air and defective. Find the probability that:

    A. In every production series of this factory, all the tubes are complete.

    B. Only 2 tubes are defective.

7. A student in the exam session is not ready for the exam at all and has to mark the answers by chance. If there are 10 four-answer questions in this exam, calculate the probability that 6 questions will be correctly marked.

8. The probability that the syrup produced by a pharmaceutical factory is not standardized and their bottle is broken or the lid is not completely closed and dripping equal to 0.002. Calculate the probability that there is no defective bottle in a 2000 shipment of this product.

9. Monthly consumption of pain relief pills in a city of 1000 households has a Poisson distribution with an average of 5. How many painkillers should be brought to the pharmacies of this city for sale each month (we know that no family consumes more than 15 pills per month).

10. There are n students in a classroom and n can take the values 5, 10, 15, 20, 25, 30, and 35. Find the probability below:

    A. Two students were born in the same day.
    B. Then plot the above probabilities against n.
    C. Two students were born today?
    D. Explain the difference between parts A and C.

11. In a laboratory, 1095 unstable isotopes are stored. If it is assumed that the probability of atomic disintegration is the same on all different days of the year, consider the following probabilities:

    A. Four isotopes today do disintegrate.
    B. Less than 4 isotopes disintegrate today.
    C. More than 4 isotopes disintegrate today.
    D. Today or tomorrow 4 isotopes disintegrate.

12. The probability of a random accident between the electron and the target material per second has a Poisson distribution with an average of 10. Find the probability that no accident occurs in one second?

13. If the probability of a baby being born with white hair (zal) is $\frac{1}{200,000}$, among the 600,000 babies born annually in a population, find the following probabilities:

    A. No infant with white-haired are born.
    B. At least one infant is born with white-haired.
    C. There are more than 5 white-haired infants.

14. Calculate the value of the expression $A = \left[1 + \frac{1}{n}\right]^n$ for $n = 2, 4, 6, 8,$ and 10. As n enlarges, does the expression A tend to the value of $e = 2.71828$? The e is the basis of the natural logarithm used in the Poisson distribution.

15. When $n = 50$ and $p = 0.1$, once, using a binomial distribution, obtain the probability that x has values of 10, 15, and 20. Then obtain the same probability using the Poisson distribution for $\lambda = n.p = 50 \times 0.1 = 5$ and compare the results.

Chapter 6
Normal Distribution

6.1 Normal Distribution

Suppose you buy a box of 16 hands disinfectious (200 ml each). When you put them together at home, you do not see the liquid on the bottle in equal level. Of course, just as you know it's very unlikely that their net volume is exactly 200 cc, you can never expect their volume to be too far from 200 cc. But you will not be surprised, if each bottle contains a few cc more or less. To ensure their quantity, you measure the volume of one. If it is only a few cc smaller in size, you will blame it on the filling machine error, but you will probably go to the next bottle. If the volume of disinfectious is more than 200 cc, you will not care. If the latter is larger and the former is smaller, you may find these two bottles sufficient to measure. But if the latter is as low as the former, your suspicious will increase and you will go to the next bottles and write down their volume. Keep doing this until you reach a certain and confidence result. If some bottles are smaller and some are larger enough to compensate each other, it is justified that the average volume is the same 200 ml "200 ± 07 ml" written on them.

However, the presence of a 120–150 ml bottle between them is not acceptable to you. The volume of disinfectant bottles is: (205–204–202–201–199–198–195–195–194–193–188–190–211–209–208–206), and their chart is a bell shape and symmetric (Fig. 6.1).

This pattern of bell-shaped distribution curves is called normal distribution. Another name for this distribution, which is used for continuous variables, is the Gaussian distribution. The larger the number of trials, the smoother, more symmetrical, and bell-like this curve will be.

Many of the events we deal with in our daily lives, such as blood pressure, marital age, maternal age, height, and weight, have an almost normal distribution. In the medical and health sciences, too, many variables can be found that have relatively normal or similar to normal distributions, or we have to assume that they are normal until their abnormality is proven.

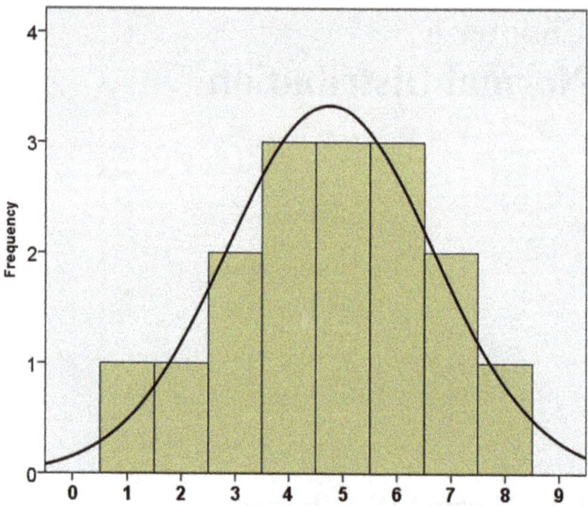

Fig. 6.1 Result of repeated weighing of disinfectant liquids

The density function of the normal distribution probability is as follows:

$$f(x) = \frac{e^{-\frac{1}{2}\left[\frac{(x-\mu)}{\sigma}\right]^2}}{\sigma\sqrt{2\pi}},$$

where $\pi = 3.14$ and $e =$. The normal distribution has two parameters μ and σ. By specifying these two parameters, the distribution pattern of the dataset is specified and it is indicated by $X \sim N(\mu.\sigma)$.

6.1.1 Distribution Function and Density Function

Typically, every distribution is characterized by two functions: the density function, denoted as $f(x)$, and the distribution function, denoted as $F(x)$. In the case of the normal distribution, the density function, $f(x)$, takes the form of a symmetrical bell-shaped curve. The integral of this function over its entire domain equals one, representing the total probability. Utilized to convey the distribution's state and structure, the density function is a fundamental aspect. On the other hand, the distribution function, $F(x)$, is primarily a cumulative probability function. It enables us to easily determine the probability of observing data up to a specific value of x, inclusive. Figure 3.5 illustrates both the distribution functions and the density function (Fig. 6.2).

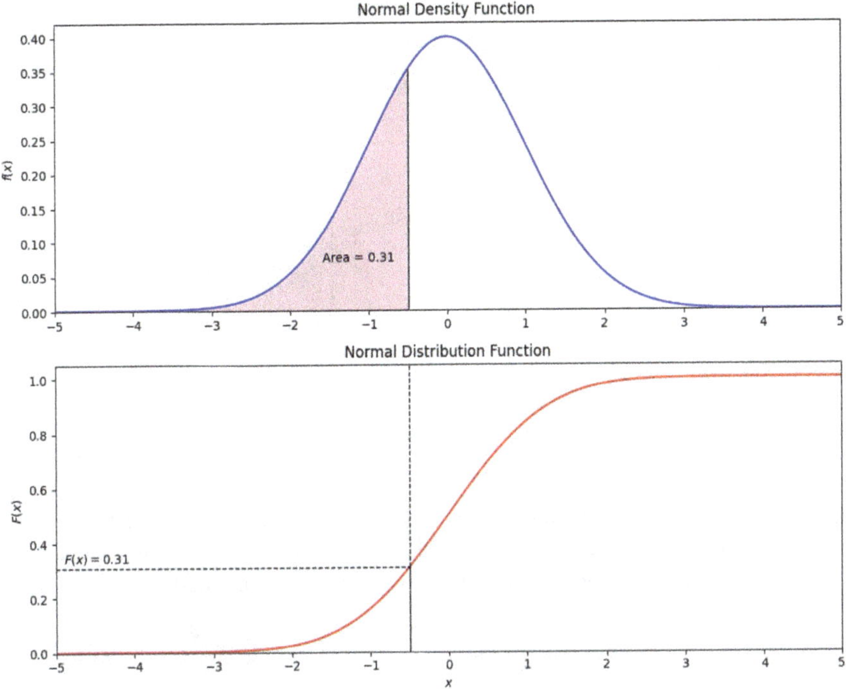

Fig. 6.2 Comparison between normal density and normal distribution functions

6.1.2 The Properties of the Normal Distribution

The normal distribution has important, obvious, and unique properties that are briefly described below:

The normal distribution curve exhibits a characteristic bell shape with a single peak. At the center of this distribution, the mean, median, and mode are all identical, showcasing a central tendency. Because of its symmetric nature, the normal distribution curve is unimodal, meaning it possesses only one mode. The total area under the normal distribution curve is considered to be 100% or 1. Half of this area (50%) lies to the right of the mean, while the remaining half (50%) is to the left. The normal distribution is perpetually asymptotic, never intersecting the X-axis but maintaining a parallel appearance to it. In essence, no matter how far the curve extends in either direction, it never reaches the X-axis, yet gets progressively closer to it. Observations occurring in the center of the distribution are more frequent than those elsewhere, with frequency decreasing the further one moves from the center. When defining the relative frequency of observations for a given class with boundary points a and b, it is equivalent to the areas under the curve between these two points. Exploiting the symmetry of this distribution and employing the sample's standard deviation estimation, we can deduce that within 1

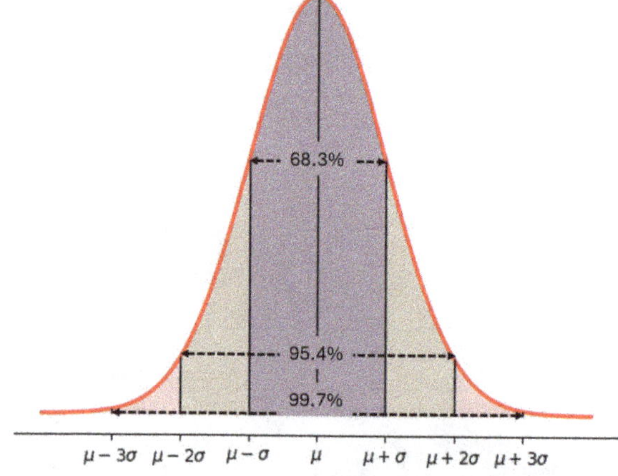

Fig. 6.3 Interpretation of normal distribution in terms of mean and standard deviation

standard deviation of the mean, approximately 68% of the area under the curve is encompassed; within 2 standard deviations, around 95% of the area is covered; and within 3 standard deviations, about 99.7% of the area is contained.

> **Tips**
>
> The empirical rule for the normal distributions:
>
> - Within one standard deviation of the mean ($\mu - \sigma$ to $\mu + \sigma$) contains approximately 68% of the population.
> - Within two standard deviations of the mean ($\mu - 2\sigma$ to $\mu + 2\sigma$) contains approximately 95% of the population.
> - Within three standard deviations of the mean ($\mu - 3\sigma$ to $\mu + 3\sigma$) contains approximately 99.7% of the population.

According to Chebyshev's theorem, in any dataset, the proportion of data will fall within K standard deviation of the mean ($\bar{x} \pm K\sigma$) which is at least $1 - \frac{1}{K^2}$, where K is any positive number greater than 1. For example, at least $1 - \frac{1}{2^2} = 0.75$ or 75% of the data will be within 2 Sd of the mean and at least $1 - \frac{1}{3^2} = 0.89$ or 89% of the data will fall between 3 Sd of the mean (Fig. 6.3).

6.2   The Standard Normal Distribution

A special situation of a normal distribution whose mean and standard deviation are 0 and 1 is called the standard normal distribution N (0 and 1). Any normal distribution can be transformed to a standard normal distribution to find the part of area under the curve that lies between two points (interval) or to obtain the ratio of an area which is more or less than a point (a certain value). This transformation is done using the following relation:

$$Z_i = \frac{X_i - \mu}{\sigma},$$

where μ is the mean, σ is the standard deviation, X_i is a certain value of the variable, and Z is the standardized variable. This procedure is called standardization of frequency distribution because it can be used to make any normal distribution equivalent to the normal standard distribution. The distance from the mean under the curve of standard normal is the criterion for comparing statistical tests and is called the Z criterion.

The standard normal distribution has another extra property (mean 0 and standard deviation 1) in addition to all normal distribution properties. Based on this feature, using the Z distribution table, the area under the curve of each normal variable can be obtained.

To find the area under the curve, we first find the probability of Z regardless of its sign from the table and then obtain it according to one of the following three basic types:

Both points are on one side of the mean (both on the left side or both on the right side). So find their probability from the standard normal table and subtract them.

Each of the two points on one side of the mean (one on the left side and the other on the right side). After finding their probabilities from the standard normal table, add them together.

When we are looking for the area under the curve to be more than one point or less than one point, after finding the probability of that point (regardless of its sign) from the standard normal table, subtract it from 0.5.

> **Tip**
>
> A normal distribution with the parameters of $\mu = 0$ and $\sigma = 1$ is named the **standard normal distribution** and denoted by $X \sim N(0, 1)$. The total area under its density curve is equal to 1 or 100%.

Example 1 Suppose the IQ has a normal distribution with a mean of 100 and a standard deviation of 15. What percentage of people has an IQ between 85 and 130?

Solution: First, we transform the above values to the standard normal variable using the $Z_i = \frac{X_i - \mu}{\sigma}$ formula.

$$p(85 \leq x \leq 130) = p\left(\frac{85 - 100}{15} \leq z \leq \frac{130 - 100}{15} \right) = p(-1 \leq z \leq +2).$$

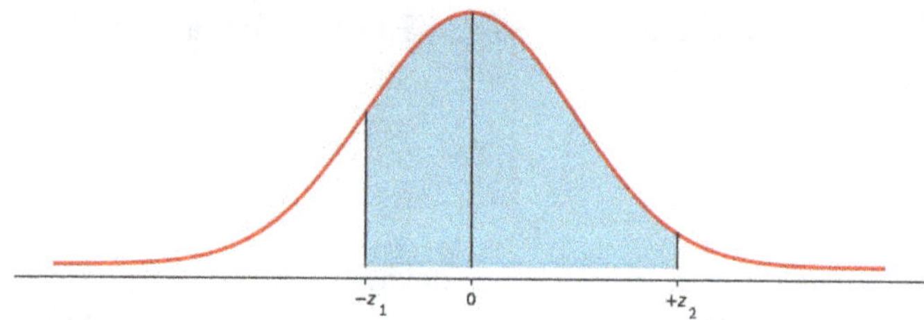

Fig. 6.4 Area between any two z values either side of the mean. Look up both corresponding values to z and add together

Due to the symmetry property of the normal curve, the following relation is established.

Due to the symmetry nature of the normal distribution curve and the fact that two values are on either side of the mean value (one smaller and the other larger), the above relation can be written as follows:

$$p(-1 \leq z \leq +2) = p(-1 < z < 0) + p(0 < z < 2).$$

Then using the standard normal area in the z table, we obtain the corresponding probabilities.

$$p(-1 \leq z \leq +2) = p(-1 < z < 0) + p(0 < z < 2) = 0.3413 + 0.4772 = 0.8185.$$

That is, approximately 81.9% of people have an IQ between 85 and 130 (Fig. 6.4).

Example 2 In Example 1, the IQ had a normal distribution: IQ $\sim N(100, 15)$. What percentage of people has an IQ greater than 130?

Solution: First, we transform the value of 130 to the standard normal variable using the $Z_i = \frac{X_i - \mu}{\sigma}$ formula.

$$p(x > 130) = p\left(z > \frac{130 - 100}{15}\right) = p(z > 2).$$

The standard normal distribution table provides the area under the curve from the mean (zero point) to a certain point. In order to obtain the area under the curve that exceeds a certain point, i.e., to the end of the distribution, we must subtract the value obtained from the table from 0.5. Because in normal distribution, the mean divides the area under the curve into two equal parts.

$$p(x > 130) = p(z > 2) = 0.5 - p(z < 2).$$

The probability of $p(z < 2)$ from the standard normal distribution table is:

Fig. 6.5 Area to the right of
any z value. Look up the
corresponding value to z and
subtract the area from 1

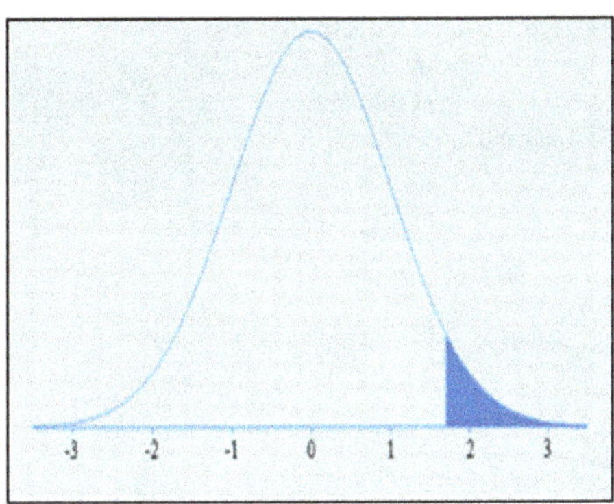

$$p(z < 2) = 0.4772.$$

And

$$p(x > 130) = p(z > 2) = 0.5 - p(z < 2) = 0.5 - 0.4772 = 0.0228.$$

That is, approximately 2.28% of people have an IQ more than 130 (Fig. 6.5).

Example 3 In Example 1, the IQ had a normal distribution: IQ $\sim N(100, 15)$. What percentage of people has an IQ less than 85?

Solution: First, we transform the value of 85 to the standard normal variable using the $Z_i = \frac{X_i - \mu}{\sigma}$ formula.

$$p(x < 85) = p\left(z < \frac{85 - 100}{15}\right) = p(z < -1).$$

Since the normal distribution curve is symmetric, we can obtain $p(z < +1)$ instead of $p(z < -1)$.

$$p(-1 < z < 0) = p(0 < z < +1) = 0.3413.$$

In order to obtain the probability of being smaller than $p(z < -1)$, we need to reduce the above value from 0.5 (half left of the curve).

$$p(z < -1) = 0.5 - p(0 < z < +1) = 0.5 - 0.3413 = 0.1587.$$

That is, approximately 15.87% of people have an IQ less than 85 (Fig. 6.6).

$$p(x < 85) = p\left(z < \frac{85 - 100}{15}\right) = p(z < -1).$$

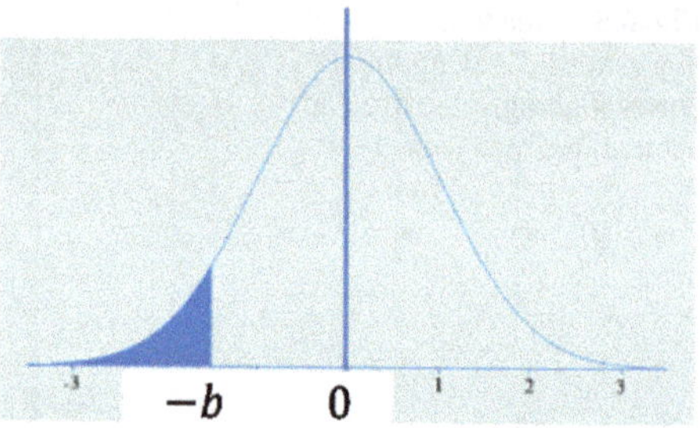

Fig. 6.6 Area to the left of any z value. Look up the value of z in the table and subtract the area from 1

Example 4 In Example 1, the IQ had a normal distribution: IQ $\sim N(100, 15)$. What percentage of people has an IQ between 115 and 130?

Solution: Likewise, to previous examples, we transform the both values of 115 and 130 to the standard normal variable using the $Z_i = \frac{X_i - \mu}{\sigma}$ formula.

$$p(115 \leq x \leq 130) = p\left(\frac{115 - 100}{15} \leq z \leq \frac{130 - 100}{15}\right) = p(+1 \leq z \leq +2).$$

The standard normal distribution table provides the area under the curve between any specified points to the mean (0). Therefore, we must find the values corresponding to the points 1 and 2.

$$p(0 < z < 2) = 0.4772,$$

$$p(0 < z < 1) = 0.3413.$$

Because both standardized values are positive, both are larger than the mean and are on one side of the mean. Therefore, we have to subtract them to achieve the area under the standard normal curve between values (1 and 2).

$$p(+1 \leq z \leq +2) = p(0 < z < 2) - p(0 < z < 1) = 0.4772 - 0.3413 = 0.1359.$$

That is, approximately 13.59% of people have an IQ between 115 and 130 (Fig. 6.7).

Example 5 In Example 1, the IQ had a normal distribution: IQ $\sim N(100, 15)$. What percentage of people has an IQ between 80 and 90?

Solution: As previous example, we transform the both values of 80 and 90 to the standard normal variable using the $Z_i = \frac{X_i - \mu}{\sigma}$ formula.

$$p(80 \leq x \leq 90) = p\left(\frac{80 - 100}{15} \leq z \leq \frac{90 - 100}{15}\right) = p(-1.33 \leq z \leq -0.67).$$

From the standard normal distribution table and due to its symmetry, we obtain the area under the curve for the values of 1.33 instead of -1.33 and 0.67 instead of -0.67.

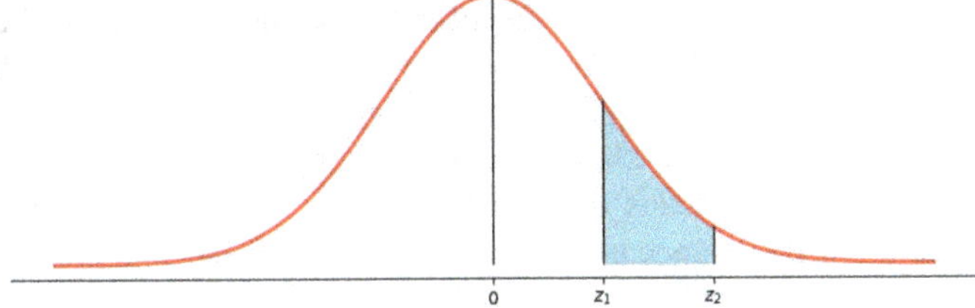

Fig. 6.7 Area between any two z values at positive side of the mean. Look up both values of z in the table and subtract corresponding areas

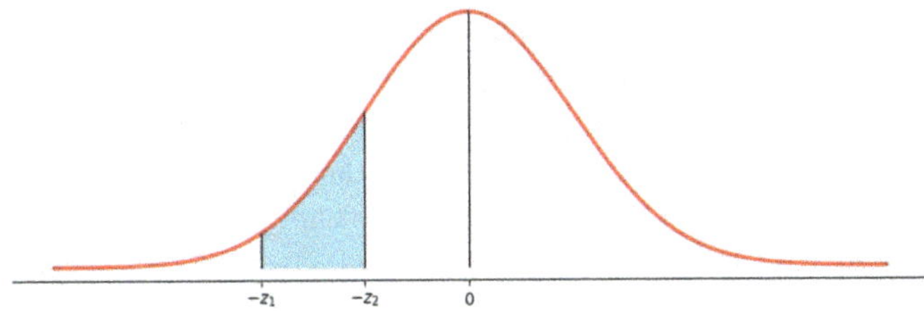

Fig. 6.8 Area between any two z values at negative side of the mean. Look up both values of z in the table and subtract corresponding areas

$$p(0 < z < -1.33) = 0.3485,$$

$$p(0 < z < -0.67) = 0.2486.$$

Because both standardized values are negative, both are less than the mean and are on one side of the mean. Therefore, we have to subtract them to achieve the area under the standard normal curve between values -0.67 and -1.33.

$$p(-1.33 \leq z \leq -0.67) = p(0 < z < -1.33) - p(0 < z < -0.67)$$
$$= 0.3485 - 0.2486 = 0.0999.$$

That is, approximately 9.99% of people have an IQ between 80 and 90 (Fig. 6.8).

6.2.1 Skewness and Kurtosis of Normal Distributions

The other two properties of the normal distribution that specify the lack of symmetry and peakedness of normal distributions are important.

Skewness: Skewness is a measure of symmetry, or more precisely the lack of symmetry. In normal distributions, in addition to being symmetric, the mean, median, and mode are very close together. But sometimes when the mean, median, and mode are not match, the distribution continued on one side (left or right) and is longer on the other. This means that the location of mean and median changes, while the location of mode appears to be unchanged in place. Therefore, the mean and the median inclined to a larger and longer tail.

When a distribution continues from the right-hand side of the X-axis toward the larger and positive values, the tail of that distribution is more elongated from the right, and the median and mean are larger than the mode of the distribution, and that distribution is called "skew to the right" or "positive skewness (such as the community income). But if the distribution continues to the left of the X-axis, the tail of that distribution extends to the left, that is, to the lesser and negative values, and that distribution is called "skew to the left" or "negative skewness" (age at death).

In addition to visually detecting of skewness, three measures for calculating the skewness of a distribution are given below, which is known as the Pearson coefficient, the Galton coefficient, and the third moment.

$$\text{Sk} = \frac{3(\mu - \text{Med})}{\sigma} \quad \text{or} \quad \text{SK} = \frac{Q_3 - 2Q_2 + Q_1}{Q_3 - Q_1}.$$

These coefficients are unitless. Because in exactly symmetric and normal distributions, the mean and the median coincide, these measures must be equal to zero. Mathematically, these coefficients do not have upper and lower limits. The closer these values to zero, the closer the distribution is to normal. They rarely even come close to $+1$ and -1. In any case, if $\text{Sk} < 0$, the skewness is negative, and if $\text{Sk} > 0$, the skewness is positive.

The G-coefficient is another measure for detecting symmetry (skewness) of the distribution:

$$G = \frac{1}{n} \times \frac{\sum (x_i - \overline{x})^3}{\sigma^3}.$$

In normal distribution, this $G = 0$. If this coefficient:

- Less than 0.1 ($|G| \leq 0.1$), the distribution is assumed to be almost normal.
- Between 0.1 and 0.5 ($0.1 \leq |G| \leq 0.5$) Low distribution skewness.
- Greater than 0.5 ($|G| \geq 0.5$), the distribution is considered to be a highly distributed skew.

It should be noted that statistical software such as SPSS and SAS calculates skewness for the sample, whose formula is slightly different (Fig. 6.9).

$$G = \frac{n}{(n-1)(n-2)} \times \frac{\sum (x_i - \overline{x})^3}{\sigma^3}.$$

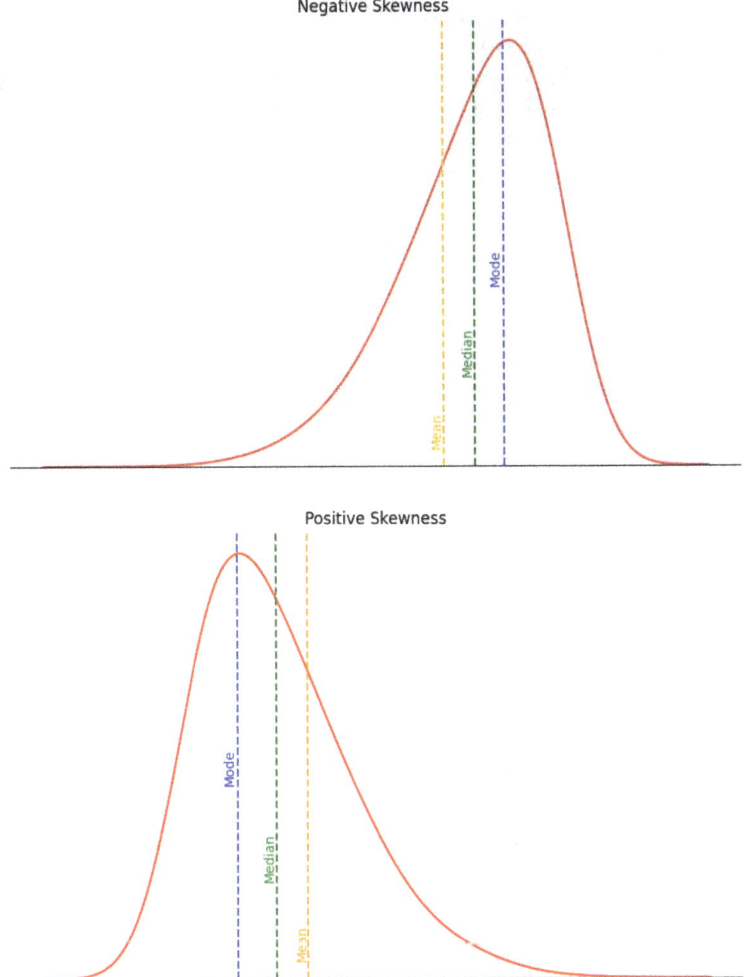

Fig. 6.9 Right (positive) and left (negative) skewnesses in distributions

Example 6 The first, second, and third quartiles of albumin level of 185 female in Table 3.2 are $Q_1 = 38.3$, $Q_2 = 41.2$ and $Q_3 = 44.3$, respectively. Find the skewness coefficient of albumin level.

Solution: Skewness coefficient for this dataset is:

$$\text{SK} = \frac{Q_3 - 2Q_2 + Q_1}{Q_3 - Q_1} = \frac{44.3 - 2 \times 41.2 + 38.3}{44.3 - 38.3} = \frac{0.2}{6.0} = 0.033.$$

The value of this coefficient is very small and close to zero. So, the distribution of albumin levels can be assumed approximately normal in terms of skewness.

Kurtosis: Some normal distributions may be taller than others. In other words, the data density around the mean is more in one distribution a is more in the tails of another distribution. The kurtosis is a measure of the thickness of the tails of a distribution and the sharpness of its peak. In other words, measure of the combined weight of a distribution's tails is relative to the center of the distribution. To measure the kurtosis of a distribution, the fourth moment is suggested by Pearson:

$$K = \frac{1}{n} \times \frac{\sum (x_i - \overline{x})^4}{\sigma^4} - 3.$$

According to this measure, the kurtosis of any normal distribution is zero and each distribution is compared to the normal distribution. This measure can take positive or negative values. Positive values (Sk > 0) indicate the elongation of the distribution and negative values (Sk < 0) indicate the flatness of the distribution. In general if:

- $K = 0$, the distribution is assumed to be mesokurtic (middle: neither high nor low).
- Less than 0 ($K < 0$) is considered to be platykurtic (broad or flat).
- Greater than 0 ($K > 0$), the distribution is considered to be leptokurtic (skinny: that are thin and tall).

It should be noted that statistical software such as SPSS and SAS calculates skewness for the sample, whose formula is slightly different (Fig. 6.10).

Skewness and kurtosis in SPSS can be calculated by choosing the path: **Analyze\Descriptive statistics\Frequencies**. After selecting the variable (s), click on the statistics option and then select the skewness and kurtosis options to calculate the skewness and kurtosis of the data.

These measures are also obtained by the path: Analyze\Descriptive statistics\descriptive and then choosing option.

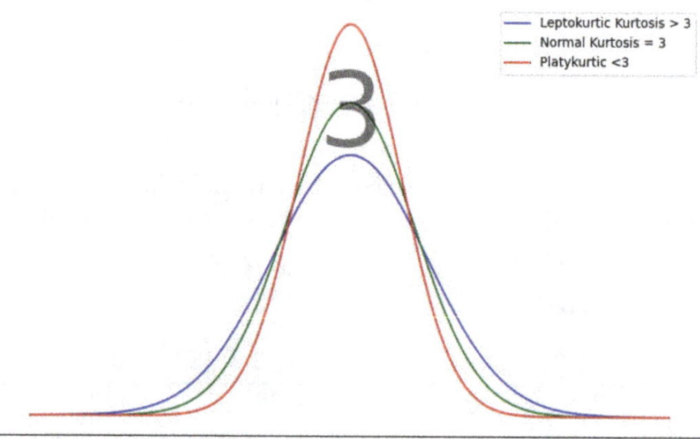

Fig. 6.10 Kurtosis and flatness of a normal distribution

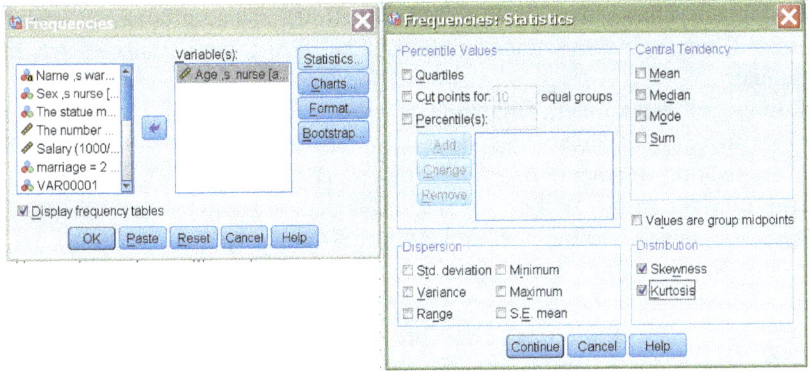

Example 7 Intraocular pressure (IOP) of the right eye of 25 people is (17–16–19–18–10–18–15–13–14–11–13–12–18–8–9–20–22–13–16–13–10–16–15–12–12). Calculate the kurtosis and skewness IOP.

Solution: The mean of this data is 14.4 and the sample standard deviation with fraction denominator $(n-1)$ is 3.594.

The skewness and kurtosis coefficients for the formulas used in the reference books are calculated as follows:

$$G = \frac{1}{n} \times \frac{\sum (x_i - \bar{x})^3}{\sigma^3} = \frac{200.4}{25 \times 3.594^3} = \frac{200.4}{1160.562} = 0.173,$$

$$K = \frac{1}{n} \times \frac{\sum (x_i - \bar{x})^4}{\sigma^4} - 3 = \frac{8863.12}{25 \times 3.594^4} - 3 = \frac{8863.12}{4171.116} - 3 = -0.875.$$

But their calculations are based on formulas used in software such as SPSS as follows:

$$G = \frac{n}{(n-1)(n-2)} \times \frac{\sum (x_i - \bar{x})^3}{\sigma^3} = \frac{25}{24 \times 23} \times \frac{200.4}{3.594^3} = \frac{5010}{25625.555} = 0.196,$$

$$K = \frac{n(n+1)}{(n-1)(n-2)(n-3)} \times \frac{\sum (x_i - \bar{x})^4}{\sigma^4} - 3\frac{(n-1)^2}{(n-2)(n-3)}$$

$$= \frac{25 \times 26}{24 \times 23 \times 22} \times \frac{8863.12}{3.594^4} - 3\frac{(24)^2}{23 \times 22} = -0.572.$$

6.2.2   Assessing Normality of a Distribution

There are various methods to ensure that a particular distribution follows an approximately normal distribution, which are described in order of simplicity below:

- **Histogram**: Draw a histogram for the data. If the histogram looks symmetrical and bell-shaped in terms of the number of data, its distribution can be assumed as normal.
- **Equality of mean, mean, and mode**: In the normal distribution, mean, mean, and mode are equal. Calculate their values. If these three measures are close together in terms of the number of data, there is a property of normal distribution.
- **Interquartile range**: In the normal distribution, the interquartile range (IQR) is approximately 1.35 times of the standard deviation (IQR $=$ 1.35Sd). If this relationship is almost established, it would be another property of a normal distribution. In the standard normal distribution (with a standard deviation of 1), the first and third quartiles are equal to -0.674, $+0.674$, and IQR $= +0.674 - (-0.674) = 1.35$.
- **Standard deviations around the mean**: Based on the characteristics of the normal distribution, if around 68% of the data values lie within 1 standard deviation of the mean, approximately 95% of the data values are within 2 standard deviations of the mean, and roughly 99.7% of the data values fall within 3 standard deviations of the mean (based on the sample size), the distribution can be deemed normal.
- **Distribution skewness and kurtosis**: We calculate the coefficients of skewness and kurtosis. As mentioned in the previous section, the normality of the distribution can be checked.
- **Kolmogorov–Smirnov test**: The more common option is normal plot, Kolmogorov–Smirnov test, and Shapiro–Wilk's test, which is easily implemented in statistical software such as SPSS.

To do this, choose the path: Analyze\Descriptive Statistics\Explore to open a window. Then click the Plots option to open another window. In this window, select Normality plots with tests. Its output program provides normal plot and Kolmogorov–Smirnov tests as well as distribution skewness and kurtosis.

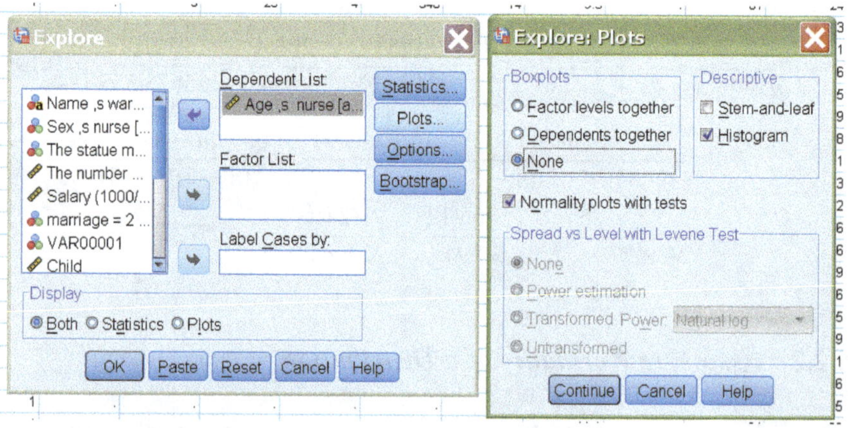

Example 8 Examine the normality of data in Example 7 (intraocular pressure (IOP) of the right eye).

Solution: The skewness coefficient and kurtosis coefficient of this data were calculated in the previous example. To facilitate the calculation of other procedures, the mean, standard deviation, median, mode and the first and third quartiles of this data are as are equal to:

$$\bar{x} = 14.40, \ \text{Med} = 14.0, \ \text{Mod} = 13.0, \ \text{Sd} = 3.594,$$

$$Q_1 = 12.0, \ Q_2 = 14.0 \ \text{and} \ Q_3 = 17.5.$$

A: The proximity of the median (14), mode (13), and mean (14.4) to each other indicates an almost normal distribution of data.

B: The interquartile is equal to $\text{IQR} = Q_3 - Q_1 = 17.5 - 12 = 5.5$ and 1.5 times the standard deviation, which is a sign of an almost normal distribution of data (in perfectly normal distributions, it is 1.35 times the standard deviation).

C: The range of one standard deviation, two standard deviations, and three standard deviations to the mean should include about 68%, 95%, and 99.7% of the data, respectively.

$$\bar{x} \pm \text{Sd} = 14.4 \pm 3.594 = (10.81 - 17.59),$$

$$\bar{x} \pm 2\text{Sd} = 14.4 \pm 2 \times 3.594 = (7.21 - 21.59),$$

$$\bar{x} \pm 3\text{Sd} = 14.4 \pm 3 \times 3.594 = (3.62 - 25.18).$$

It is observed that the interval of one standard deviation to the mean includes 17 observations (68%), the interval of two standard deviations to the mean includes 24 observations (96%), and the interval of three standard deviations to the mean includes 25 observations (100%), which testifies to the almost normal distribution of data.

D: The histogram diagram also shows the nearly normal distribution of data.

E: The p value (Sig column) in both Kolmogorov–Smirnov test and Shapiro–Wilk test is more than 0.05 which does not reject the normality of the data distribution Fig. 6.11.

Tests of normality

	Kolmogorov–Smirnov[a]			Shapiro–Wilk		
	Statistic	df	Sig.	Statistic	df	Sig.
IOP	0.132	25	0.200[*]	0.980	25	0.890

[a] Lilliefors Significance Correction
[*] This is a lower bound of the true significance

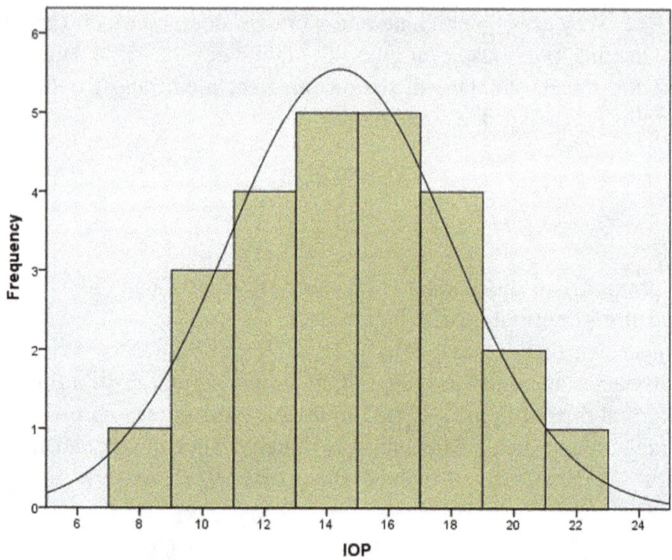

Fig. 6.11 Histogram with approximately normal curve

6.3   t Distribution

Many of statistical analyzes are based on the assumption of normality distribution of the variable under study and that the variance of the sample population is known. According to the central limit theorem, when the sample is large enough and its variance is known (regardless of the population distribution), the sample means are assumed to be normal. But in practice, this assumption is not true, and usually, the number of samples is small and the variance of the population is unknown. In these samples, instead of the variance of the population, its sample estimation should be used, i.e., $S^2 = \frac{\sum(x_i - \bar{x})^2}{n-1}$. Under such conditions, the statistic $z = \frac{\bar{x} - \mu}{\frac{\sigma}{\sqrt{n}}}$ will become the statistic $t = \frac{\bar{x} - \mu}{\frac{S}{\sqrt{n}}}$, which has a distribution t.

The shape of this distribution is not exactly normal, but it is very similar to the standard normal distribution and has similar properties.

- It is bell-shaped and symmetrical.
- The mean of this distribution is zero and its variance is greater than one. But the larger the sample size, the closer the variance gets to one.
- It is flattered compared to normal distribution and its tails are ticker. That is, its variance is greater than the normal distribution.
- The tails of distribution are asymptotic. That is, they can vary from $-\infty$ to $+\infty$.

This distribution has a parameter called degree of freedom (df $= n - 1$). The greater the degree of freedom, the closer the t distribution to the normal distribution.

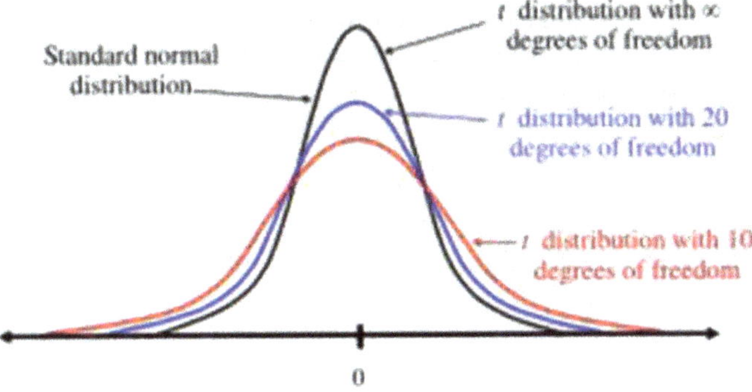

Fig. 6.12 Student t distributions for $n = 10$ and $n = 20$ compared to standard normal distribution

Figure 6.12 shows the distribution of t compared to the standard normal distribution.

Tips

Similarity of t Distribution to the Standard Normal Distribution:

(1) It is bell-shaped and symmetric about the mean.
(2) The mean, median, and mode are equal to 0 and are located at the center of the distribution.
(3) The curve is asymptotic and never touches the X-axis.

Dissimilarity of t Distribution from the Standard Normal Distribution:

(1) Its variance is greater than 1.
(2) The t distribution curves are actually different based on the *degrees of freedom* $(n - 1)$.
(3) As the sample size increases, the t distribution approaches the standard normal distribution.

One of the applications of t distribution is to test the hypothesis and construct the confidence interval of the mean sample, which we will discuss in Chaps. 7 and 8. In this section, it is enough to present the confidence interval formula and solve a couple of example.

$$\bar{x} \pm t_{\left(1-\frac{\alpha}{2},\, n-1\right)} \times \frac{S}{\sqrt{n}}.$$

Example 9 The accommodation (ACC) of right eye for 25 patients is equal to (15–14–10–9.5–11–12–11–10–10.5–9–11.5–9–8.5–7–6.5–7.5–8–5–6–5.5–5–2.5–3.5–1–11.5). These patients were randomly selected from a large population. Construct the 95% confidence interval for the population mean.

Solution: In order to construct the confidence interval for the mean, we must first estimate the variance and the standard error and find the critical t value with df = 25–1 = 24 from the student t-table distribution and then substitute them on the formula.

$$S^2 = \frac{\sum (x_i - \bar{x})^2}{n - 1} = \frac{289.5}{25 - 1} = 12.06,$$

$$Se(x) = S_{\bar{x}} = \frac{S}{\sqrt{n}} = \frac{\sqrt{12.06}}{\sqrt{25}} = 0.695,$$

$$\bar{x} = \frac{\sum x_i}{n} = \frac{210}{25} = 8.4,$$

$$t_{(0.025, 24)} = 2.064,$$

$$\bar{x} \pm t_{(1 - \frac{\alpha}{2}, n-1)} \times \frac{S}{\sqrt{n}} = 8.4 \pm 2.064 \times 0.695$$

$$= 8.4 \pm 1.43 = (6.97 \text{ to } 9.83).$$

In Chap. 7, we will interpret the confidence interval and its details in more detail.

Example 10 In Example 9, the mean and variance of accommodation (ACC) of right eye calculated from a sample of 25 subjects were 8.4 and 12.06, respectively. If the same mean and variance are obtained from a sample of 81, construct the 95% confidence interval for the population mean.

Solution: After calculating the standard error, we find the critical t value with df = 81–1 = 80 from the student t-table distribution and then substitute them on the formula.

$$\bar{x} = 8.4, \ S^2 = 12.06 \text{ and } df = 81 - 1 = 80,$$

$$t_{(0.025, 80)} = 1.990,$$

$$Se(x) = S_{\bar{x}} = \frac{S}{\sqrt{n}} = \frac{\sqrt{12.06}}{\sqrt{81}} = 0.386,$$

$$\bar{x} \pm t_{(1 - \frac{\alpha}{2}, n-1)} \times \frac{S}{\sqrt{n}} = 8.4 \pm 1.990 \times 0.386$$

$$= 8.4 \pm 0.768 = (7.63 \text{ to } 9.17).$$

As can be seen, as the sample size increases, the standard error and the critical value of t become smaller and the confidence interval for the mean becomes narrower.

6.4 Chi-Square Distribution

Many of important probabilistic distributions in statistics are related to the normal distribution. One of these important distributions that is widely used in the analysis of categorical data, goodness of fit of the observed distributions, and the confidence interval of the variance of the population when the distribution of the main population is normal is the Chi-square distribution.

Suppose k standard normal random variable $(Z_1, Z_2, Z_3, \cdots, Z_k)$, their sum of squares $(Z_1^2 + Z_2^2 + Z_3^2 + \cdots + Z_k^2)$ has the Chi-square distribution with degree of freedom k and is denoted by the Greek letter χ^2. The degree of freedom plays an important role in this distribution. The greater the degree of freedom of Chi-square distribution, the more symmetrical the distribution shape will be. In this book, only Chi-square distribution with the degree of freedom 1 will explain and we denote it by $\chi_{(1)}^2$.

The standardized normal variable z follows the distribution $N(1, 0)$. Its square (Z^2) is also a non-negative random variable and can change from 0 to $+$ (skewed to the right). The distribution of the random variable Z^2 is called the distribution of $\chi_{(1)}^2$ and its graph is shown in Fig. 5.10.

This distribution was first proposed in 1900 by the famous statistical scientist Karl Pearson (Fig. 6.13).

The percentages of this distribution (critical values) are given in Table VI of the appendix for different degrees of freedom. This statistic is always executed as one tailed. In the distribution of $\chi_{(1)}^2$ with 1 degree of freedom, one or two points are important.

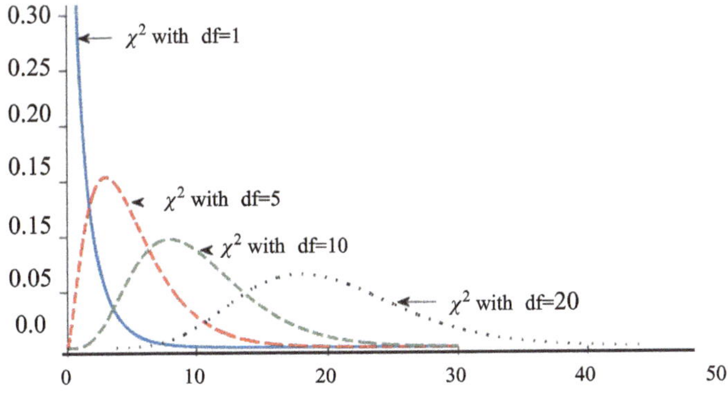

Fig. 6.13 Chi-square distribution with different degrees of freedom

Mean distribution is 1, because $E(z^2) = \frac{E(x-\mu)^2}{\sigma^2} = \frac{\sigma^2}{\sigma^2} = 1$.

Percentages of this distribution can also be obtained from the standard normal distribution. For example, the percentile corresponding to the 0.05 probability is $+1.96$ or -1.96 and their square is $(1.96)^2 = 3.841$. The distribution of $\chi^2_{(1)}$ at the same 5% level equals to 3.841. The same is true for other percentile values of this distribution. This same relation is true for other percentile points of the distribution.

Due to the equivalency between the standard normal distribution $N(1, 0)$ and the Chi-square distribution of $\chi^2_{(1)}$, many issues of the standardized random variable can be expressed by both distributions. However, it should be noted that using Z^2 (or $\chi^2_{(1)}$) eliminates the z direction. Therefore, if the direction of the deviations from the mean is important, the $N(1, 0)$ distribution should be used.

Chi-square computational statistic is defined as follows:

$$\chi^2 = \sum_{i=1}^{k} \frac{(O_i - E_i)^2}{E_i},$$

where O_i is the ith observed value and E_i is the expected value for the ith subject.

The expected frequency (E_i) is also obtained by multiplying the "sum of corresponding column for the cell (C_i)" by the "sum of the related row for the same cell (R_i)" divided by the "grand total (N)":

$$E_i = \frac{C_i \times R_i}{N}.$$

Example 11 The last digit of the weight of a sample of 96 adults (in kilograms) is recorded in Table 6.1. We want to investigate and judge the accuracy of weight in this study.

Solution: If the weighing and recording of people are accurate and there is no tendency to round the last digits to 0 or 0.5, all the unit's digits on the right place (0, 1, 2, … and 9) should be repeated equally. Because the probability of observing each one is equal to 0.1. Therefore, the expected number of each unit's digit is obtained by multiplying this probability by the total number of observations (96). That is, each units' digit ($96 \times 0.1 = 9.6$) is repeated, which is given in the third row (E_i) of the table above. The set of all values in this row is called theoretical distribution.

It should be noted that the sum of the expected frequencies and the sum of the observed frequencies must always be equal.

Table 6.1 Last digit of the weight of individuals in a sample of 96

Last digit	0	1	2	3	4	5	6	7	8	9	Total
Freq	13	8	10	9	10	14	5	12	11	4	96
E_i	9.6	9.6	9.6	9.6	9.6	9.6	9.6	9.6	9.6	9.6	96

Now to judge the accuracy of the weights, we must use the goodness of fit to compare the observed distribution with the theoretical distribution that is completely solved in Chap. 8.

Example 12 Example: In a survey, the opinions of a sample of 200 people (50 men and 150 women) about the use of social media were asked and are shown in Table 6.2. Are people's genders and their opinions about the social media independent?

Solution: If the gender of subjects is independent of their opinions about social media, the proportion of women and men should be equal between those who agree and those who disagree. That is, we expect that the proportion of disagree women and the proportion of disagree men are to be equal to the proportion of total disagrees (30%). Therefore, $(45 = 150 \times 30\%)$ women and $(50 \times 30\% = 15)$ men are disagree. We also expect that the proportion of women who agree with the proportion of men who agree is equal to the proportion of all those who agree (50%), i.e., $(150 \times 50\% = 75)$ women and $(50 \times 50\% = 25)$ men agree with social media. It is also expected that the number $(150 \times 20\% = 30)$ of women and $(50 \times 20\% = 10)$ of men will be neural. These calculated values are called the expected frequency, which means that when the respondents' opinions are independent of their gender (hypothesis zero), we should have Table 6.3.

Now, using Chi-square statistic, it can be tested whether there is a significant difference between observed frequency and expected frequency in this sample. Its full solution will be presented in Chap. 8.

It is widely believed that "cigarettes are health-threatening". Does the data of this study confirm public belief? In other words, is the cause of death related to smoking habits?

Table 6.2 Opinions of 200 samples re social media in terms of gender

	Male	Female	Total
Agree	17	43	60
Disagree	22	78	100
Neural	11	29	40
Total	50	150	200

Table. 6.3 Expected values of Table 6.2

	Male	Female	Total
Agree	15	45	60
Disagree	25	75	100
No idea	10	30	40
Total	50	150	200

6.5 Exercises

1. The label of a Decongestant syrup shows that it has 1.4% of alcohol. The pharmaceutical factory has announced that the production process of the factory is such that the alcohol in these syrups has a normal distribution with mean of 1.4% and a standard deviation of 0.1%. Find out the probability that:

   A. Is the alcohol in a bottle more than 1.5%?
   B. In a box of 24, only one bottle has more than 1.5% alcohol?

2. In the biostatistics exam, a student is asked to explain the concept of a "central limit theorem". "The more the sample size increases, the closer the sample average ($\overline{x}$) gets to normal", he replied. Do you think that the student's answer was correct?

3. When X is following a normal distribution with mean μ and variance σ^2, find the following values:

   A. $p(\mu - 3\sigma < x < \mu + 3\sigma)$.
   B. $p(\mu - 2\sigma < x < \mu + 2\sigma)$.
   C. $p(\mu - \sigma < x < \mu + \sigma)$.
   D. $p(\mu - 3\sigma < x < \mu + 2\sigma)$.
   E. $p(\mu - \sigma < x < \mu + 3\sigma)$.

4. IQ with Wechsler criteria has a normal distribution with mean of 100 and a standard deviation of 15. If a 10-year-old boy has an IQ of 126. What percentage of people in the population has a higher IQ than him?

5. Find the following probabilities using the standard normal distribution table:

   A. $p(Z > 1.5)$.
   B. $p(Z > 1.645)$.
   C. $p(Z < 1.645)$.
   D. $p(-1.645 < Z < +1.645)$.
   E. $p(-1.96 < Z < +1.96)$.
   F. $p(Z > 0)$.

6. If $\sigma = 15$ and $\mu = 120$, obtain a constant value of C in the relation $p(-C < Z < +C) = 0.5$.

7. If $\sigma = 15$ and $\mu = -150$, obtain a constant value of C in the relation $p(-C < Z < +C) = 0.5$.

8. Using Table 2.4 about the hemoglobin level of 70 female samples and assuming the normality of hemoglobin distribution, determine the hemoglobin level to be used for the instructions of laboratories and physicians (i.e., 95% of population are in that range).

9. Audiometric scores have a distribution of $N(600, 100)$. A patient is selected at random. What is the probability that his score is between 650 and 750?

Table 6.4 Causes of death in 1000 men over the age of 45 and smoking habits

Causes of death	Cancers	Cardiovascular diseases	Other	Total
Smoker	135	310	205	650
Non-smoker	55	155	140	350
Total	190	465	345	1000

10. Men's brain weight has a normal distribution with a mean and standard deviation of 1400 and 110 g. The brain weight of a how many percent of men is between 1500 and 1700 kg.

11. The mean and standard deviation of the shelf life of the battery (X) for a medical device have been reported with a normal distribution of 30 and 5.5 h. How many percent of the batteries in this device have a shelf life more than 20 h?

12. The specific radiation energy in a particular atom has a normal distribution with mean and standard deviation of 270 and 10 keV. How many of the 100,000 atoms of the same type have more radiant energy than 287 keV?

13. In a specialized hospital, Interactive operative Radiation Therapy (IORT) method is used to treat tumors. In this hospital, about 1.4% of the radiation is gamma radiation. Radiated gamma rays have a normal distribution with mean and standard deviation of (1.4% of 0.1%). Find the following probabilities:

    A. More than 1.5% of radiation is gamma type?
    B. Among 100 patients, only one person received more than 1.5% gamma?

14. The causes of death in 1000 men over the age of 45, along with their smoking habits, are summarized in Table 6.4:

Chapter 7
Sampling and Sample Size

7.1 Introduction

A full survey and data collection (census) of all patients referred to a rehabilitation center provides only reliable information about that center, but the results cannot be generalized to all rehabilitation centers. If we want to have information about rehabilitation centers in the country, there are two approaches: to consider all rehabilitation clinics and centers (without exception), or to select a sample of the centers and generalize the result to all rehabilitation centers. The method and number of selected cases are called sampling techniques. The obtained result is named estimation and how to generalize the results to the reference population is called judgment and statistical test, which are all three components of statistical inference.

7.2 Definitions

In this section, only the three keywords: population, sample, and sampling, which are widely used in the sampling process, are defined.

Population

The term "population" in ordinary conversations means a set of people living in a geographical area/region or a country. But the term "population" in statistics has a broader meaning and is different from the general concept and does not refer necessary only to individuals or living organisms (for instance, the population of school students, the patient population by Covid-19).

In statistics, a **population** is a set of similar items or events which is of interest for some research and survey. Statisticical population may include of a population of objects, or events, or procedures, observations or clinics. Thus, a population is "an aggregate of creatures, things, cases, and so on with the similar and common

© The Author(s), under exclusive license to Springer Nature Singapore Pte Ltd. 2024

S. H. Saneii and H. Doosti, *Practical Biostatistics for Medical and Health Sciences*,

https://doi.org/10.1007/978-981-97-3083-4_7

characteristics". Therefore, any set of individuals grouped together by a common feature can be said to be a population. Then a statistician should clearly define the features of population. For example, the set of all cards in a deck or the set of all cars within the county.

Population can be finite or infinite. The more common features for individual in a population, the more finited (closed) population becomes: for example, **"Women's population"** is wider than **"Educated women's population"** and this is wider than **"Married educated women's population"** and so on.

In statistics, a population is the pool of individuals from which a statistical sample is drawn for a research study.

Remember that: the process of data collection from all members of population requires a lot of money, time, facilities, and human resources. On the other hand, for various reasons, all members of the population are not accessible and the samples under study are in motion and changeable, or they may be destroyed by testing and experiment. Therefore, the process of observing information and data collection is problematic. Hence, the study of a relatively small portion of population as a "sample" is inevitable, especially in clinical and medical studies. For this reason, sampling is very common in the medical profession. For example, in order to determine the patient's fat or blood sugar, not all of his blood is tested and only a few drops of the patient's blood are enough.

Sample

In statistics, the term "sample" refers to a small section of population that is randomly selected and looks like population in terms of the feature under study. By selecting a smaller number of individuals in the population, the interested variables are measured in a shorter time and at a lower cost, with less facilities and resources, and the required information is collected. What emerges from such an example can be generalized to "main population", "mother population" or "reference population". Because such sample is representative of the population. Thus, it should be chosen in such a way that all subgroups of population have a proportionate role in it. That is, the sample selection method should be based on chance and random so that all members of the population have a proportional chance and probability of selection, which is called random sampling.

For example, if we want to select 6 students from a class of 30, we have to write the names of all 30 students separately on small pieces of paper and put them in an urn, and after mixing thoroughly, 6 people will be selected by chance and randomly. Or assign a number to each student and then, using the table of random numbers, select 6 numbers (students) as a sample. In this way, all students have an equal chance of $\frac{6}{30}$ to be selected. But if 6 students are selected from the first rows of the class, it will not be a good sample of the class. This is because students who have poorer vision or are more interested in the lecture may be sitting in the front rows of the classroom. So, their opinions are not generalizable to the whole class.

For the interpretation of sample data to be valid, the sample must be representative of the population. It means **"You may know by a handful the whole sack"** and is large enough to be able to discover the truths of the reference population.

Sampling

Sampling involves selecting a smaller subset of participants, known as a "sample", from the larger population of interest. This approach is necessary because it may not be feasible, either logically or logistically, to involve the entire population in the research. Sampling is a fundamental process in statistical analysis, whereby a predetermined number of observations are drawn from the larger population. By studying a representative subset obtained from the population of interest, researchers can generalize their findings and make inferences about the entire population. The methodology employed for sampling from a larger population varies depending on the type of analysis being conducted, but it can include probabilistic (random) sampling or non-probabilistic (non-random) sampling methods.

> **Tip**
> Only the obtained results from probabilistic (random) sampling are generalizable to the reference population. The results obtained from non-probabilistic (non-random) sampling cannot generalized, even if sample size is large.

The most important goal of sampling is to have the best estimate of the variable in question in the population so that the results can be generalized to the whole population. For example, when we conclude in a study that mortality from Covid-19 is higher in people with the background disease than in others, or that in a laboratory experiment, cells grow better under certain temperatures and light conditions, we expect that the same is true in the reference population.

7.3 Sampling Techniques

Sampling can be done randomly (probabilistic) and non-random (non-probabilistic). In this book, only random (probabilistic) methods are explained. Because it has more advantages than non-random sampling. The main advantages of probability sampling are:

- The biasness of the results is less.
- The absence of systematic error and sampling bias.
- Better representation of the population under study.
- Allows accurate statistical inference to be made.
- The results can be generalized to the reference population.
- There is no systematic and biased sampling error.
- Everyone in the population has a known and equal chance of being elected.

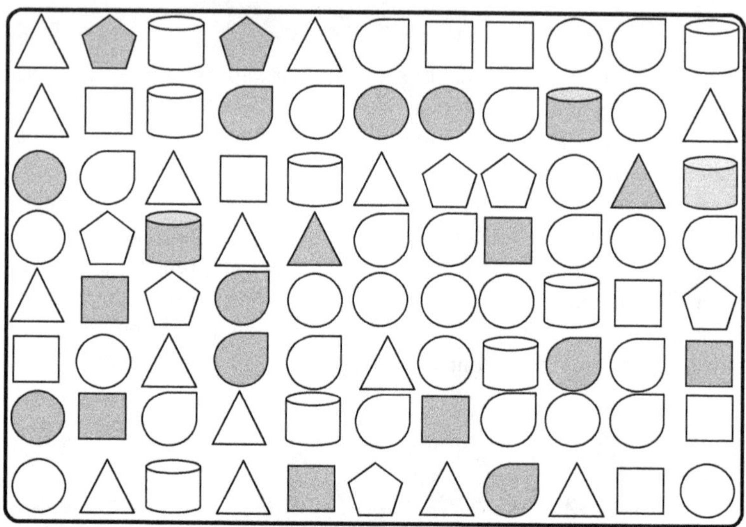

Fig. 7.1 Schematic representation of simple random sampling

Simple Random Sampling (SRS)

Simple random sampling (SRS) stands out as the simplest and most widely used form of probabilistic sampling. It represents an impartial surveying approach and serves as a foundational sampling method, often integrated into more intricate sampling strategies. At the core of SRS lies the principle that each subject within the population holds an equal likelihood of being chosen. Consequently, a simple random sample constitutes a subset of individuals selected from a population in which each individual is chosen entirely at random, devoid of bias or predetermined criteria. In essence, every individual maintains an equal chance of selection at any given stage throughout the sampling procedure, with no prescribed guidelines governing the selection process (Fig. 7.1).

The estimated mean and standard deviation by this method are unbiased estimates of parameters in the reference population and also are very easy to calculate.

While this sampling method is highly representative of the population, this method may not be practical if the units of the population are very small. Because preparing a sampling framework for large-scale research is time consuming, tedious, and costly.

To perform SRS, the researcher must ensure that all members of the population are included in a master list, and that subjects are then selected randomly from this master list.

Systematic Sampling

Systematic sampling is a random method that is employed by most researchers due to its simplicity and its periodic quality. Systematic sampling is a method in which first a random starting point is determined and then the rest of the sample members are selected based on that random point with a fixed periodic interval, every kth (e.g., every 5th) element from a pre-provided list, where k, is defined the sampling interval as $k = \frac{N}{n}$.

Advantages of this methodology include: simplicity of selecting cases, coverage entire population and eliminating the phenomenon of clustered selection, because it ensures that each selected sample has a fixed distance apart from its former and later one. This means the sample spreads through the entire population.

Disadvantages include following a particular pattern (showing over-or under-representation) and increasing the likelihood of data manipulation. That is, in special circumstances where the list of individuals in the population is followed a certain pattern and the sampling interval is the same as the special conditions, the reliability of systematic sampling is greatly reduced. For example, when the sampling interval is 4 and we want to use systematic sampling from a list of families of 4 sorted by age, the reliability of sampling is severely reduced.

In order to perform systematic sampling, the researcher must prepare a list of everyone in the population and, by calculating the sampling interval ($k = \frac{N}{n}$), select the first sample from 1 to k and continuously add "k" to the previous sample number. For example, if the first sample is the mth member of the list, the second sample is the $(m + k)$th member of the list, the third is the $(m + 2k)$th member of the list, ... and the last one is the $[m + (n - 1)\,k]$th member of the list (Fig. 7.2).

Cluster Sampling

Cluster sampling is a method of probability sampling and a cost-effective method compared to other sampling methods. When the full and update sampling frame of all population elements is not available, we can resort only to the cluster sampling.

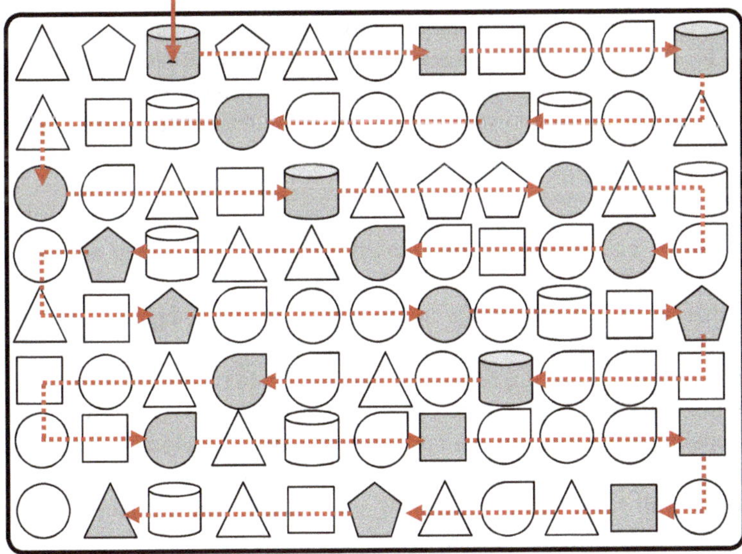

Fig. 7.2 Schematic representation of systematic sampling

Cluster sampling is often used to study large populations, particularly those that are widely geographically dispersed. Cluster sampling is preferred when There is not available a reliable list of elements and it is expensive to prepare it or even there is a list of elements, the location or identification of the units may be difficult.

Researchers typically opt to employ pre-existing units such as schools or cities as their clusters in cluster sampling methodology. In cluster sampling, the entire population is subdivided into smaller groups, referred to as clusters, and a simple random sample of these clusters is selected. Subsequently, all elements within each selected cluster are sampled, constituting a "one-stage" cluster sampling plan.

Cluster sampling offers the advantage of significantly lower operating costs. Instead of having to commute to each unit in different locations "n" times, researchers need only travel once to the location of all "n" units. Consequently, travel costs between sample units are markedly reduced, and the expenses associated with preparing a sampling framework are also minimized.

A primary motivation behind utilizing cluster sampling is to achieve cost efficiency while maintaining the desired level of accuracy. With a fixed sample size, if a substantial portion of the variation in the population occurs internally within the groups rather than between the groups, employing cluster sampling results in a smaller expected random error.

The sampling error of this design is higher than the SRS. However, the smaller the size of the clusters, the closer this method will be to simple random method and has less sampling error. When there is an intraclass correlation for the concerned variable (e.g., infectious diseases and Covid-19), the sample size should be increased compared to other methods.

An example of cluster sampling is area sampling or geographical cluster sampling. Because any geographical area is assumed as a cluster. Since surveying in geographically dispersed population is expensive, grouping several respondents within a local area into a cluster has lower costs than simple random sampling. It is usually necessary to increase the total sample size to achieve equivalent confidence and precision, but saving cost compensates for the increase in the sample size (Fig. 7.3).

Stratified Sampling

When dealing with a grand population characterized by varying subpopulations (heterogeneity), it can be statistically advantageous to independently sample each subpopulation, thereby enhancing sample precision and reducing sampling error. This process of categorizing population members into homogeneous subgroups prior to sampling is termed stratification, and the strata must form a partition of the population. For stratification to be effective, the strata must collectively cover the entire population while also being mutually exclusive—ensuring that each element belongs to precisely one stratum, with no element remaining unallocated. Additionally, all strata must be sampled, even if only one sample is taken. Each stratum is sampled separately, and the resulting estimates from each stratum are combined to form an overall estimate for the entire population.

In stratified random sampling, or stratification, the population is divided based on shared attributes or characteristics of its members (e.g., income, education levels,

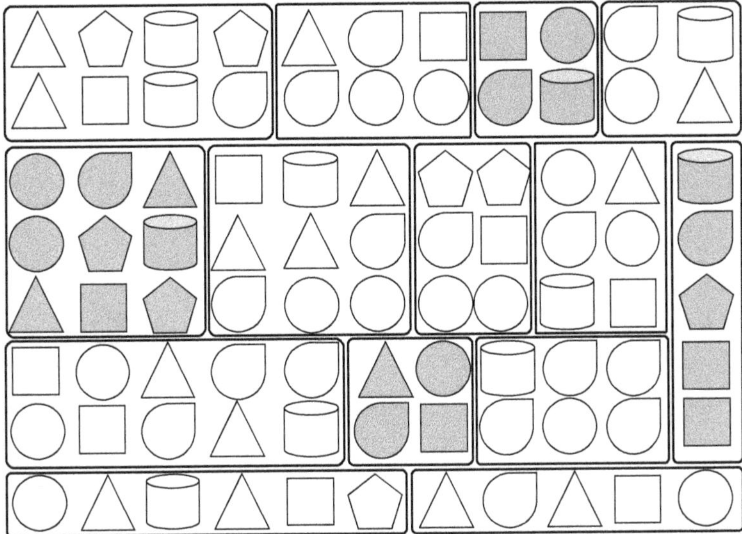

Fig. 7.3 Schematic representation of cluster sampling

race, gender, location) into homogeneous subpopulations known as strata. However, in practical applications, strata are often determined by field conditions, such as geographic regions or administrative structures. The researcher then selects samples independently from each stratum using simple random sampling, allowing him/her to estimate statistical measures for each sub-population. Stratified random sampling is also called proportional random sampling or quota random sampling (Fig. 7.4).

In this method, the probability of selecting individuals between different strata is different. But the probability of choosing sample within the same stratum is equal. The sampling ratio in all strata is the same and equal to the sampling ratio in the whole population. Therefore, the number of samples of each class is proportional to the size of that stratum. Sampling error of this method is less than simple random sampling even without increasing the sample size. For this reason, the use of this sampling method is recommended when population stratification is possible.

The advantage of this technique is that ensures that observations from all relevant strata are included in the sample.

Multi-stage Sampling

Multi-stage sampling is a kind of random sampling which is done sequentially at least two hierarchical levels. Multi-stage sampling involves selecting the sample in stages, typically considering the hierarchical or nested structure of the population. In this method, sample units are not directly chosen; rather, they are selected through larger units, taking into account the population's hierarchical organization. Multistage sampling is actually a combination form of other random sampling methods, because it divides the population into strata or clusters. It takes samples at

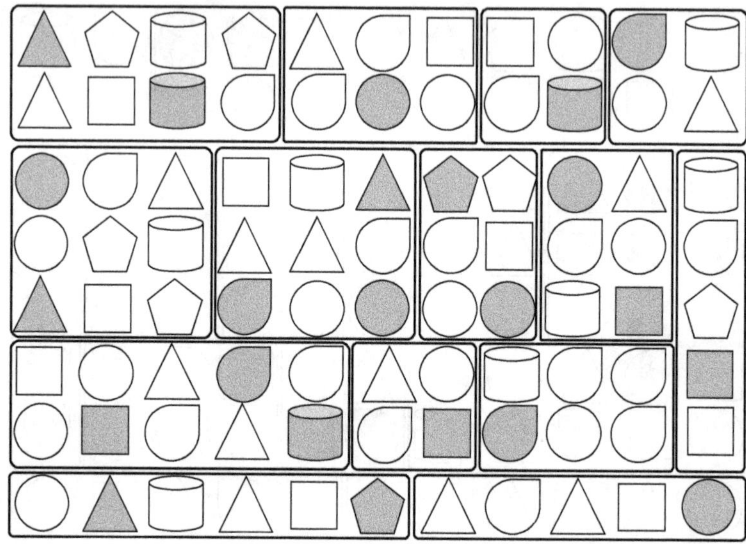

Fig. 7.4 Schematic representation of stratified sampling

each stage using smaller and smaller sampling units. In the last stage, the researcher randomly selects samples from each last layer.

The multi-stage sampling process is as follows:

In the first stage, a population is divided into several large clusters of elements, called **primary sampling units (PSU)**.

In the second stage, several primary sampling units must be selected randomly. The other unselected units are ignored.

In the third stage, from those selected primary sampling units, some smaller units, called **secondary sampling units (SSU)**, are selected again randomly. Again, unselected units are ignored at this stage.

The process of partitioning continues frequently until an appropriate list of all elements of the population is available. Sometimes, several levels of cluster selection may be applied before the ultimate sample elements are reached.

Finally, units are selected randomly within the available list of the last layers.

Usually, the primary sampling unit (PSU) is a compact geographical area, for instance, counties and the second stage unit (SSU), can be a local unit, such as cities in each county. The next stage may involve the selection of Covid-19 patients or hospitals units.

When the primary sampling units were selected in the first stage, the next step is to repeat the random sampling process of sub-sampling from the chosen units at the immediate previous stage. Therefore, it is emphasized that the sample units of the second stage are chosen from the first stage sample of units. The third stage samples of units are chosen from the second stage sample of units, and so on.

In multi-stage sampling, the fewer the stages of selection, the better and more efficient the sample would be. Because in the narrowing down, you're going to lose a little bit of accuracy and be a little bit less precise, which is the disadvantage of this method. The advantage of this method is that you can really narrow down a large population appropriately.

7.4 Sample Size

The sample size is the minimum number of cases that must be participated in each empirical study in order to gain a satisfactory answer with sufficient accuracy and confidence to the purpose and subject of the study. Because the important goal of any research study is to make inferences about the main population from a sample. Sample size is denoted by lowercase of letter n.

Tips

The sample size is calculated based on the variance of a given variable or the proportion of the occurrence of events in the population, which are usually unknown quantities. Therefore, it requires a prior knowledge of the population parameters (mean, standard deviation, and proportion). Investigators often determine these parameters estimates approximately by following practicable and acceptable ways:

(1) *Use of results of previous studies*: You can utilize the results and information of previous investigations to determine the size of the sample.
(2) *An intelligent guess*: An experienced investigator should be familiarized with population structure, be updated and aware with similar carried out researches and also be experienced with the topic of enquiry. Therefore, s/he should be able to make an intelligent guess about the standard deviation of the variable or the proportion of outcome in the geographical area of research.
(3) *Carry out a pilot study*: A small-scale pilot study can lead the researcher to the estimating mean, standard deviation, proportion (or prevalence) and so on.

Larger sample sizes indeed yield more reliable results with increased precision and statistical power. However, they also entail greater time and financial investments. Furthermore, excessively large sample sizes can be unnecessary and unethical, as they may involve subjecting more individuals to research procedures than is warranted or justified. The small sample size is not true representative of the target population then it may lead to unreliable conclusions and is unscientific and also unethical. Therefore, the necessary sample size must be determined before conducting any research project to ensure that we have a sufficient large sample size to be able to draw meaningful conclusions, without wasting resources on sampling more than we really need. Then an appropriate sample size is also important for economic and ethical reasons.

> **Tip**
>
> The test power represents the ability to detect an effect that is present in a population using a test based on a sample from that population (true negative). The test power is the complement of beta $(1 - \beta)$ and represents the probability of avoiding a false-negative conclusion.

Determination of sample size depends on a number of factors including: dispersion of the variable, confidence level or significant level, effect size (error of estimation), power of the test, population size, and sampling method. In this book, we will focus on the first four factors, and the last two factors are beyond the scope of this book.

Dispersion of the variable: The dispersion of the population is shown by variance (σ^2). The larger the variance, the larger the sample size. In studies that aim to estimate the ratio of an event, we use the value $\sigma^2 = p(1 - p)$ instead of variance. Obviously, the closer the value of p is to 0.50, the higher the product of $p(1 - p)$.

> **Tips**
>
> The effect size or the error value can be determined in two situations:
>
> (1) **The purpose is to estimate populations mean**: The error value (d) is the minimum difference between the estimated value by the research and its actual value in the population that the researcher can detect it. For example, if the purpose of the study is to estimate blood pressure mean and the researcher chooses $d = 15$, it means that the sample size of the study is able to estimate the average of blood pressure with an accuracy of 15 mm Hg (15 mm Hg. Less to 15 mm Hg.) from the actual value.
>
> (2) **The purpose is to compare the mean of the two populations**: The value of error (d) is the minimum clinical difference between two populations that is clinically relevant and acceptable to the researcher. For example, the aim of a study is to compare the effect of two treatments on the blood pressure mean, and for the researcher, the difference of $d = 15$ mm is clinically significant $(d = \mu_1 - \mu_2 = 15)$. This means that his sample size must be able to detect a difference of more than 15 mm Hg. between the blood pressures of the two populations.

Confidence level or significant level: In most health, medical, and epidemiological studies, the significance level of $\alpha = 0.05$ is sufficient. Unless in a certain circumstance, a study emphasizes a lower level of significance (e.g., 0.01) or a higher level of significance (e.g., 0.10). Significance levels of 0.01, 0.05 and 0.10

are equivalent to confidence levels of 0.99, 0.95, and 0.90, respectively. Their corresponding values, denoted by the symbol $z_{1-\frac{\alpha}{2}}$, are equal to 1.64, 1.96, and 2.58 from the normal distribution table. Obviously, the higher the confidence level, the larger the sample size.

Effect size (error of estimation): The estimation error, also called the estimation precision, depends on the objectives, facilities, and resources of the research, as well as the mean (or ratio) of the variable. The error value in the formula is denoted by the symbol d and is placed in the denominator of the fraction. The greater the error, the smaller the sample size.

> **Tip**
>
> Mostly, the calculated sample size is not a whole number and has decimals. For conservativeness, always round up the sample size to make sure that the sample size is representative of the population (whether decimals are less than 0.5 or more). Because these formulas calculate the minimum sample size required. For example, round both 245/09 and 245/87 to get the whole number of 256.

Power of the test: The test power is usually assumed to be 0.80 and the corresponding value, denoted by the symbol $z_{1-\beta}$, from the normal distribution table is 0.84. The higher the power, the larger the sample size. The researcher may increase or decrease the test power according to the research resources.

In this book, sample size calculations are explained only for the 4 situations that are most useful and beneficial for researches.

7.4.1 Sample Size Determination for Estimating a Single Population Mean

When the purpose of a research study is to estimate the mean population at the confidence level of %100 $(1 - \alpha)$ and with a predefined error of "d", the minimum number of subjects who must participate in the study (n) is obtained from the following formula:

$$z = \frac{z_{1-\frac{\alpha}{2}}^2 \times \sigma^2}{d^2}.$$

Example 1 To estimate the mean of serum cholesterol in men of a population, how many samples are needed to estimate this parameter at 95% confidence level and 5 mg/100 ml difference from its true value? According to studies conducted in similar populations, the standard deviation of serum cholesterol is 40 mg/100 ml.

Solution: In this example $d = 5$, $\sigma^2 = 40$, $1 - \alpha = 0.95$ and from the standard normal distribution table $z_{1-\frac{\alpha}{2}} = 1.96$. By replacing these values in the above formula, the minimum sample size required to estimate the mean serum cholesterol is equal to:

$$n = \frac{z_{1-\frac{\alpha}{2}}^2 \times \sigma^2}{d^2} = \frac{1.96^2 \times 40^2}{5^2} = 245.9 \cong 246.$$

That is, at least 246 samples must be included in the study.

In such cases, if we include the power of the test $(1 - \beta)$ in sample size determination, the sample size will increase and its formula will be as follows:

$$z = \frac{\left(z_{1-\frac{\alpha}{2}} + z_{1-\beta}\right)^2 \times \sigma^2}{d^2}.$$

Note that the corresponding value of the power of the test must be found as one-tailed value from the standard normal distribution table.

Example 2 If in example 1 we want to obtain the sample size with 95% confidence and 80% power of the test, what should be the minimum required sample size?

Solution: The corresponding value of 80% for power of the test can be found from the standard normal distribution table $(z_{1-\beta} = 0.84)$. By replacing these values in the above formula, the minimum sample size required to estimate the mean serum cholesterol is equal to:

$$z = \frac{\left(z_{1-\frac{\alpha}{2}} + z_{1-\beta}\right)^2 \times \sigma^2}{d^2} = \frac{(1.96 + 0.84)^2 \times 40^2}{5^2} = 501.76 \approx 502.$$

That is, at least 502 samples must be included in the study.

7.4.2   Sample Size Determination to Estimate the Proportion

When the aim of the study is to estimate the prevalence or incidence of a disease and generally to estimate the proportion (or percentage) of an event in the population, given that the variance of the p proportion is equal to $\sigma_p^2 = p(1 - p)$, the minimum sample size with the above similar formula is obtained from the following formula:

$$z = \frac{z_{1-\frac{\alpha}{2}}^2 \times p(1 - p)}{d^2}.$$

> **Tip**
>
> If a wide range for p is guessed, a value closer to 0.5 should be included in the formula. Because the closer the p-value to 0.5, the higher the product $[p\,(1-p)]$ and the sample size. But if it is not possible to guess a value for p at all, then it is necessary to put the value of $p = 0.5$ in the formula to maximize the sample size with caution.

If we include the power of the test $(1 - \beta)$ in sample size determination, the sample size will increase and its formula will be as follows:

$$z = \frac{\left(z_{1-\frac{\alpha}{2}} + z_{1-\beta}\right)^2 \times p(1-p)}{d^2}.$$

In these formulae, p is an estimate of the proportion or rate of studied events in the reference population. This estimate can be obtained from similar studies (or previous studies) or according to experts or from a pilot study.

> **Tip**
>
> The proportion is a special situation of the mean. Because if the members of the population only take the values 0 and 1, the population mean will be the same proportion.

Example 3 To estimate the prevalence of pulmonary tuberculosis (TB) in Kermanshah province, a study has been planned. No information is available on the prevalence of TB in this province or similar studies. But according to experts, the prevalence of TB in this province cannot be more than 3 per thousand (0.003). If this study wants to estimate the prevalence of TB in this province with a precision of a quarter of the true prevalence. At the 95% confidence level, how many people in this province should be considered in this project? If the power of the test is 85%, how much will the sample size be?

Solution: The researcher could not access to any information in the literature about the prevalence of TB, but infectious disease experts believe that its prevalence is not higher than 0.003. Therefore, the maximum disease prevalence rate ($p = 0.003$) that is closer to 0.5 which should be replaced in the formula.

$$n = \frac{(1.96)^2 \times 0.003 \times (1 - 0.003)}{\left(\frac{1}{4} \times 0.003\right)^2} = 20427.067 \approx 20428.$$

Therefore, at least 20,427 individuals from this province should be participated in this study in order to estimate the prevalence of pulmonary tuberculosis in Kermanshah province with 95% confidence and within a quarter of it (i.e., from 2.25 per 1000 to 3.75 per 1000).

The corresponding value of 85% for power of the test from the standard normal distribution table ($z_{1-\beta} = 1.036$). By replacing these values in the above formula, the minimum sample size required to estimate the mean serum cholesterol is equal to:

$$z = \frac{\left(z_{1-\frac{\alpha}{2}} + z_{1-\beta}\right)^2 \times p(1-p)}{d^2}$$

$$= \frac{(1.96 + 1.036)^2 \times 0.003 \times (1 - 0.003)}{\left(\frac{1}{4} \times 0.003\right)^2} = 47729.$$

7.4.3 Sample Size Determination for Comparing the Mean of Two Populations

When the purpose of the study is to evaluate and compare the mean of two independent populations together, the minimum required sample size in each population is equal to:

$$n = \frac{\left(z_{1-\frac{\alpha}{2}} + z_{1-\beta}\right)^2 \times \left(\sigma_1^2 + \sigma_2^2\right)}{(\mu_1 - \mu_2)^2},$$

where μ_1 and μ_2 are the mean of the two populations, which their difference is used as the precision or error in the denominator of the fraction. Also, σ_1^2 and σ_2^2 are the variance of the two populations, which their sum is used as the dispersion in nominator of fraction.

Example 4 It is generally believed that the weights of infants whose mothers are smoking are less than infants' weight whose mothers do not smoke. For this purpose, a study has been designed to investigate the effect of smoking in pregnant women during pregnancy on birth weight. If the average weight difference between the two groups of 300 g is clinically significant and obvious, at least how many samples should be taken from each group to reveal the weight difference between the two groups with 99% confidence and 90% of power of the test?

Solution: In this example, the following elements have been specified:

(A) The weight difference between the two groups, which is clinically important for the researcher and cannot be ignored, is 0.3 kg (300 g), i.e., ($d = \mu_1 - \mu_2 = 0.3$).

(B) Standard deviation of infant's weight in two groups. Similar studies indicate that the standard deviation of infants weight in the smoking mothers is $\sigma_1 = 0.49$ and in non-smoking mothers is $\sigma_2 = 0.37$.

(C) 90% for test power is requested and its corresponding value from the table is $z_{1-\beta} = 1.282$.

(D) The desired significance level is 99% and its corresponding value from the table is $z_{1-\frac{\alpha}{2}} = 1.28$.

By replacing the above values in the relevant formula, we have:

$$n = \frac{\left(z_{1-\frac{\alpha}{2}} + z_{1-\beta}\right)^2 \times \left(\sigma_1^2 + \sigma_2^2\right)}{(\mu_1 - \mu_2)^2} = \frac{(2.58 + 1.28)^2 \times \left(0.49^2 + 0.37^2\right)}{(0.3)^2} = 62.5 \approx 63$$

Therefore, at least 63 random samples should be taken from each group to reveal the average weight difference of 0.3 kg in these two groups of infants with a 99% probability.

In order to facilitate the calculation of the sample size, the resultant of the expression $\left(z_{1-\frac{\alpha}{2}} + z_{1-\beta}\right)^2$ is calculated by different values of α and $1 - \beta$ and is presented in Table 7.1.

> **Tip**
>
> In the sample size formula, the standard deviations and means measurement unit should be the same, both in grams, both in kilograms, both in meters or both in centimeters.

Table 7.1 Coefficients of $\left(z_{1-\frac{\alpha}{2}} + z_{1-\beta}\right)^2$ for sample size calculations

Significant level of α						
The value for power of test $(1 - \beta)$ is one-tailed		Level	0.10	0.05	0.01	0.001
	Level	Value	(1.645)	(1.96)	(2.58)	(3.29)
	0.50	(0.000)	2.69	3.84	6.66	10.82
	0.60	(0.253)	3.58	4.90	8.03	12.56
	0.70	(0.524)	4.68	6.17	9.64	14.56
	0.80	(0.842)	6.16	7.85	11.71	17.08
	0.85	(1.036)	7.19	8.99	13.08	18.71
	0.90	(1.282)	8.54	10.51	14.92	20.91
	0.95	(1.645)	10.76	12.96	17.81	24.31
	0.99	(2.58)	17.81	20.61	26.63	34.46

7.4.4 Sample Size Determination to Compare the Proportion of Two Populations

When the purpose of the study is to evaluate and compare the observed proportion of two independent populations together, the minimum required sample size in each population is equal to:

$$n = \frac{\left[z_{1-\beta}\sqrt{p_1 \cdot q_1 + p_2 \cdot q_2} + z_{1-\frac{\alpha}{2}}\sqrt{2 \times \bar{p}(1-\bar{p})}\right]^2}{(p_1 - p_2)^2}.$$

In above formula, p_1 and $q_1 = 1 - p_1$ are diseased and undiseased proportions in the first population, p_2 and $q_2 = 1 - p_2$ are diseased and undiseased proportion s in the second population, $\bar{p} = \frac{p_1+p_2}{2}$ and $\bar{q} = 1 - \bar{p}$.

Where p_1 and $q_1 = 1 - p_1$ are the proportion of individuals with and without characteristics in the first population and p_2 and $q_2 = 1-p_2$ proportion of individuals with and without characteristics in the second population. The difference between p_1 and p_2 is used on the denominator of the fraction as precision or error. Also, the average of $\bar{p} = \frac{p_1+p_2}{2}$ and $\bar{q} = 1 - \bar{p}$ is used in nominator of fraction.

In some statistical books, applying the relations of $\sigma_1^2 = p_1(1 - p_1)$ and $\sigma_2^2 = p_2(1 - p_2)$, the sample size formula is calculated as follows:

$$n = \frac{\left(z_{1-\frac{\alpha}{2}} + z_{1-\beta}\right)^2 \times \left[p_1(1 - p_1) + p_2(1 - p_2)\right]}{(p_1 - p_2)^2}.$$

Or more simply, they use the following formula.

$$n = \frac{\left(z_{1-\frac{\alpha}{2}} + z_{1-\beta}\right)^2 \times 2 \times \bar{p}(1 - \bar{p})}{(p_1 - p_2)^2}.$$

The results of the calculations of all three formulas are very close to each other.

Example 5 The prevalence rate of Covid-19 is thought to be lower in rural areas (due to fewer crowds) than in urban areas. At least how many samples should be taken in each area so that with 90% power and at a significance level of 0.05, we can detect the difference in the prevalence of Covid-19 between urban and rural areas? Similar studies indicate that the prevalence of Covid-19 in rural and urban areas is 5 and 11 per 1000, respectively.

Solution: The following information is provided in this example:

$$p_1 = 0.005, \ p_2 = 0.011, \ q_1 = 1 - 0.005 = 0.995,$$

$$q_2 = 1 - 0.01 = 0.989, \ \bar{p} = \frac{0.005 + 0.011}{2} = 0.008, \ \bar{q} = 1 - 0.008 = 0.982$$

$$z_{1-\frac{\alpha}{2}} = 1.96 \quad \text{and} \quad z_{1-\beta} = 1.282$$

$$n = \frac{\left[1.282\sqrt{0.005 \times 0.995 + 0.011 \times 0.989} + 1.96\sqrt{2 \times 0.008 \times 0.992}\right]^2}{(0.011 - 0.005)^2}$$

$$= \frac{\left[1.282\sqrt{0.015854} + 1.96\sqrt{0.015872}\right]^2}{(0.006)^2} = \frac{0.16675}{0.000036} = 4631.92$$

$$\approx 4632.$$

The required sample size for rural and urban areas is 4632 samples. Sample size calculation by another two suggested formulae is:

$$n = \frac{\left(z_{1-\frac{\alpha}{2}} + z_{1-\beta}\right)^2 \times \bar{p}(1 - \bar{p})}{(p_1 - p_2)^2}$$

$$= \frac{(1.96 + 1.28)^2 \times (0.005 \times 0.995 + 0.011 \times 0.989)}{(0.011 - 0.005)^2}$$

$$= \frac{10.51 \times 0.015854}{0.000036} = 4629$$

$$n = \frac{\left(z_{1-\frac{\alpha}{2}} + z_{1-\beta}\right)^2 \times 2 \times \bar{p}(1 - \bar{p})}{(p_1 - p_2)^2}$$

$$= \frac{(1.96 + 1.28)^2 \times 2 \times 0.008 \times 0.992}{(0.011 - 0.005)^2} = 4634.$$

Example 6 It seems that the mortality rate of Covid-19 patients among men is higher than that of women. What is the minimum number of subjects who need to be sampled in each group of men and women in order to be able to achieve the difference in mortality rates of Covid-19 patients among men and women with a significant level of 0.05% and 90% power of the test? Previous studies show that mortality rates among men and women affected by care are 8.8% and 4.4% of patients and hospitalized.

Solution: The following information has been given on the example:

$$p_1 = 0.088, \ p_2 = 0.044, \ q_1 = 1 - p_1 = 1 - 0.088 = 0.912,$$
$$q_2 = 1 - q_2 = 1 - 0.044 = 0.956,$$
$$\bar{p} = \frac{p_1 + p_2}{2} = \frac{0.088 + 0.044}{2} = 0.066 \quad \text{and} \quad \bar{q} = 1 - \bar{p}$$
$$= 1 - 0.066 = 0.934.$$

By substituting them in each of the three given formulas, the required sample size is obtained.

$$n = \frac{\left[1.282\sqrt{0.088 \times 0.912 + 0.044 \times 0.956} + 1.96\sqrt{2 \times 0.066 \times 0.934}\right]^2}{(0.088 - 0.044)^2}$$

$$= \frac{\left[1.282\sqrt{0.12232} + 1.96\sqrt{0.12329}\right]^2}{(0.044)^2} = \frac{1.292}{0.001936} = 667.26$$

$$\approx 668$$

$$n = \frac{\left(z_{1-\frac{\alpha}{2}} + z_{1-\beta}\right)^2 \times 2 \times \bar{p}(1-\bar{p})}{(p_1 - p_2)^2} = \frac{(1.96 + 1.28)^2 \times 2 \times 0.066 \times 0.934}{(0.088 - 0.044)^2}$$

$$= 668.51 \approx 669$$

$$n = \frac{\left(z_{1-\frac{\alpha}{2}} + z_{1-\beta}\right)^2 \times \left[p_1(1-p_1) + p_2(1-p_2)\right]}{(p_1 - p_2)^2}$$

$$= \frac{(1.96 + 1.28)^2 \times (0.088 \times 0.912 + 0.044 \times 0.956)}{(0.088 - 0.044)^2} = \frac{10.51 \times 0.12232}{0.001936}$$

$$= 664.04 \approx 665.$$

Therefore, at least 669 samples in each group should participate in order to be able to achieve an actual difference in mortality rate between male and female patients with a 90% probability and a significant level of 0.05.

7.5   Exercises

1. Based on previous studies, the variance of height in men and women is 160 and 140 cm respectively. We want to estimate the height of people in population with an error of 4 cm and at a significance level of 5%:

    A. How many samples are needed to estimate women's height?
    B. How many samples are needed to estimate omen's height?
    C. Explain the reason for the difference in the sample sizes required for men and women.

2. Should the relation $\alpha + \beta = 1$ be considered in estimating the sample size? Why?

3. To compare the height of men and women in the population with 90% probability and 95% confidence, how many samples are needed to find out the difference in height of 10 cm between men and women. However, we know that the variance of height in these two groups is 155 and 165 cm, respectively.

4. To make X-ray tubes, a tungsten disk known as anode disk is used. The mean and standard deviations of the radius of this disk are 26 and 4 cm, respectively. To prevent disk damage, its radius should not exceed 34 cm. How many samples are needed so that with a probability of 90% the disk radius is not more than the standard.

5. The researcher wants to estimate the proportion of pregnant women seeking prenatal care in the first trimester of pregnancy. How many samples should be taken to make this ratio at the level of significance of 5% and the actual difference between the proportion of applicants and subsequent pregnancies is 20%. Previous similar studies show that this proportion is not more than 40%.

6. In the design of dental radiography devices, the aluminum thickness in the radiographic device is lower than the panorex device. If the average thickness difference between the filters of the two devices is 2.5 mm, what is the minimum required sample in each group to reveal the above thickness difference with 95% confidence and 90% power?

7. Assuming the heights of students in a college campus are normally distributed with a Sd = 5 inch, find the required minimum sample size to construct a 95% confidence interval for mean with a maximum error = 0.5 inch.

8. We are interesting to compare cholesterol concentration between male and female students in one health sciences faculty. Since cholesterol level is numerical kind of data basically, you will be comparing their mean between two groups of people (here is gender, either male or female). Search for standard deviation and mean of the two groups from reliable sources and find the minimum sample size required for this comparison at the significance levels of 5% and 10%.

9. To detect a statistically significant mean difference between the treatment and placebo groups (1:1 allocation), how many samples are needed in each group? We assume that the antiviral effect was 87% and 93%, respectively.

10. Here you are going to find smoking habit (yes/no) in one university population. Smoking habit is categorical data, and in a pilot study, 10 out of 30 university students experienced smoking. How many samples are needed in this university campus?

Chapter 8
Estimation and Statistical Inference

8.1 Introduction

In the preceding chapter, we discussed how sampling serves the purpose of obtaining results from a random sample of variables under study, enabling us to generalize those results to the larger population. The fundamental principle underlying all probability (random) sampling methods is that data is collected from a random sample of subjects, and this information is then used to draw inferences about the entire population. However, it's important to note that different samples may yield different conclusions. Each sample obtained is not an exact replica of the original population, leading to deviations from the true population values. Consequently, we cannot assert the absolute correctness of our conclusions. Instead, the conclusion derived from a sample represents our perception and inference of the true population value, albeit with a certain degree of uncertainty. This process is known as statistical inference or inferential statistics. Statistical inference employs the language of probability to quantify the reliability of our conclusions. Essentially, statistical inference involves using statistical methods to draw conclusions about unknown aspects of a population (population parameters) based on a random sample drawn from that population. The outcomes derived from this random sample are referred to as "estimations". The two primary approaches to statistical inference are estimation and hypothesis testing, which we will delve into in the subsequent chapters.

8.2   Estimation

When evaluating a new treatment on a random sample of individuals with a specific disease within a population, it's improbable to obtain identical results when replicating the study with another random sample from the same population. Even if multiple random samples are drawn from the population of individuals with the disease and the same test statistic (such as the mean) is calculated for each sample, it's highly unlikely that all sample means will match precisely. Variations will inevitably occur, even in the absence of errors, due to chance factors. Collating all the means obtained from different samples forms a sampling distribution of the mean. The standard deviation of this sampling distribution is referred to as the standard error, which represents the variability in the mean obtained from all random samples taken from the population. The standard error aids in estimating the true population parameter.

To better understand the sampling variations, we consider 7 numbers (8–7–6–5–4–3–2) as the reference population. The mean and variance of this hypothetical population are 5.00 and 5.60, respectively.

We now estimate the mean of this population by drawing a random sample of 3. Using a simple random sampling method, the first sample consists of numbers 4, 6 and 1 with the mean of 3.67. Another sample of 3 includes numbers 2, 3, and 8 with the mean of 4.33 will be obtained. If sampling is repeated up to 20 times, the extracted samples 3 from this population and their means are given in Table 8.1.

The estimated means in the above table show the sample variations of how the estimation of a parameter will vary from sample to sample. As can be seen, none of these means are exactly equal to the mean of the reference population. For this reason, calculating interval estimate is necessary.

If we extract all 210 possible samples (3 out of 7, without replacement) from this population and calculate their means, these means themselves will have a distribution named the "sample distribution of the mean".

The mean and variance of these 20-sample means are 4.4335 and 0.7177 respectively. The mean of the reference population, which is 5.00, is close to the mean of the samples, but the variance of the reference population, which is 5.60, is significantly larger than the variance of the sample means. As mentioned, if all 210

Table 8.1   Extracted samples of 3 from a population of 7 elements

	x_i	$\bar{x}$	x_i	$\bar{x}$	x_i	$\bar{x}$	x_i	$\bar{x}$
Samples	1–6–4	**3.67**	2–7–8	**5.67**	6–8–2	**5.33**	3–7–5	**5.00**
	8–2–3	**4.33**	7–8–3	**6.00**	3–2–4	**3.00**	6–2–4	**4.00**
	2–6–7	**5.00**	6–5–2	**4.33**	3–6–4	**4.33**	7–8–4	**6.33**
	4–8–5	**5.67**	7–2–8	**5.67**	2–8–5	**5.00**	3–6–5	**4.67**
	2–6–5	**4.33**	4–7–3	**4.67**	6–1–4	**3.67**	2–3–7	**4.00**

possible samples are taken and their means are calculated, it is proved that the average of these means will be exactly equal to 5.00 and their variance will be equal to 1.867 which is exactly proportional to the inverse of the sample size (3). That is, its variance is equal to one-third of the variance of the reference population, and it can be said that the sample mean is an unbiased estimate of the population mean, i.e.:

$$\mu_{\bar{x}} = \mu_x.$$

If we plot the sample means and the reference population mean, it is observed that the central location of both is the same. But the dispersion of the reference population will be greater than the dispersion of the samples, because:

$$\sigma_{\bar{x}}^2 = \frac{\sigma^2}{n}.$$

The square root of variance of the sample means is called the standard error and is denoted by Se, i.e.:

$$Se_{\bar{x}} = \frac{\sigma}{\sqrt{n}}.$$

Usually, because the population standard deviation is not known, an estimate of $S = \sqrt{\frac{(x-\bar{x})^2}{n-1}}$ is used. So:

$$Se = \frac{S}{\sqrt{n}}.$$

Here, we will explain another example:

Example 1 The mean and standard deviations of diastolic blood pressure of 250 nurses are $\sigma = 9.4\,$mm/HG, $\mu = 78.2\,$mm/HG. We assume these 250 nurses as the reference population and write their blood pressure measurements on similar papers and put them in an urn. Bring this urn into a class of 30 students and ask them to draw 10 samples each one. After recording the blood pressure, return it to the urn. The next student, after mixing thoroughly, takes 10 samples and records the blood pressure. All 30 students do this and calculate their sample mean of 10 ($\bar{x}$). Finally, we have 30 different means obtained from 10 different samples. Each of these is an estimate of the reference population mean of 250 flight attendants. But if we take the average of these entire 30 sample means, the value of $\mu = 78.23\,$mm/HG is obtained, which is very close to the mean of the reference population μ. The standard deviation of these 30 means is 3.01 which is close to $\frac{\sigma}{\sqrt{n}} = \frac{9.4}{\sqrt{10}} = 2.97$. Now if students take 20 samples instead of 10 samples, and we repeat the same cycle, then the mean of the

20s is $\mu = 78.14\,\text{mm/HG}$, which is very close to the mean of the main population $\mu = 78.20$. The standard deviation of the means of 20 is equal to 2.07, which has a good agreement with $\sigma/\sqrt{n} = 9.4/\sqrt{20} = 2.10$.

For all the measures and indicators presented in the previous chapters, two types of estimation can be calculated, point estimate and interval estimate.

> **Tips**
>
> Any estimator should have two important properties:
>
> - **Unbiasedness**: The unbiasedness property of the estimators means that, if we have many samples for the random variable and we calculate the estimated value corresponding to each sample, the average of these estimated values approaches the unknown parameter. In other words, an expected value equal to a population parameter is estimated.
> - **Minimum variance**: It means that it has lower **variance** than any other **unbiased estimator** for all possible values of the parameter.

8.3   Estimation of the Mean

8.3.1   Point Estimate of the Mean

In statistics, the process of finding an approximate value of the parameter in a population (such as the mean or proportion), from random samples taken of the population is called point estimation. In other words, a point estimate is a single numerical value that estimates of an unknown parameter. The best point estimate of the population mean μ is the sample mean $(\overline{x})$, and the best estimate of proportion population (p) is the sample proportion $(\hat{p})$, which are estimated by the following formulae:

$$\hat{p} = \frac{r}{n}, \ \overline{x} = \frac{\sum_{i=1}^{n} f_i.x_i}{n}.$$

In Table 6.2, the values 3.67, 3.00, 6.00, 4.33, and… each are point estimates of the mean. This estimate varies by changing the sample size and even drawing another sample. Sometimes, this deviation is too high. As can be seen in Table 6.2, the means of the 3 from the same population varies from sample to sample. In this example, it fluctuated between 3.00 and 6.33. These fluctuations are due to sampling changes and are always present. This fluctuation will be less in larger samples and more in smaller samples. As can be seen, none of the means of these samples is exactly equal to the reference population mean. However, some are closer and some are farther away. There is no guarantee that the mean of subsequent samples will fully match to the mean of the reference population. Therefore, the mean of the population will not be exactly equal to its estimate. But as the sample size increases, the probability of the two becoming closer increases. In terms of probabilities, the chances of matching these two values with each other are very low.

8.3.2 *Interval Estimate of the Mean*

The goal of statistical analysis is to estimate a true population parameter, which is a constant but unknown value, using sample data. However, a point estimate derived from sample data represents the most probable value of the true population parameter, yet it tends to be imprecise. Interval estimation serves as a method to address and enhance this imprecision by furnishing a range of values within which the true population parameter is likely to lie. For instance, the 95% interval estimation includes the range within which we are 95% certain that the true population parameter will lie. On the other words, an interval estimate displays the probability that a population parameter will fall between a pair of values around the estimated value and measures the degree of uncertainty in a sampling method.

An interval estimate is obtained by adding and subtracting a value to the point estimate of the mean (or proportion). So, using standard error, we create a range called "interval estimate" or "confidence interval". This range is expanded on both sides of the mean (or proportion) and is a multiplication of the standard error and a certain probability includes the true value of the population parameter.

$$\hat{p} - d \quad \text{to} \quad \hat{p} + d$$
$$d = \text{Se} \times z_{1-\frac{\alpha}{2}}.$$

Interval estimates, also known as confidence intervals (CIs), offer a range of plausible values for a population parameter, providing insight into the precision of the measured variable. Additionally, CIs may offer valuable information regarding the clinical significance of results. With the interval estimate, the probability of matching this interval with the population parameter (proportion or mean) increases. In terms of probabilities, the chances of a point being at an interval also increase.

Tips

If a measure is taken from entire population, calculating the confidence interval or a statistical test has no advantage. Because the confidence interval and hypothesis test are to provide uncertainty due to sampling from population. Therefore, if the entire population is measured, there is no need to worry about sampling error and there is no need to calculate the confidence interval and statistical tests.

This argument has another side. Others argue that even if the measures are taken from entire population, that population represents only a sample of subjects from a hypothetical super-population. In other words, the study population, even if enumerated completely without any sampling, represents merely a biologic sample of a larger set of subjects; therefore, a confidence interval and statistical tests are justified.

The validity of each argument depends on the context. If consider voter priorities, it is the actual population they are interested in, and the first argument is sound. But for biologic phenomena, what happens in an actual population may be of less interest than the biologic norm that describes the super-population. Thus, for biologic phenomena, the second argument is more convincing.

The interval estimate is influenced by the following 3 factors:

Standard deviation of the variable: The more dispersed the variable, the less homogeneity the data and the larger the standard deviation and the interval estimate will be wider and broader.

The confidence level or probability of containing the true mean: The more likely that the true parameter (e.g., mean) will fall within the interval estimate, the estimated interval should be wider. For instance, 99% intervals, which are from $-3 \times$ Sd to $\mu +3 \times$ Sd, are wider than 95% intervals, which are from $\mu - 2 \times$ Sd to $\mu + 2 \times$ Sd, and are wider than 90% intervals, which are from $\mu - 1 \times$ Sd to $\mu + 1 \times$ Sd. If you are willing to be only 50% sure that an interval contains the true value, then it can be much narrower.

The sample size of the study: The larger the sample size, the narrower and closer the interval estimates will be.

Therefore, d can be defined as follows:

$$d = z_{1-\frac{\alpha}{2}} \frac{\sigma}{\sqrt{n}} = z_{1-\frac{\alpha}{2}} \times \text{Se},$$

where $\text{Se} = \frac{\sigma}{\sqrt{n}}$ is the standard error or standard deviation of the mean. By adding and subtracting the above value from the sample point estimated mean, an interval estimate or confidence interval is obtained.

$$\bar{x} - z_{1-\frac{\alpha}{2}} \frac{\sigma}{\sqrt{n}} \quad \text{and} \quad \bar{x} + z_{1-\frac{\alpha}{2}} \frac{\sigma}{\sqrt{n}}.$$

The confidence interval of $\bar{x} + z_{1-\frac{\alpha}{2}} \frac{\sigma}{\sqrt{n}}$ to $\bar{x} - z_{1-\frac{\alpha}{2}} \frac{\sigma}{\sqrt{n}}$ is interpreted that this interval contains the true mean of the population with probability of $(1 - \alpha) \times 100\%$. That is $(1 - \alpha) \times 100\%$ we are sure that the true mean of the population (μ) is at this interval.

The smaller value is named the lower limit of the estimate, the larger value upper limit of the estimate, and both are called "confidence limits".

In epidemiological studies, interval estimates of the mean are usually calculated with 95% confidence (sometimes 90 or 99%). Their corresponding values of $z_{1-\frac{\alpha}{2}}$ are 2.58 for 99%, 1.96 for 95%, and 1.64 for 90%. If another confidence level is considered by the researcher, the corresponding value can be obtained from the standard normal distribution table.

Example 2 Male diastolic blood pressure in a population has an approximately normal distribution with a variance of $\sigma^2 = 10.4$. Based on a sample of 25, the mean diastolic blood pressure of men was estimated to be $\bar{x} = 8.5$:

Calculate the 90% interval estimate of the mean.

Obtain 95% confidence interval of the mean.

Determine limits that can contain the true mean with the 99% probability.

Solution: The three terms "interval estimate", "confidence interval", and "limits determination" are synonymous, and similar methods are used to calculate. Based on given information, $\bar{x} = 8.5$ and $Se = \frac{\sigma}{\sqrt{n}} = \frac{\sqrt{10.4}}{\sqrt{25}} = \frac{3.22}{5} = 0.645$. Assuming two-tailed estimation, $z_{0.90} = 1.64$, $z_{0.95} = 1.96$, $z_{0.99} = 2.58$

$$8.5 \pm 1.64 \times 0.645 = 8.5 \pm 1.06 = (7.44 \quad \text{to} \quad 9.56)$$
$$8.5 \pm 1.96 \times 0.645 = 8.5 \pm 1.26 = (7.24 \quad \text{to} \quad 9.76)$$
$$8.5 \pm 2.58 \times 0.645 = 8.5 \pm 1.66 = (6.84 \quad \text{to} \quad 10.16).$$

It can be seen that with increasing confidence level, the confidence interval has become wider.

Example 3 If in the Example 2, the mean is based on a sample of 256. Recalculate the mentioned estimates.

Solution: According to the given information: $\bar{x} = 8.5$ and $Se = \frac{\sigma}{\sqrt{n}} = \frac{\sqrt{10.4}}{\sqrt{256}} = \frac{3.22}{16} = 0.201$. Assuming two-tailed estimation, $z_{0.90} = 1.64$, $z_{0.95} = 1.96$, $z_{0.99} = 2.58$

$$8.5 \pm 1.64 \times 0.201 = 8.5 \pm 0.33 = (8.17 \quad \text{to} \quad 8.83)$$
$$8.5 \pm 1.96 \times 0.201 = 8.5 \pm 0.39 = (8.11 \quad \text{to} \quad 8.89)$$
$$8.5 \pm 2.58 \times 0.201 = 8.5 \pm 0.52 = (7.98 \quad \text{to} \quad 9.02).$$

It is observed that with increasing the sample size, the standard error becomes smaller and the confidence interval becomes narrower and closer.

Example 4 If in the Example 2 the mean is based on a sample of 25, but the population variance is $\sigma^2 = 15.8$. Recalculate the mentioned estimates.

Solution: According to the given information: $\bar{x} = 8.5$ and $Se = \frac{\sigma}{\sqrt{n}} = \frac{\sqrt{15.8}}{\sqrt{25}} = \frac{3.97}{5} = 0.795$. Assuming two-tailed estimation, $z_{0.90} = 1.64$, $z_{0.95} = 1.96$, $z_{0.99} = 2.58$

$$8.5 \pm 1.64 \times 0.795 = 8.5 \pm 1.30 = (7.20 \quad \text{to} \quad 9.80)$$
$$8.5 \pm 1.96 \times 0.795 = 8.5 \pm 1.56 = (6.94 \quad \text{to} \quad 10.06)$$
$$8.5 \pm 2.58 \times 0.795 = 8.5 \pm 2.05 = (6.45 \quad \text{to} \quad 10.55).$$

It can be seen that when the dispersion of the population is greater, the standard error is larger and the confidence interval widens.

By comparing the results of these three examples, the role of the above three components in estimating the confidence interval can be realized.

It should be noted that the true value of variance of the population is usually unknown to the researchers and they should use the estimated variance of S^2 instead. In this case, if the sample size is small (less than 30), another coefficient called t is used instead of $z_{1-\frac{\alpha}{2}}$.

$$d = t_{(df,1-\frac{\alpha}{2})} \times \frac{S}{\sqrt{n}}.$$

The degree of freedom ($df = n - 1$) plays a role in this coefficient, the value of which is obtained from the t distribution table at the end of statistical books.

Therefore, when the variance of the population is unknown and estimated from the sample, the interval estimate of the mean is calculated as follows:

$$\bar{x} + t_{(df,\ 1-\frac{\alpha}{2})} \times \frac{S}{\sqrt{n}} \quad \text{to} \quad \bar{x} - t_{(df,\ 1-\frac{\alpha}{2})} \times \frac{S}{\sqrt{n}}.$$

Example 5 The pain relief time of a new drug for 6 patients with arthritis is: 4.3–3.7–2.5–4.9–2.4–2–2 h. Calculate a 95% confidence interval of mean.

Solution: In this example, the mean is $\bar{x} = \frac{4.3+3.7+\cdots+2.2}{6} = 3.3$, standard deviation $Sd = 1.13$, $n = 6$ and $df = 1-6 = 5$, and standard error $Se = \frac{Sd}{\sqrt{n}} = \frac{1.13}{\sqrt{6}} = 0.46$. The critical value of **t** with a $df = 5$ is equal to $t_{(5,\ 1-\frac{0.05}{2})} = t_{(5,\ 0.975)=2.57}$. Therefore, the 95% confidence interval is calculated as follows:

$$3.3 \pm 2.57 \times 0.46 = 3.3 \pm 1.18 = (2.12 \quad \text{to} \quad 4.48)\,\text{h}.$$

8.4   Point and Interval Estimation of a Population Proportion

In epidemiology and health sciences, we are usually dealing with the proportion of subjects with a feature or the rate of a disease in the population. Because many attributes and variables are used in binomial scale (yes, no). That is, they indicate "occurrence or non-occurrence" or "existence or non-existence" in population. Variables whose responses were "positive and negative", "dead and alive", "healthy and sick", "failure and success", "yes and good", "vaccinated and not vaccinated", "under 15 years and over 15", "male and female", etc. are examples of binomial variables. For example, what proportion of infants are boys? What is the prevalence rate of depression in the population? What percentages of patients recover with a particular treatment? Or what is the immunity of a vaccine? These questions have different forms and shapes that require another answer.

Tip

When the sample size (n) is large, the critical values of t and z are approximately the same and the z distribution can be used instead of t.

In fact, the proportion itself is a kind of mean that can be reached by making changes in the formula of estimating the mean to estimate the proportion in population. Therefore, the same logic expressed for estimating the mean can be generalized to estimate the proportion.

Tip

If we assign a value of 1 to all subjects who have a feature or an attribute and a value of 0 to people who do not have that feature, their mean will be equal to the proportion.

If a sample of n is selected from the study population and r subjects have the interested characteristic, the p proportion estimate is the number of individuals who have the characteristic (r) divided by the total number of samples (n), i.e.:

$$\hat{p} = \frac{r}{n}.$$

This is the relative frequency or point estimate of the proportion in the population. But as mentioned about the mean, the proportion estimate is also subject to sampling variations, and its value varies from sample to sample and its confidence interval should be calculated.

Like any other estimator, the confidence interval is obtained by the following equation.

$$\text{Estimator} \pm (\text{Constant coefficient}) \times (\text{Standard error}).$$

The standard error of proportion is equal to:

$$\sigma_p = \sqrt{\frac{p(1-p)}{n}} = \sqrt{\frac{r(n-r)}{n^3}}.$$

Interval estimates or confidence intervals of the proportion can be obtained.

$$\frac{r}{n} \pm z_{1-\frac{\alpha}{2}} \sqrt{\frac{r(n-r)}{n^3}} \quad \text{or} \quad \frac{r}{n} \pm z_{1-\frac{\alpha}{2}} \sqrt{\frac{\hat{p}(1-\hat{p})}{n}}.$$

This range means that the above interval with a probability of $\%(1-\alpha) \times 100$ includes the true value of the proportion in the population.

Example 6 In the only maternity hospital in a city, 84 babies were born in one month, of which 37 were girls. Calculate the point estimate and 95% confidence interval of the proportion of girl birth rate in this city.

Solution: Point estimate of girl birth rate is:

$$\hat{p} = \frac{r}{n} = \frac{37}{84} = 0.44.$$

The 95% confidence interval or interval estimate for girl birth rate is:

$$\frac{r}{n} \pm z_{1-\frac{\alpha}{2}} \sqrt{\frac{r(n-r)}{n^3}} = \frac{37}{84} \pm 1.96 \sqrt{\frac{37(84-37)}{84^3}} = 0.44 \pm 0.106$$
$$= (0.334 \quad \text{to} \quad 0.546).$$

Therefore, the 95% interval estimate for girl birth proportion in this city is between 0.334 and 0.546, and this means that with 95% confidence, the above interval covers the true proportion of girl birth in this city.

Example 7 In an area of 800 patients with Covid-19 have been hospitalized and 32 out of 800 are dead. Find 90% confidence interval for the mortality rate in this area.

Solution: Point estimate of mortality rate is:

$$\hat{p} = \frac{r}{n} = \frac{32}{800} = 0.04.$$

The 90% confidence interval or interval estimate for mortality rate is:

$$\frac{r}{n} \pm z_{1-\frac{\alpha}{2}} \sqrt{\frac{r(n-r)}{n^3}} = \frac{32}{800} \pm 1.645 \sqrt{\frac{32(800-32)}{800^3}} = 0.04 \pm 0.011$$
$$= (0.029 \quad \text{to} \quad 0.051).$$

Therefore, the 90% interval estimate for mortality rate in this area is between 0.029 to 0.051, and this means that with 90% confidence, the above interval covers the true rate of mortality by Covid-19 in this area.

If n is large enough such that both $n.p$ and $n(1-p)$ are greater than 10, or both n and $(n-r)$ are greater than 10, the distribution of $\hat{p} = \frac{r}{n}$ can be assumed as normal distribution, i.e.:

$$\hat{p} = \frac{r}{n} \sim N\left(\hat{p}, \sqrt{\frac{\hat{p}(1-\hat{p})}{n}}\right) \approx N\left(\frac{r}{n}, \sqrt{\frac{r(n-r)}{n^3}}\right).$$

In a sample of n, the probability of the sample proportion is greater than p_1 is:

$$p\left(\frac{r}{n} > p_1\right) = \frac{p\left(\frac{r}{n} - p\right)}{\sigma_p} > \frac{(p_1 - p)}{\sigma_p} = p\left(z > \frac{p_1 - p}{\sigma_p}\right).$$

Similarly, the probability of the sample proportion is less than p_2 is:

$$p\left(\frac{r}{n} < p_2\right) = p\left(z < \frac{p_2 - p}{\sigma_p}\right).$$

As a result, the probability of the sample proportion is between p_1 and p_2 is:

$$p\left(p_1 < \frac{r}{n} < p_2\right) = p\left(\frac{p_1 - p}{\sigma_p} < z < \frac{p_2 - p}{\sigma_p}\right).$$

The corresponding value of z can be obtained from the standard normal distribution table.

Example 8 The proportion of girl birth in Example 6 was 0.44; calculate the following probabilities in a sample of 75.
The probability of girl birth is less than a third.
The probability of girl birth is greater than two-thirds.
The probability of girl birth is between 0.50 and 0.65.

Solution: According to the provided information in the example, proportion estimate and estimation of its standard deviation are:

$$p = 0.44 \quad \text{and} \quad \sigma_p = \sqrt{\frac{p(1-p)}{n}} = \sqrt{\frac{0.44(1-0.44)}{75}} = 0.0573.$$

The probability of girl birth less than a third is:

$$p\left(\frac{r}{n} < p_2\right) = p\left(z < \frac{\frac{1}{3} - 0.45}{0.0573}\right) = p(z < -2.033).$$

According to the standard normal distribution table, we will have:

$$p(z < -2.033) = 0.5 - 0.4778 = 0.0212.$$

The probability of girl birth greater than two-thirds is:

$$p\left(\frac{r}{n} > p_1\right) = p\left(z > \frac{\frac{2}{3} - 0.45}{0.0573}\right) = p(z > 0.217).$$

According to the standard normal distribution table, we will have:

$$p(z > 0.217) = 0.5 - 0.086 = 0.414.$$

The probability of girl birth between 0.50 and 0.65 is:

$$p\left(p_1 < \frac{r}{n} < p_2\right) = p\left(\frac{p_1 - p}{\sigma_p} < z < \frac{p_2 - p}{\sigma_p}\right)$$
$$= p\left(\frac{0.50 - 0.45}{0.0573} < z < \frac{0.65 - 0.45}{0.0573}\right)$$
$$= p(0.871 < z < 3.484).$$

According to the standard normal distribution table, we will have:

$$p(0.871 < z < 3.484) = p(z < 3.484) - p(z < 0.871)$$
$$= 0.4988 - 0.308 = 0.1908.$$

8.5 Central Limit Theorem

The central limit theorem (CLT) is a fundamental theorem in probability and statistics, with broad applications across various fields such as medicine, health, epidemiology, psychology, and social sciences. Specifically, the CLT concerning the distribution of means enables us to validate the assumption of normal distribution, thereby facilitating the utilization of statistical formulas that necessitate normality in our datasets. Furthermore, the CLT permits the application of a normal distribution in numerous significant and impactful scenarios. Additionally, the grand mean and standard deviations tend to approximate the population mean and standard deviations, further highlighting the utility of the CLT in statistical analysis.

Tips

Whatever the population distribution is (normal or non-normal) with mean μ and standard deviation σ, the sampling distribution of $\bar{x}$ for any large sample ($n \geq 30$), is approximately normal with the same mean μ and same standard deviation $\frac{\sigma}{\sqrt{n}}$.

This means that a sufficiently large sample size can predict the characteristics of a population accurately.

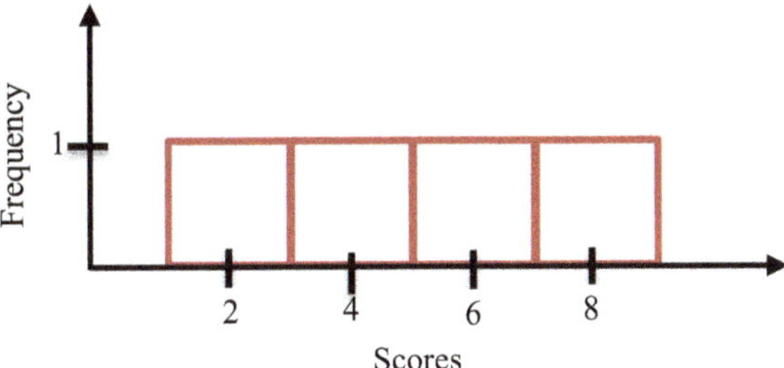

Fig. 8.1 Uniform distribution of rolling a tetrahedral die

Central limit theorem (CLT) states that the distribution of sample means ($\bar{x}$) from a population with *any* distribution (uniform, skewed, whatever), regardless of the distribution shape in a main population, can be approximated by a normal distribution with mean μ and standard deviation $\frac{\sigma}{\sqrt{n}}$, given a sufficiently large sample size ($n \geq 30$). The theorem is a key concept in probability theory because it shows that probabilistic and statistical methods that work for normal distributions can be applied to many problems related to other types of distributions. The larger the sample, the better the approximation will be.

Based on the central limit theorem, the probability that the sample mean $\bar{x}$ is between the values a and b is obtained by the following formula:

$$p(a < \bar{x} < b) = p\left(\frac{a - \mu}{\sigma_{\bar{x}}} < \frac{\bar{x} - \mu}{\sigma_{\bar{x}}} < \frac{b - \mu}{\sigma_{\bar{x}}} \right) = p(z_1 < z < z_2).$$

Example 9 Suppose a tetrahedral four-sided dice, which is shaped like a pyramid, is used in role-playing games. It has four triangular faces corresponding to numbers of 2, 4, 6, and 8. So, when the player tosses a tetrahedral die, it comes up with one of the numbers 2, 4, 6, or 8, which constructs our population. Tossing a tetrahedral die is like sampling from the above population.

The graph of the original distribution is shown in Fig. 8.1. This is called a uniform distribution.

The mean, variance, and standard deviation of the hypothetical population are:

$$\text{Population mean: } \mu = \frac{2 + 4 + 6 + 8}{4} = \frac{20}{4} = 5.$$

$$\text{Population variance: } \sigma^2 = \frac{(2-5)^2 + (4-5)^2 + (6-5)^2 + (8-5)^2}{4} = \frac{9+1+1+9}{4} = 5.$$

Population standard deviation: $\sigma = \sqrt{\sigma^2} = \sqrt{5} = 2.236$.

Now, we are rolling two tetrahedral dice simultaneously. The resulting sequence of two observed numbers represents a sample of size 2 drawn with replacement from the specified population. There are 16 possible samples, and they are all equally likely. Table 8.2 lists them and provides the value of the sample mean $\bar{x}$ for each.

A frequency distribution of 16 sample means is shown in Table 8.3:

The histogram of the 16 samples means appears to be approximately normal.

Table 8.2 16 possible samples of size 2 and their sample means

Samples	Mean ($\bar{x}$)
2, 2	2
2, 4	3
2, 6	4
2, 8	5
4, 2	3
4, 4	4
4, 6	5
4, 8	6
6, 2	4
6, 4	5
6, 6	6
6, 8	7
8, 2	5
8, 4	6
8, 6	7
8, 8	8

Table 8.3 Frequency distribution of sample means of rolling tetrahedral dice together

$\bar{x}$	f
2	1
3	2
4	3
5	4
6	3
7	2
8	1

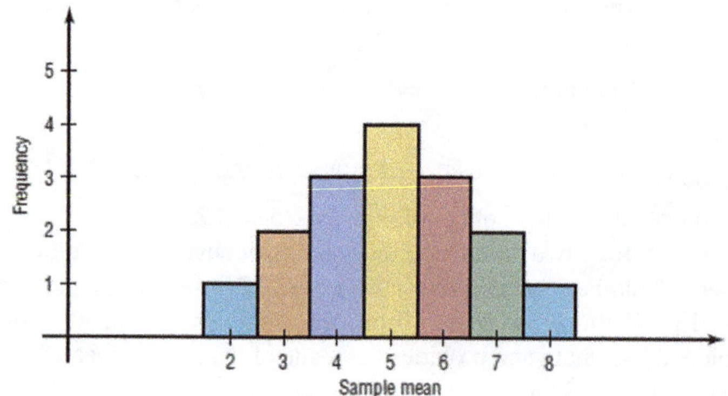

The mean and standard deviations of the sample means are denoted by $\mu_{\bar{x}}$ and $\sigma_{\bar{x}}$, respectively.

$$\mu_{\bar{x}} = \frac{(2 \times 1) + (3 \times 2) + \cdots + (7 \times 2) + (8 \times 1)}{16} = \frac{80}{16} = 5$$

$$\sigma_{\bar{x}} = \sqrt{\frac{\left[(2-5)^2 \times 1\right] + \left[(3-5)^2 \times 2\right] + \cdots + \left[(7-5)^2 \times 1\right] + \left[(8-5)^2 \times 1\right]}{16}}$$

$$= \sqrt{\frac{40}{16}} = \sqrt{2.5} = 1.581$$

If we divided the population standard deviation by $\sqrt{n} = 2 = 1.414$.

$$1.414 \times 1.581 = 2.236.$$

Conclusion is made that:

$$\mu_{\bar{x}} = \mu_x \quad \text{and} \quad \sigma_{\bar{x}} = \frac{\sigma_x}{\sqrt{n}}.$$

In statistics, $\sigma_{\bar{x}} = \frac{\sigma_x}{\sqrt{n}}$ is named standard error, which is denoted by Se.

Example 10 In the example of pain relief in patients with arthritis (Example 5), assume that the pain relieved time in the population has normal distribution with mean and standard deviations of 3.3 and 1.3 h. In a sample of 6, find the following probabilities:

(a) The mean is less than 2 h.
(b) The mean is between 2 to 4 h.
(c) The maximum distance between $\bar{x}$ and population mean is one.

Solution: In the example, the information of $\bar{x} = 3.3$, $S = 1.13$, $n = 6$, $df = 6 - 1 = 5$ and Se $= \frac{S}{\sqrt{n}} = \frac{1.13}{\sqrt{6}} = 0.46$ is provided.

(a) The probability that the pain relief time is less than 2 h is equal to:

$$p(\bar{x} < 2) = p\left(\frac{\bar{x} - \mu}{\sigma_{\bar{x}}} < \frac{b - \mu}{\sigma_{\bar{x}}}\right) = p\left(z < \frac{2 - 3.3}{0.46}\right) = p(z < -2.83)$$

According to the normal distribution table, we have:

$$p(\bar{x} < 2) = p(z < -2.83) = 0.5 - 0.4997 = 0.0023.$$

(b) The probability that the pain relief time is between 2 and 4 h is equal to:

$$p(2 < \bar{x} < 4) = p\left(\frac{a - \mu}{\sigma_{\bar{x}}} < \frac{\bar{x} - \mu}{\sigma_{\bar{x}}} < \frac{b - \mu}{\sigma_{\bar{x}}}\right) = p\left(\frac{2 - 3.3}{0.46} < z < \frac{4 - 3.3}{0.46}\right)$$

$$= p(2.83 < z < 1.52) = p(0 < z < 2.83) + p(0 < z < 1.52).$$

According to the normal distribution table, we have:

$$p(0 < z < 2.83) + p(0 < z < 1.52) = 0.4977 + 0.4357 = 0.9334.$$

(c) The probability that the sample mean of pain relief time to the population mean of up to one hour is equal to:

$$p(\mu - 1 < \bar{x} < \mu + 1) = p\left(\frac{\mu - 1 - \mu}{\sigma_{\bar{x}}} < \frac{\bar{x} - \mu}{\sigma_{\bar{x}}} < \frac{\mu + 1 - \mu}{\sigma_{\bar{x}}}\right)$$

$$= \left(\frac{-1}{\sigma_{\bar{x}}} < z < \frac{+1}{\sigma_{\bar{x}}}\right) = \left(\frac{-1}{0.46} < z < \frac{+1}{0.46}\right)$$

$$= (-2.17 < z \leq 2.17) = 2 \times (0 < z \leq 2.17).$$

According to the standard normal distribution table, we have:

$$2 \times (0 < z \leq 2.17) = 2 \times 0.4850 = 0.97.$$

8.6 Estimation of Variance and Standard Deviation

Aside from the population mean, the variance and standard deviation of the population may also be of interest. Because the standard deviation (often SD) is a measure of variability and indicates how close the data is to the mean. The variance is used as an estimate of the variability of the population from which the sample was drawn. The estimation of variance is important because it can monitor the status of quantities and is highly indicative of how one clinical trial or experiment is performing. It can be used to measure the confidence in statistical data and is used to compare different sets of data.

On the other hand, uniformity in research work as well as in most production processes is an important criterion for quality control. That is, it must be ensured that variations in measurements do not exceed a certain value, because having similar outputs is important. Therefore, it is important to calculate the confidence intervals for the standard deviation and variance of a population, assuming that the distribution of the population is normal.

Point estimate of variance and standard deviation from a sample of size n is obtained from the following formula:

$$S^2 = \hat{\sigma}^2 = \frac{\sum_{i=1}^{n}(x_i - \bar{x})^2}{n - 1} \quad \text{and} \quad S = \hat{\sigma} = \sqrt{\frac{\sum_{i=1}^{n}(x_i - \bar{x})^2}{n - 1}}.$$

Because S^2 has a Chi-square distribution with degree of freedom $(n-1)$, the interval estimate of the variance is calculated using Chi-square:

$$\chi^2 = \frac{(n-1)S^2}{\sigma^2}$$

$$\frac{(n-1)S^2}{\chi^2_{df,\,\frac{\alpha}{2}}} \quad \text{to} \quad \frac{(n-1)S^2}{\chi^2_{df,\,1-\frac{\alpha}{2}}}.$$

The $(1-\alpha)\%$ interval estimate of standard deviation is also obtained by square root from the above limits.

$$\sqrt{\frac{(n-1)S^2}{\chi^2_{df,\,\frac{\alpha}{2}}}}, \quad \sqrt{\frac{(n-1)S^2}{\chi^2_{df,\,1-\frac{\alpha}{2}}}}.$$

For ease of 95% confidence interval is equal to:

$$S\sqrt{\frac{(n-1)}{\chi^2_{df,\,0.025}}}, \quad S\sqrt{\frac{(n-1)}{\chi^2_{df,\,0.975}}}.$$

Example 11 The standard deviation of pain relief time in patients with arthritis in a sample of 6 is 1.3 h. Find the 95% confidence interval for variance and standard deviation of pain relief time.

Solution: The following information is provided, $S^2 = 1.69$, $n = 6$, $df = n-1 = 5$ and from the Chi-square distribution table $\chi^2_{0.025} = 14.449$, $\chi^2_{0.975} = 0.831$ replacing on the previous formula, gives interval estimates of variance and standard deviation.

$$\frac{(6-1)1.69}{14.449} \quad \text{to} \quad \frac{(6-1)1.69}{0.831} = (0.585,\ 10.168)$$

$$\left(\sqrt{0.585},\ \sqrt{10.168}\right) = (0.765,\ 3.189).$$

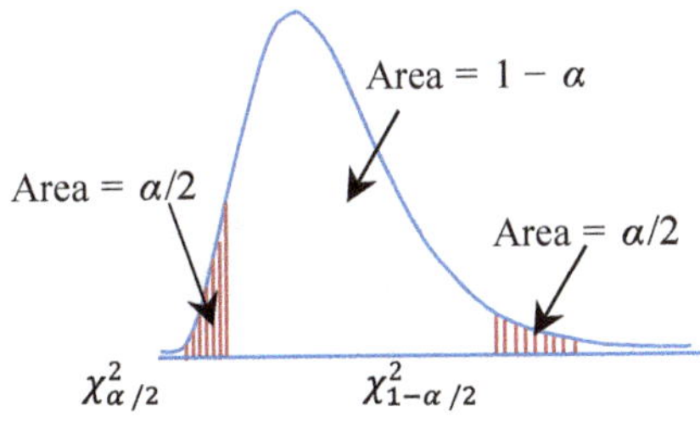

8.7   Exercises

1. Specify the answers of the three examples of interval estimation (diastolic blood pressure) on the axis and interpret the role of factors affecting confidence interval.
2. In a sample of 225 people with asthma, 40% had a positive reaction to dust. Calculate a 95% confidence interval for the proportion of people with positive reactions and interpret it.
3. Based on a sample of 25 men, the mean and variance of their height are 165 and 160 cm, respectively. Based on another sample of 36 women, the mean and variance of their height were 155 and 140 cm, respectively.

    A. Calculate and interpret the 95% confidence interval for the height of men and women in population.
    B. Is it possible to judge the mean height of men and women based on the confidence interval? Why and how?

4. As part of the malaria control plan, 10,000 rural homes must be sprayed. For this purpose, cost estimation is essential. But measuring all houses is not possible. A sample of 100 is selected and the area to be sprayed is measured. The average spraying area in these 100 houses is 23.2 square meters with a standard deviation of 5.9.

    A. Find an interval estimation that covers 95% of the total mean.
    B. If this probability increases to 99%, what will be the above interval?
    C. If the above estimation is from a 20 sample, what will happen in parts A and B.

5. The distribution of blood types in a study is included in the table below. Can the results of this table be generalized to the entire population? This means that can we calculate the proportion of the blood types in the population corresponding to the percentages of this table? If not, how do you interpret these data?

Blood types	A	B	AB	O	Unknown	Sum
Frequency	30	32	8	34	96	200
Percentage	28.84	30.77	7.69	32.69	.	100
Relative percentage	15.0	16.0	4.0	17.0	48.0	100

6. The mean and standard deviations of diastolic blood pressure in 35-year-old men in a 64 sample were 105 and 12 mmHg, respectively. Based on this information, what range of diastolic blood pressure includes the actual blood pressure of the population.
7. In a sample of 225 medical diagnostic devices, 40% had piezoelectric transducers. Calculate and interpret the 95% interval estimate for the proportion of devices with piezoelectric transducers.
8. Eighty compound generators are set up simultaneously in a factory. After checking the output voltage of the generators, it was found that 11 of them were defective compounds and the rest were additional compounds. If a company buys 30 generators a year from this factory, what percentage of the companies are defective.

Chapter 9
Health and Disease Measures

9.1 Introduction

Epidemiology and clinical research always deal with strong fundamental knowledge about probability and statistical theory. Estimation of disease indicators and health measurement, analysis of rates and ratios (comparison and confidence interval) are common applications of statistics in epidemiology. Because sound understanding of statistics can provide valuable insights about the proper epidemiological and research projects. This chapter of the book explains the conceptuality of health indicators and the occurrence of events, rates, coefficient of agreement and how to calculate their confidence intervals. However, since the terms ratio, rate, and proportion are used extensively, it is preferred to first distinguish between these three terms in epidemiology.

9.2 Difference Between Ratios, Rates, and Proportions

A ratio represents the relative magnitude or comparison of two quantities, expressed as one measure against another. It is calculated by dividing one phenomenon, typically with an interval or ratio scale, by another within the same unit. The numerator and denominator do not necessarily need to have a direct relationship. Thus, one could compare unrelated variables, such as patients with beds or nurses with the number of physicians in a city. Ratios are used to express relative magnitudes and are commonly employed to compare similar quantities. For instance, in a classroom with 12 males and 18 females, the ratio of males to females is 12 to 18 or 2 to 3. Ratios can be utilized to compare quantities of the same type of objects as well as quantities of different types.

S. H. Saneii and H. Doosti, *Practical Biostatistics for Medical and Health Sciences*,
https://doi.org/10.1007/978-981-97-3083-4_9

> **Tips**
>
> Ratio can be used to compare quantities of the same type of objects and of different types.
>
> A rate measures the frequency of an event in a defined population over a specified period of time.
>
> A proportion is a type of ratio in which the denominator includes the numerator.

Rates are utilized to compare dissimilar quantities with distinct units, where the numerator and denominator possess different units. Similar to ratios, rates can be simplified to their smallest terms. Given that rates contextualize disease frequency relative to the population size, they are valuable for comparing disease occurrence across various locations, time periods, or demographic groups with differing population sizes. Therefore, rates represent a specialized form of ratio that integrates the dimension of time into the denominator. In epidemiology, a rate quantifies the frequency of an event within a defined population over a specified time frame. Examples include disease prevalence rates per population, mortality rates for Covid-19 per 100,000 individuals, and measurements of speed (e.g., kilometers per hour) or water flow (e.g., gallons per minute).

A proportion is a type of ratio in which the denominator includes the numerator. In other words, a proportion is a ratio that relates a part to a whole in order to compare with the whole. A proportion might be used to describe what fraction of medical laboratory patients tested positive for Covid-19. A proportion may be expressed as a decimal, a fraction, or a percentage. For example, in the class with 12 males and 18 females, the total class size is 30, and the proportion of males is 12/30 or 40%.

9.3   Rates

A rate serves as a metric for the frequency of occurrence of a phenomenon relative to the population, taking into consideration the number of events transpiring within a specified period per unit of time (e.g., per month, per year). Given that this frequency can fluctuate over time, the rate is expressed in terms of time, with time being an integral component of this concept. Rates are fundamental measures of disease occurrence in populations and provide an estimate of the speed at which disease or mortality is occurring within a particular population. In epidemiology, the term "rate" is commonly used in conjunction with the prevalence and incidence of diseases. The integral components of a rate are as follows:

> **Tip**
>
> Rate describes risk of disease among a population at risk during a given time interval, while proportion shows the likelihood of disease among the population at a specific point in time. The difference is in the denominator.

1. Numerator: This represents the count or frequency of the desired events observed within a specified period of time among the population. It serves as the numerator of the fraction.
2. Denominator: The denominator typically comprises the number of individuals in the group at risk, whose events are included in the numerator. In some cases, the denominator may need to be estimated as the average population size over the duration of follow-up.
3. Specified duration (period of time): This refers to the time frame during which the events of the study population are tallied. It commonly encompasses calendar periods such as months, seasons, or years.
4. Multiplier: This factor, such as percent or per thousand, aids in converting the fraction into a more convenient and easily interpretable numeric value, facilitating better comprehension of the rate.

The denominator of a rate encompasses entities such as population, patients, or subjects, which can fluctuate both temporally and geographically. The primary function of computing a rate is to standardize the denominator, enabling valid comparisons between different groups, times, or locations. However, it's crucial to note that a rate is a reliable tool only under stable conditions throughout the period for which it is calculated. In cases where the population undergoes variation within this period, such as at the beginning and end of a year, it is customary to utilize the average population (typically the middle of the year) to ensure consistency and accuracy in the calculation.

Rates can only be averaged if the denominator of the fractions is the same. For example, if the prevalence rate of a disease in a county is 3 per 1000 per year and in another county is 5 per 1000 per year, the rate for both counties together will not be 4 per 1000 per year unless both counties have equal population. In these cases, the weighted mean is recommended, which takes into account the relative frequencies.

Rates describe risk of disease among a population at risk during a given time interval, while proportions describe the likelihood of disease among the population at a specific point in time.

9.3.1 Incidence

In epidemiology, incidence refers to the occurrence of new cases of a disease, disorder, or specific health-related event within a population during a defined time period, such as a month or a year. In the calculation of incidence, the numerator represents the number of new events that arise within the specified time frame, while the denominator signifies the population at risk of experiencing the event during this period. Achieving an accurate incidence, whether as a proportion or rate, necessitates a precise definition of the denominator, which comprises the population at risk. Additionally, two other critical elements include a well-defined numerator, representing the number or frequency of new events occurring without any prior history of the disease in question, and a clearly defined duration or period of time over which the incidence is calculated.

> **Tip**
>
> The incidence of disease indicates how quickly subjects are developing the disease, disorder or incident.

Incidence can typically be computed as either a proportion or a rate, each serving distinct purposes. The proportion indicates the risk of an event occurring within a given time frame, while the rate quantifies the number of new cases in a population over time.

Incidence proportion (IP), also known as proportional incidence, measures the likelihood that an individual will develop a disease or disorder during a specified period. This proportion is calculated by dividing the number of new events (disease or disorder) occurring during the period by the total population at risk. Incidence proportion is commonly utilized in studies with a short follow-up duration, as it requires complete follow-up and observation of all subjects at risk for the outcome throughout the study period, or at least until they experience the outcome. It assumes that individuals were initially free of the disease. Incidence proportion is also referred to as cumulative incidence.

Incidence rate (IR) represents the number of new cases divided by the total person-time at risk. The incidence rate is typically more precise than incidence proportion for long-term studies, as it accommodates individuals entering and leaving study populations.

Incidence rate

$$= \frac{\text{Number of new cases in a period of time}}{\text{Number of person} - \text{years when people were at risk of getting the disease}}$$

Incidence proportion

$$= \frac{\text{Number of new cases in a period of time}}{\text{Number of people at risk of getting the disease at the start of the period}}$$

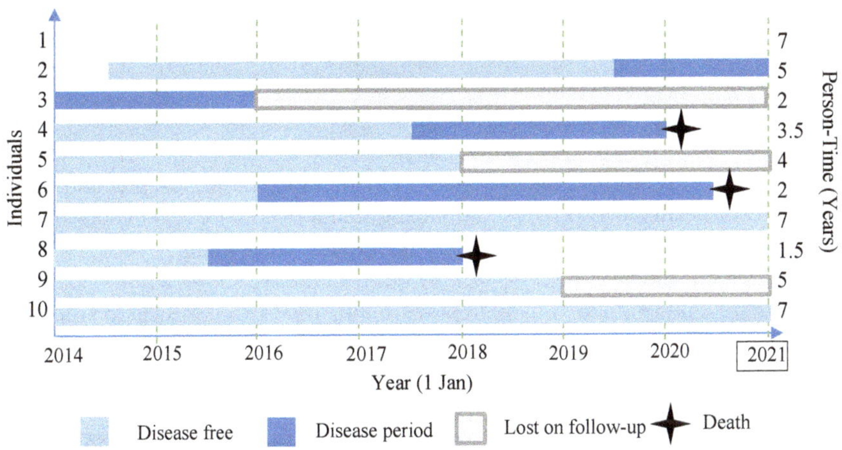

On June 30, 2020, there were 3 people who were sick out of the remaining 5 people in the group. This yields a prevalence of 3 out of 5, which is equivalent to 60%. The total number of person-years followed up in this study is obtained by summing the numbers to the right of the graph, which equals 44.0 person-years.

$$7 + 5 + 2 + 3.5 + 4 + 2 + 7 + 1.5 + 5 + 7 = 44$$

Incidence rate

$$= \frac{\text{number of new cases in a period of time}}{\text{Number of person} - \text{years when people were at risk of getting the disease}}$$

Incidence proportion

$$= \frac{\text{number of new cases in a period of time}}{\text{Number of people at risk of getting the disease at the start of the period}}$$

Example 1 In a sample population of a city with 50,000 people, we want to determine the incidence rate of developing Covid-19 over a period of 6 months. At the beginning of the study ($t = 0$), 30 cases of Covid-19 were identified. Assuming these people cannot get Covid-19 for the second time, they will not be counted in the population. In the follow-up 2 months later ($t = 2$), 1000 new cases of Covid-19, and in the second follow-up at the end of the study ($t = 6$), 2000 new cases were identified. Find the incidence and prevalence rate of Covid-19.

Solution: To measure the prevalence rate, we simply divide the total number of cases ($30 + 1000 + 2000 = 3030$) by the total number of sample population (50,000). So, the prevalence is equal to:

$$\text{PR} = \frac{x}{n} = \frac{3030}{50000} = 0.0606.$$

This measure tells the widespread of Covid-19 for any person in our sample population but provides little information about the actual risk of developing of Covid-19 in coming month.

To calculate the incidence rate, we must take into account how many months each person participated in the study and when they developed Covid-19. When it is not clear exactly when a person develops the disease in question, epidemiologists frequently use actuarial method **interpolating,** and it is assumed that the person was developed half-way point between follow-ups. In this calculation:

In the first 2 months, 1000 new cases were identified, so we assume that they developed Covid-19 at half-way (one month), thus 1000 (1000 × 1) person-months of disease-free life.

Two thousand new cases identified in the sixth month were free of Covid-19 disease for the first two months. Therefore, we assume that they were infected half-way between two and 6 months and had disease-free life for 4 (2 + 2) months. Thus contributing 8000 (4 × 2000) person-months of disease-free life. That is a total of (8000 + 1000) = 9000 person-months so far.

On the other hand, 46,970 people (50,000 − 2000 − 1000 − 30) who have never developed Covid-19 over the 6-month period should be accounted as 281,820 (46,970 × 6) contributing person-months of disease-free life.

That is a total of 290,820 (281,820 + 9000) person-months of life. Now, by dividing 3000 new cases of Covid-19 by 290,820, the rate of 0.0103 or 103 cases of Covid-19 per 10,000 populations per month can be obtained. In other words, if you were to follow 10,000 people for one month, you would see 103 new cases of Covid-19. This is a much more accurate measure of risk than prevalence rate.

$$IR = \frac{3000}{29,820} = 0.0103.$$

Confidence interval for incidence rate

First, we obtain the standard error of the incidence rate using the normal approximation for the situation with x new cases existing in n person-time:

$$Se\ (Ir) = \sqrt{\frac{1 - Ir}{x}}.$$

Then the 95% confidence interval of the incidence rate is calculated by the following formula.

$$95\%CI = \left(e^{\ln Ir - 1.96 \times Se(Ir)}\ \text{to}\ e^{\ln Ir + 1.96 \times Se(Ir)}\right).$$

However, due to the ease of calculation, the confidence interval for "the number of new cases" can be obtained based on the Poisson distribution from the simpler relation below.

$$CI = \left[\left(\frac{z_{1-\frac{\alpha}{2}}}{2} - \sqrt{x} \right)^2 \text{ to } \left(\frac{z_{1-\frac{\alpha}{2}}}{2} + \sqrt{x+1} \right)^2 \right].$$

Example 2 In example 1, the incidence of Covid-19 in a 6-month period was 0.0103 with 3000 new cases. Calculate a 95% confidence interval for the incidence rate.

Solution: In this example $x = 3000$ and $z_{1-\frac{\alpha}{2}} = 1.96$. The confidence interval for incidence rate is:

$$Se\,(Ir) = \sqrt{\frac{1-Ir}{x}} = \sqrt{\frac{1-0.0103}{3000}} = 0.01816$$

$$95\%CI = \left(e^{\ln Ir - 1.96 \times Se(Ir)} \text{ to } e^{\ln Ir + 1.96 \times Se(Ir)} \right)$$
$$= \left(e^{-4.576 - 1.96 \times 0.01816} \text{ to } e^{-4.576 + 1.96 \times 0.01816} \right)$$
$$= \left(e^{-4.6116} \text{ to } e^{-4.5404} \right) = (0.0099 \text{ to } 0.01067).$$

The confidence interval for the number of new cases of the disease is:

$$CI = \left[\left(\frac{z_{1-\frac{\alpha}{2}}}{2} - \sqrt{x} \right)^2 \text{ to } \left(\frac{z_{1-\frac{\alpha}{2}}}{2} + \sqrt{x+1} \right)^2 \right]$$
$$= \left[\left(\frac{1.96}{2} - \sqrt{3000} \right)^2 \text{ to } \left(\frac{1.96}{2} + \sqrt{3001} \right)^2 \right]$$
$$= 2893 \text{ to } 3109).$$

9.3.2 Prevalence

In epidemiology, prevalence, also known as point prevalence, is determined by dividing the total number of individuals in a population who have a disease or disorder at a specific point in time by the total population, typically expressed as a percentage. Prevalence encompasses both new and existing cases of the disease. It provides a snapshot of the situation at a single moment, hence its alternative name, "point" prevalence. In contrast, periodic prevalence refers to the proportion of the population affected by the disease at any given time period.

The prevalence can be defined as follows:

$$\text{prevalence rate} = \frac{\text{Total number of exixting cases at a given time}}{\text{Total number of people in the population}}.$$

Confidence interval for prevalence rate.

First, the standard error of the prevalence rate must be calculated.

$$Se\ (Pr) = \sqrt{\frac{x(n-x)}{n^3}}.$$

Then obtain the confidence interval for prevalence rate as the confidence interval of the rates from the following formula:

$$95\%\ CI = Pr \pm z_{1-\frac{\alpha}{2}}.Se(Pr).$$

Example 3 In the population of 50,000 people, for example 1, obtain a 95% confidence interval for the prevalence rate.

Solution: In this example $x = 3030$ and $z_{1-\frac{\alpha}{2}} = 1.96$. The confidence interval for prevalence rate is:

$$Se\ (Pr) = \sqrt{\frac{3030(50,000 - 3030)}{50,000^3}} = 0.00107$$

$$95\%CI = Pr \pm z_{1-\frac{\alpha}{2}}.Se(Pr)$$
$$= 0.0606 \pm 1.96 \times 0.00107$$
$$= (0.0585\ to\ 0.0627).$$

9.3.3   Mortality Rate

Mortality rate quantifies and depicts the trend and tendency of a particular population to die (or due to a specific cause) during a given time period. Mortality rate is typically expressed in units of deaths per 1000 individuals and always refers to a time period, usually 1 year. However, sometimes mortality rates are calculated for shorter time periods such as monthly or even daily. The mortality rate is major (inverse) health indicators, because death is the ultimate situation of "disease", or "unhealthiness". An important specific mortality measure is the crude death rate, which looks at mortality from all causes in a given time interval for a given population. The annual crude death rate (m) can be calculated as a proportion:

$$m = \frac{d}{P} \times 10^k,$$

where d is the count of deaths during the year and p is the total population or "population at risk", which is generally approximated by the population at midyear. If $k = 3$, the result is per thousand or when $k = 4$ the result is per ten thousand.

We can compute different types of mortality rates such as: age-specific, sex-specific, or age-and-sex-specific mortality rates to gauge health conditions in

Table 9.1 Formulas of frequently used mortality measures

Measure	Numerator	Denominator	Multiplier (10^k)
Crude death rate	Total number of deaths during a given time interval	Mid-interval population	1000 or 100,000
Cause-specific death rate	Number of deaths assigned to a specific cause during a given time interval	Mid-interval population	100,000
Proportionate mortality	Number of deaths assigned to a specific cause during a given time interval	Total number of deaths from all causes during the same time interval	100 or 1000
Death-to-case ratio	Number of deaths assigned to a specific cause during a given time interval	Number of new cases of same disease reported during the same time interval	100
Neonatal mortality rate	Number of deaths among children < 28 days of age during a given time interval	Number of live births during the same time interval	1000
Postneonatal mortality rate	Number of deaths among children 28–364 days of age during a given time interval	Number of live births during the same time interval	1000
Infant mortality rate	Number of deaths among children <1 year of age during a given time interval	Number of live births during the same time interval	1000
Maternal mortality rate	Number of deaths assigned to pregnancy-related causes during a given time interval	Number of live births during the same time interval	100,000

specific demographic groups. Table 9.1 summarizes the most commonly used mortality measures formulas.

9.4 Standardization

Comparison of health indicators is common in epidemiological studies, but this comparison is misleading regardless of differences in population's characteristics. Because the populations are different in terms of demographic, age and gender structure. For example, age is one of the most common and important confounding factors in health studies. When the age distribution of the groups being compared is different and the age is related to the outcomes of interest (such as death or prevalence of disease), comparisons can be disrupted. Therefore, a method should be employed to eliminate the confounding effect of age so that we can make a sound and logical comparison. Standardization is a classic epidemiological method for comparing populations disease or death experiences of rates. There are two

Fig. 9.1 Direct standardazition

main methods for standardization to characterize whether the population distribution used is standard (direct method) or uses a set of specific measures (indirect method). These two methods are presented below.

9.4.1   Procedure for Direct Standardization

Compute the mortality rates specific to each age group within both populations. Select one population to serve as the standard (reference) population. Typically, this is the national population when comparing mortality rates of a specific community.

For the other population being studied, determine the expected number of deaths by multiplying its age-specific mortality rates by the number of individuals in each age group of the standard population.

Sum up the expected deaths across all age groups.

Subsequently, calculate the age-adjusted mortality rates by dividing the total expected deaths by the standard population size.

Finally, compare the age-standardized mortality rates of the two populations to draw conclusions (Fig. 9.1).

9.4.2   Procedure for Indirect Standardization

We propose the following method for computing indirect standardization:

- Select a reference or standard population.
- Determine the observed number of deaths in the population(s) under investigation.
- Utilize the age-specific mortality rates from the chosen reference population and apply them to the population(s) of interest.
- Multiply the population count in each age group of the population(s) of interest by the age-specific mortality rate corresponding to the equivalent age group in the reference population.
- Aggregate the total expected deaths for each population of interest.
- Calculate the standardized mortality ratio (SMR) by dividing the total observed deaths in the population(s) of interest by the expected deaths (Fig. 9.2).

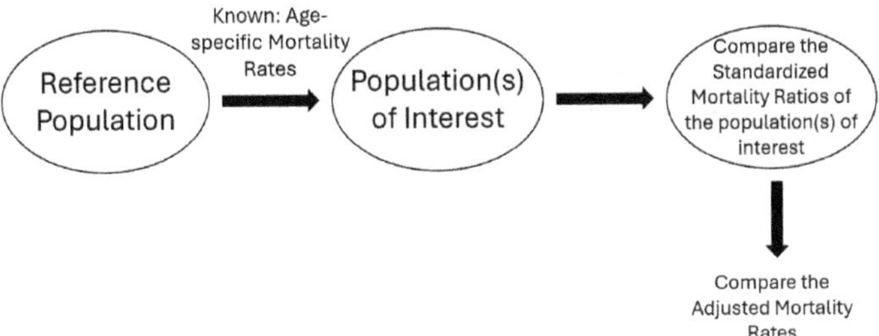

Fig. 9.2 Indirect standardization

Direct and indirect standardization methods often yield similar outcomes, but direct standardization is generally preferred over the indirect method. The preference for direct standardization stems from the fact that it applies the age-specific rates of the study populations to a single standard population, ensuring consistent weighting across age groups. Conversely, in indirect standardization, the weighting applied to the standard age-specific rates is contingent upon the age distribution of the study populations. Consequently, if there are substantial variations in age distribution among the study populations, the resulting standardized mortality ratios (SMRs) may be based on disparate weightings, rendering them incomparable.

Another advantage of direct standardization lies in its transitive property, as it employs a consistent denominator across all datasets. For instance, if direct standardization is employed on three datasets (C, D, and E), and the compressed mortality file (CMF) between C and D is 2, while that between D and E is 0.5, it implies that the CMF between C and E can be derived as $2/0.5 = 4$.

9.5 Screening Tests

At times, there is a need to assess the efficacy of a particular test. This involves comparing its outcomes with those of a widely recognized benchmark test, often referred to as the "gold standard", to ascertain the level of concordance and compatibility. The gold standard test typically delivers the most accurate and reliable results and serves as the benchmark against which the performance of other tests is measured. However, this gold standard test may be costlier, more time-intensive, or less accessible to all individuals. Alternatively, it may involve a combination of diagnostic procedures typically conducted in a hospital setting, rendering it unsuitable for routine or widespread use in screening.

For instance, consider the scenario where students undergo a basic hearing assessment during their initial year of schooling. Those who do not pass this initial screening test are subjected to retesting or are referred to a specialized hearing clinic for more extensive evaluations to determine the presence of an actual hearing impairment. Let's assume a school with a student population of 600, out of which

60 students indeed have a genuine hearing impairment. Among these, 50 students and 40 students with normal hearing (possibly affected by temporary conditions like a cold during the test) failed the school hearing test. The outcomes of this hearing test are outlined in Table 9.2.

As depicted in Table 9.3, each student in the school can experience one of four possible outcomes. A student with an actual hearing impairment may either be identified as having a problem during the test, which is a true positive outcome (group "a" in Table 9.3), or may incorrectly pass the test, erroneously suggesting the absence of a problem, resulting in a false negative outcome (group "c"). Similarly, a student without a hearing impairment may correctly pass the test and be accurately diagnosed as not having a problem, representing a true negative outcome (group "d"), or may be erroneously identified as having a problem and fail the test, resulting in a false positive outcome (group "b").

In assessing the accuracy of the school hearing test, we evaluate two key aspects: its ability to correctly identify students with hearing problems and its accuracy in classifying students with normal hearing.

Referring to Table 9.3, vertically, we observe that 83% (50 out of 60) of students with hearing problems were correctly identified as having issues during the school test (cell "a"), while 93% (500 out of 540) of students with normal hearing were accurately classified as such (cell "d"). These measures are known as sensitivity and specificity, respectively.

Examining the table horizontally, we find that 56% (50 out of 90) of those diagnosed with a positive result actually had a hearing problem (cell "a"), and 98% (500 out of 510) of those diagnosed with a negative result were correctly identified as having normal hearing (cell "d"). These metrics are termed positive predictive value (PPV) and negative predictive value (NPV), respectively.

Table 9.2 Results from a school hearing test program

School hearing test	True hearing status		
	Have problem	Normal	Total
Reject (positive test result)	50	40	90
Pass (negative test result)	10	500	510
Total	60	540	600

Table 9.3 Possible outcomes for a screening test

		Disease status	
		Positive	Negative
Test result	Positive	*True Positive* (a)	False Positive (b)
	Negative	False Negative (c)	*True Negative* (d)

Sensitivity indicates the test's ability to detect true positive cases, calculated by dividing the number of true positive results (cell "a") by the total number of subjects with the disease (sum of cells "$a + c$"). In this example, the sensitivity of the school hearing test is 83%.

Specificity, on the other hand, denotes the test's capacity to correctly identify disease-free subjects, obtained by dividing the number of true negative results (cell "d") by the total number of subjects truly without the disease (sum of cells "$b + d$"). The specificity of the school hearing test is 93%.

Positive predictive value (PPV) gauges the probability of correctness of a positive test, indicating the proportion of patients with positive results correctly identified as having the disease. It is computed by dividing the number of true positive results (cell "a") by the total number of subjects with a positive test result (sum of cells "$a + b$"). In this scenario, the PPV of the school hearing test is 56%.

Negative predictive value (NPV) assesses the probability of correctness of a negative test, reflecting the proportion of patients with negative results accurately identified as not having the disease. NPV is calculated by dividing the number of true negative results (cell "d") by the total number of subjects with a negative test (sum of cells "$c + d$"). The NPV of the school hearing test is 98%.

Example 4 A sample of 1000 people has been examined at a communicable disease clinic. 50 of them were diagnosed with Covid-19 and the rest were healthy and asymptomatic. Then the same people have been sent for PCR testing, and the results of the test and the experts' opinions are given in Table 9.4.

Estimate sensitivity, specificity, positive predictive value, and negative predictive value.

Solution: In this $a = 44$, $b = 9$, $c = 6$ and $d = 941$.

$$\text{Sen} = \frac{a}{a + c} = \frac{44}{44 + 6} = 0.880$$

$$\text{Spe} = \frac{d}{b + d} = \frac{941}{941 + 9} = 0.991$$

$$\text{PPV} = \frac{a}{a + b} = \frac{44}{44 + 9} = 0.830$$

Table 9.4 Results of clinical diagnosis and PCR test for 1000 people

		Clinic		
		+	−	
PCR	+	44	9	53
	−	6	941	947
		50	950	1000

$$\text{NPV} = \frac{d}{c+d} = \frac{941}{941+6} = 0.994.$$

Screening tests can be expressed as conditional probabilities in the following relationships:

$$\text{Prevalence} = P(\text{disease})$$

$$\text{Sen} = P(\text{test} + |\text{disease})$$

$$\text{Spe} = (\text{test} - |\text{no disease})$$

$$\text{PPV} = P(\text{disease}|\text{test}+)$$

$$\text{NPV} = P(\text{no disease}|\text{test}-)$$

$$1 - \text{Spe} = P(\text{test} + |\text{no disease})$$

$$1 - \text{Sen} = P(\text{test} - |\text{disease}).$$

Using the above relations, the following equations can be derived between positive predictive value, negative predictive value, sensitivity, and specificity:

$$\text{PPV} = \frac{\text{Sen} \times \text{Pr}}{(\text{Sen} \times \text{Pr}) + (1 - \text{Spe}) \times (1 - \text{Pre})}$$

$$\text{NPV} = \frac{\text{Spe} \times (1 - \text{Pre})}{\text{Spe} \times (1 - \text{Pre}) + \text{Pre} \times (1 - \text{Sen})}.$$

The other two rates, called "false positive rate" and "false negative rate", along with screening tests are also used. When the patient's true situation is positive and the test result is negative, it is called a false negative. If the patient's true situation is negative and the test result is positive, it is called a false positive. Table 9.5 can provide the concepts of these two rates more clearly.

Table 9.5 Concepts of true and false positive and negative

Name of rate	Disease situation	Test result
True positive	+	+
False positive	-	+
False negative	+	-
True negative	−	−

9.5.1 Confidence Interval for Screening Tests

To estimate the confidence interval for sensitivity (Sen) and specificity (Spe), it is necessary to calculate the standard error of each one. Given that sensitivity and specificity are considered a kind of ratio (p), their standard error is equal to:

$$Se_{Sen} = \sqrt{\frac{p(1-p)}{n}} = \sqrt{\frac{Sen(1-Sen)}{a+c}}$$

$$Se_{Spe} = \sqrt{\frac{p(1-p)}{n}} = \sqrt{\frac{Spe(1-Spe)}{b+d}}.$$

Hence, the confidence interval for sensitivity (Sen) is equal to:

$$CI = \left(Sen \pm z_{\frac{1-\alpha}{2}} \times Se_{Sen}\right).$$

The confidence interval for specificity (Spe) is equal to:

$$CI = \left(Spe \pm z_{\frac{1-\alpha}{2}} \times Se_{Spe}\right).$$

Similarly, the standard error and the confidence interval for positive predictive value (PPV) are obtained from the following equation:

$$Se_{PPV} = \sqrt{\frac{p(1-p)}{n}} = \sqrt{\frac{PPV(1-PPV)}{a+b}}$$

$$CI = \left(PPV \pm z_{\frac{1-\alpha}{2}} \times Se_{PPV}\right).$$

Also, the standard error and the confidence interval for negative predictive value (NPV) are obtained from the following equation:

$$Se_{NPV} = \sqrt{\frac{p(1-p)}{n}} = \sqrt{\frac{NPV(1-NPV)}{c+d}}$$

$$CI = \left(NPV \pm z_{\frac{1-\alpha}{2}} \times Se_{NPV}\right).$$

Example 5 In Example 4, the sensitivity was estimated to be 0.88 and the positive predictive value was to be 0.83. In that example, $a = 44$, $b = 9$, $c = 6$ and $d = 941$. Find the confidence interval for sensitivity and positive predictive value.

Solution: The standard error of sensitivity and positive predictive values are:

$$\mathrm{Se}_{\mathrm{Sen}} = \sqrt{\frac{\mathrm{Sen}\,(1-\mathrm{Sen})}{a+c}} = \sqrt{\frac{0.88(1-0.88)}{44+6}} = 0.0460$$

$$\mathrm{Se}_{\mathrm{PPV}} = \sqrt{\frac{\mathrm{PPV}\,(1-\mathrm{PPV})}{a+b}} = \sqrt{\frac{0.83(1-0.83)}{44+9}} = 0.0516.$$

The confidence intervals for them are:

$$\mathrm{CI} = \left(\mathrm{Sen} \pm z_{\frac{1-\alpha}{2}} \times \mathrm{Se}_{\mathrm{Sen}}\right) = (0.88 \pm 1.96 \times 0.0460) = (0.790\,\mathrm{to}\,0.970)$$

$$\mathrm{CI} = \left(\mathrm{PPV} \pm z_{\frac{1-\alpha}{2}} \times \mathrm{Se}_{\mathrm{PPV}}\right) = (0.83 \pm 1.96 \times 0.0516) = (0.729\,\mathrm{to}\,0.931.$$

9.6   Cohen's Kappa Coefficient

Cohen's kappa coefficient (κ) serves as a statistic used to assess inter-rater reliability, particularly when observations are categorized as a qualitative dichotomous variable. It offers a stronger and more robust measure compared to a simple percent agreement calculation, as it considers the probability of agreement occurring by chance. It's imperative to note that both raters evaluate the same set of subjects. Hence, Cohen's kappa measures the agreement between two raters who each classify a group of n subjects into two mutually exclusive categories.

> **Tip**
>
> Cohen's kappa is only valid for nominal categories. If used for ordinal and metric categories, it ignores their order and considers them as nominal.

To calculate Cohen's kappa coefficient, the results of two experts should be presented in a table, such as Table 9.6:

Table 9.6 General contingency table for kappa coefficient

		Expert1		
		Yes	No	
Expert 2	Yes	a	b	r_1
	No	c	d	r_0
		c_1	c_0	n

> **Tip**
>
> Kappa values from different studies may not be comparable as the value also varies with prevalence, which is the number of subjects in categories relative to the total.

where r_1 and r_0 are the sum of rows of positives and negatives and c_1 and c_0 are the column sum of positives and negatives. Given the cells in the table above, we first obtain the probability of the observed agreement or the observed relative agreement p_o and the theoretical probability of a chance agreement p_e between the two raters:

$$p_e = \frac{(r_1 \times c_1) + (r_0 \times c_0)}{n^2}$$

$$p_o = \frac{a + d}{n}.$$

Cohen's kappa coefficient is calculated from the following equation:

$$\kappa = \frac{p_o - p_e}{1 - p_e}.$$

The kappa value is not like a correlation between -1 and $+1$ and may take any negative values. But only values 0 to 1 are important for interpretation and have useful meaning. All negative values are interpreted such as zero, which means that there is no agreement at all between the two experts. A value of 1 indicates the perfect agreement of the two experts, while the negative and zero values indicate that there is no agreement between the two experts at all. Therefore, the closer kappa coefficient is to 1, the greater the agreement, and the closer it is to zero, the lower the agreement. Table 9.7 is a good guide for interpreting Cohen's Kappa coefficient.

Table 9.7 Guideline for interpreting Cohen's Kappa coefficient

Cohen's Kappa	Interpretation
0.0–0.19	Poor agreement
0.20–0.39	Fair agreement
0.40–0.59	Moderate agreement
0.60–0.79	Good agreement
0.80–1.0	Very good agreement

> **Tips**
>
> The following simple and practical assumptions must be fulfilled for Cohen's Kappa:
>
> - Independence of subjects,
> - Independence of raters, observers, and laboratories from each other,
> - The rating categories to which individuals are assigned are mutually exclusive and exhaustive.

The assumptions for Cohen's Kappa, which are easily fulfilled in most practical situations, are that the categories are nominal and only valid for nominal categories. This measure ignores the order and value of ordinal and metric categories and considers them nominal. The other assumptions of this statistic are: the independence of subjects, the independence of rater, laboratories and measurement methods from each other, and the rating categories are mutually exclusive and exhaustive.

9.6.1 Confidence Interval for Cohen's Kappa

The formula for a kappa confidence interval is:

$$\kappa - 1.96 \times \mathrm{Se}_\kappa \text{ to } \kappa + 1.96 \times \mathrm{Se}_\kappa,$$

where is Se_κ the standard error of kappa obtained by:

$$\mathrm{Se}_\kappa = \sqrt{\frac{p_o(1 - p_o)}{n(1 - p_e)}}.$$

Example 6 A sample of 1000 people has been examined at a communicable disease clinic. 50 of them were diagnosed with Covid-19 and the rest were healthy and asymptomatic. Then the same people have been sent for PCR testing, which is the result of the test and the experts' opinions are given in Table 9.8.

Find kappa coefficient and its 95% confidence interval.

Solution: The given information in this example is $a = 44$, $d = 941$, $r_1 = 53$, $r_0 = 947$, $c_1 = 50$, $c_0 = 950$, and we substitute them in the formulas:

Table 9.8 Results of clinical diagnosis and PCR test for 1000 people

		Clinic		
		+	−	
PCR	+	44	9	53
	−	6	941	947
		50	950	1000

$$p_e = \frac{(50 \times 53) + (947 \times 950)}{1000^2} = 0.9023$$

$$p_o = \frac{44 + 941}{1000} = 0.985$$

$$\kappa = \frac{p_o - p_e}{1 - p_e} = \frac{0.985 - 0.9023}{1 - 0.9023} = 0.846.$$

The agreement between clinical diagnosis and PCR test is %84.6 and very good and nearly perfect.

The standard error of kappa is:

$$Se_\kappa = \sqrt{\frac{p_o(1 - p_o)}{n(1 - p_e)}} = \sqrt{\frac{0.985(1 - 0.985)}{1000(1 - 0.9023)}} = 0.0249.$$

The 95% confidence interval for Cohen's kappa is

$$\kappa - 1.96 \times Se_\kappa \text{ to } \kappa + 1.96 \times Se_\kappa$$
$$= 0.846 - 1.96 \times 0.0249 \text{ to } 0.846 + 1.96 \times 0.0249$$
$$= (0.797 \text{ to } 0.895).$$

If the two experts are in perfect agreement with each other, it means that in Table 9.10, the b and c cells should be 0, i.e., the table should be like Table 9.9.

In case the two raters have negative agreement with each other at all, it means that in Table 9.9, a and d cells should be 0, i.e., the table should be like Table 9.10.

Table 9.9 Contingency table for two experts with perfect agreement

		Expert1		
		Yes	No	
Expert 2	Yes	a	0	a
	No	0	d	d
		a	d	n

Table 9.10 Contingency table for two experts with negative agreement

		Expert1		
		Yes	No	
Expert 2	Yes	0	b	b
	No	c	0	c
		c	b	n

Example 7 Suppose that in Example 1, the result of clinical diagnosis and PCR test is exactly the same, i.e., $b = 0$ and $c = 0$. Find the kappa coefficient.

Solution: If the clinical diagnosis and PCR test are exactly the same, we will have Table 9.11.

The probability of observed agreement, agreement by chance and kappa coefficient is:

$$p_e = \frac{(50 \times 50) + (950 \times 950)}{1000^2} = 0.905$$

$$p_o = \frac{50 + 950}{1000} = 1.00$$

$$\kappa = \frac{p_o - p_e}{1 - p_e} = \frac{1.00 - 0.905}{1.00 - 0.905} = 1.00.$$

This kappa means that the two raters are in perfect agreement.

Example 8 If in Example 6, the result of clinical diagnosis and PCR test is not the same at all, i.e., $a = 0$ and $d = 0$. Find the kappa coefficient.

Solution: If the clinical diagnosis and PCR test are exactly the same, we will have Table 9.12.

The probability of observed agreement, agreement by chance, and kappa coefficient is:

$$p_e = \frac{(50 \times 50) + (950 \times 950)}{1000^2} = 0.905$$

Table 9.11 Results of clinical diagnosis and PCR test with perfect agreement

		Expert1		
		Yes	No	
Expert 2	Yes	50	0	50
	No	0	950	950
		50	950	1000

Table 9.12 Results of clinical diagnosis and PCR test with no agreement at all

		Expert1		
		Yes	No	
Expert 2	Yes	0	50	50
	No	950	0	950
		950	50	1000

$$p_o = \frac{0+0}{1000} = 0.00$$

$$\kappa = \frac{p_o - p_e}{1 - p_e} = \frac{0.00 - 0.905}{1.00 - 0.905} = -9.53.$$

The sample size does not affect the calculation of kappa coefficient and increasing or decreasing the sample size does not change it. Because the kappa coefficient uses relative frequencies. For example, if in Table 9.12, all cells are reduced 10 times, i.e., $d = 0$, $c = 95$ instead of 950, $b = 5$ instead of 50, $a = 0$, the kappa coefficient will still be $\kappa = -9.53$.

Example 9 If in example 6, the sample size is 100 and $d = 0$, $c = 90$, $b = 10$, $0 = a$ (Table 9.13). Find the kappa coefficient.

Solution: The purpose of this example is not to show the effect of increasing or decreasing the sample size on the kappa coefficient, but we want to express the effect of changes in the cells within the table or the probability of observations (prevalence).

The probability of observed agreement, agreement by chance, and kappa coefficient is:

$$p_e = \frac{(10 \times 10) + (90 \times 90)}{100^2} = 0.82$$

$$p_o = \frac{0+0}{100} = 0.00$$

$$\kappa = \frac{p_o - p_e}{1 - p_e} = \frac{0.00 - 0.82}{1.00 - 0.82} = -4.56.$$

Comparison of examples 8 and 9 shows that when the cells in the table change, the probability of chance and random agreement decreases from 0.905 to 0.82, and as a result, the kappa coefficient will change from -9.53 to -4.56.

Example 10 If in Example 9, the sample size is 100 and $d = 0$, $c = 50$, $b = 50$, $0 = a$ (Table 9.14). Find the kappa coefficient.

Table 9.13 Results of clinical and laboratory diagnosis with different agreement

		Expert1		
		Yes	No	
Expert 2	Yes	0	10	5
	No	90	0	95
		90	10	100

Table 9.14 Results of clinical and laboratory diagnosis with different agreement

		Expert1		
		Yes	No	
Expert 2	Yes	0	50	50
	No	50	0	50
		50	50	100

Solution: The purpose of this example is to show the kappa coefficient is equal -1.

The probability of observed agreement, agreement by chance, and kappa coefficient is:

$$p_e = \frac{(50 \times 50) + (50 \times 50)}{100^2} = 0.50$$

$$p_o = \frac{50 + 50}{100} = 1.00$$

$$\kappa = \frac{p_o - p_e}{1 - p_e} = \frac{1.00 - 0.50}{1.00 - 0.50} = -1.0.$$

Understanding the concept of zero is very important for the Kappa coefficient.

The zero coefficient occurs when one of the experts assigns all observations to only two cells in a row (or column) of a table and does not assign anything to two cells in another row (or column).

> **Tip**
>
> The kappa coefficient will be zero when all observations are assigned to only two cells of one row (column) of the table and nothing is assigned to the other two cells of the row (column).

Example 11 If in Example 10, the sample size is 100 and the number of items in the cells is equal to $d = 50$, $c = $, $b = 50$, $a = 0$ (Table 9.15), the kappa coefficient will be equal to zero.

Table 9.15 Hypothetical results for zero agreement

		Expert1		
		Yes	No	
Expert 2	Yes	0	50	50
	No	0	50	50
		0	100	100

$$p_e = \frac{(0 \times 50) + (100 \times 50)}{100^2} = 0.50$$

$$p_o = \frac{0 + 50}{100} = 0.50$$

$$\kappa = \frac{p_o - p_e}{1 - p_e} = \frac{0.50 - 0.50}{1.00 - 0.50} = 0.0.$$

Also, if the number of items in the row cells is equal to in another example are equal to $a = 85$, $b = 15$, $c = 0$, $d = 0$, (Table 9.16), the kappa coefficient will be equal to zero.

$$p_e = \frac{(100 \times 85) + (0 \times 1550)}{100^2} = 0.85$$

$$p_o = \frac{0 + 85}{100} = 0.85$$

$$\kappa = \frac{p_o - p_e}{1 - p_e} = \frac{0.85 - 0.85}{1.00 - 0.85} = 0.0.$$

When raters classify observations into n classes instead of 2, the kappa coefficient is calculated from the following formula

$$\kappa = \frac{\sum O_{kk} - \sum \left(\frac{O_{k.} \times O_{.k}}{n}\right)}{n - \sum \left(\frac{O_{k.} \times O_{.k}}{n}\right)},$$

where O_{kk} is the cell frequency in the kth row and the kth column (diagonal element) of the $K \times K$ table, n is the grand total, $O_{k.}$ is the marginal total in the kth row, $O_{.k}$ is the marginal total in the kth column. Obviously, this formula can also be used for 2 $\times$ 2 tables.

Table 9.16 Another hypothetical results for zero agreement

		Expert1		
		Yes	No	
Expert 2	Yes	85	15	100
	No	0	0	0
		85	15	100

Table 9.17 Hypothetical assessment of intrathecal synthesis by two laboratories

Laboratory 1	Laboratory 2			
	Positive	Doubtful	Negative	Total
Positive	**40**	6	4	**50**
Doubtful	7	**10**	8	**25**
Negative	3	4	**58**	**65**
Total	**50**	**20**	**70**	**140**

Example 12 In the hypothetical example, following the intrathecal assessment of 130 patients have sent to two laboratories and the results were reported positive, negative, and doubtful in Table 9.17. Calculate the kappa agreement coefficient.

Solution: In this example, $O_{11} = 40, O_{22} = 10, O_{33} = 58$ and $n = 140$. By substituting them in the formula, we have:

$$\kappa = \frac{(40 + 10 + 58) - \left(\frac{50 \times 50}{130} + \frac{20 \times 25}{130} + \frac{70 \times 65}{130}\right)}{140 - \left(\frac{50 \times 50}{130} + \frac{20 \times 25}{130} + \frac{70 \times 65}{130}\right)} = \frac{49.92}{81.92} = 0.609.$$

This kappa coefficient means that the agreement between two laboratories is moderate.

9.7 Relative Risk (RR)

When the population is divided into two groups: those exposed to a factor (E) and those not exposed to that factor ($\overline{E}$), and the occurrence of a certain disease in them is denoted by D, the relative risk indicates how much more likely it is for the disease to occur among the exposed (E) group compared to the unexposed individuals.

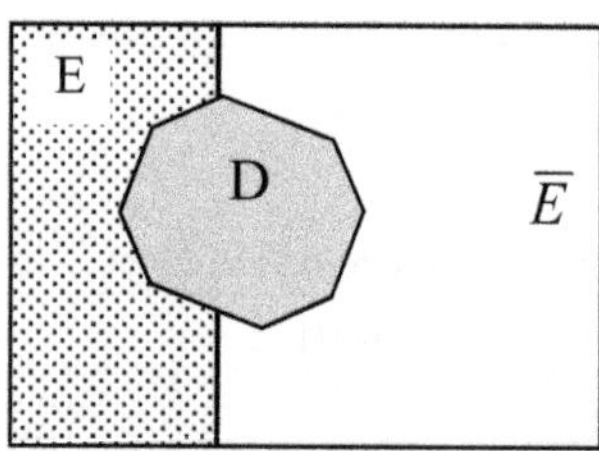

In epidemiology, the relative risk (RR) provides insights into the increased likelihood of an exposed group developing a specific event (such as a disease, injury, risk factor, or death) compared to a non-exposed group. By understanding the exposure and disease status of a research population, the corresponding numbers can be filled in a table, such as Table 9.18.

Table 9.18 Disease status in terms of exposure status

		Disease	
		Yes	No
Exposure	Yes	Diseased + Exposed	Healthy + Exposed
	No	Diseased + Unexposed	Healthy + Unexposed

Thus, the probability or risk of disease among the exposed ($p_1 = p(D|E)$) is equal to:

$$\text{probability (risk) of disease}$$
$$= \frac{\text{No. of diseased \& Exposed}}{\text{No. of diseased \& Exposed + No. of Healthy \& Exposed}}.$$

And the probability or risk of disease among the unexposed ($p_0 = p(D|\overline{E})$) is equal to:

$$\text{probability (risk)disease}$$
$$= \frac{\text{No. of diseased \& Unexposed}}{\text{No. of diseased \& Unexposed + No. of Healthy \& Unexposed}}.$$

The relative risk (RR) is obtained by dividing the probability of disease among the exposed by the probability of disease among unexposed.

$$\text{RR} = \frac{\text{probability of getting disease in exposed group}}{\text{probability of getting disease in unexposed group}}$$

$$\text{RR} = \frac{p_1}{p_0} = \frac{p(D|E)}{p(D|\overline{E})}.$$

The relative risk points out that the occurrence of disease (D) in the presence and exposure to a factor (E) is how many times greater or lesser than the occurrence of disease (D) in the absence of factor E. This measure is also called the incidence relative risk or "risk ratio" of the disease (mostly related to prospective and cohort studies). Relative risk is compared with 1, because the expected value of relative risk (RR) under the null hypothesis is 1 and means there is no difference between groups.

Tip

The relative risk or risk ratio (RR) of an event is the probability of its occurrence in exposed group to a risk factor as compared with the probability of its occurrence in an unexposed or control or reference group. The RR is estimated as the absolute risk with the risk factor divided by the absolute risk in the unexposed group.

We must remind that in a prospective or cohort study, groups of subjects are followed up for predetermined time to observe whether the outcome of interest occurs; we can use the relative risk (RR).

In prospective and cohort studies, we look at the table in this (horizontal) way (Table 9.19).

And then:

$$RR = \frac{\mathrm{pr}(D|E)}{\mathrm{pr}(D|\overline{E})} = \frac{{}^{a}/{}_{(a+c)}}{{}^{b}/{}_{(b+d)}}.$$

When $\left(\frac{\mathrm{pr}(D|E)}{\mathrm{pr}(D|\overline{E})}\right)$ is interpreted as the ratio of the prevalence of the disease in the two groups, it is the "relative risk of disease" and shows that the prevalence of a particular disease in exposed group is much higher than the prevalence of the same disease in unexposed people.

Example 13 Out of 951 mothers who drank alcohol at least once a day during pregnancy, 7 babies were born with impairment. Of the 31,530 mothers who drank less than once a day during pregnancy, 86 had birth impairment babies. Does drinking alcohol during pregnancy increase the risk of birth impairment in the newborns?

Solution: In this example, event $I =$ impairment and event $E =$ daily consumption of alcohol at least once during pregnancy is defined. The conditional probability of having a impairment is $\mathrm{pr}(I|E) = \frac{7}{951} = 0.00736$ provided the pregnant mother has consumed alcohol at least once a day equals and the conditional probability of having a impairment is $\mathrm{pr}(I|\overline{E}) = \frac{86}{31530} = 0.00273$ given the pregnant mother has consumed alcohol less than once a day. Then according Table 9.20, $a = 7$, $b = 944$, $c = 86$, $d = 31,444$, $a + c = 93$, and $b + d = 32,388$.

Dividing these two probabilities gives the relative risk, and we have to compare RR with 1.

$$RR = \frac{\mathrm{pr}(I|E)}{\mathrm{pr}(I|\overline{E})} = \frac{0.00736}{0.00273} = 2.70.$$

Table 9.19 Disease status and exposure in cohort studies

Prospective Cohort study	Disease	
	D	$\overline{D}$
Exposure E	a	b
Exposure $\overline{E}$	c	d
total	$a + c$	$b + d$

Table 9.20 Impairment of newborns in terms of mother drinking

Cohort study		Impairment	
		I	$\bar{I}$
Alcohol drinking	E	7	944
	$\bar{E}$	86	31,444
	Total	93	32,388

This means that impairment in newborns who's their mothers drank alcohol at least once a day during pregnancy is 2.70 times more likely than newborns whose mothers drank less than once a day.

Confidence interval for relative risk.

Like other statistical estimations, we can confidence interval for RR. But because its logarithm follows the normal distribution, we first find the standard error of $\log_e$ RR using the following formula:

$$Se\left(\log_e RR\right) = \sqrt{\frac{1}{a} + \frac{1}{b} - \left(\frac{1}{a+c} + \frac{1}{b+d}\right)}.$$

So, the lower and upper boundaries of $(1 - \alpha)\%$ confidence interval for $\log_e$ RR are:

$$\left(\log_e RR - Z_{1-\frac{\alpha}{2}} \times Se\left(\log_e E\right) \text{ to } \log_e RR + Z_{1-\frac{\alpha}{2}} \times Se\left(\log_e RR\right)\right),$$

where $Z_{1-\frac{\alpha}{2}}$ is corresponding value obtained from standard normal distribution. Finally, using back transformation (antilogarithm) from these boundaries the confidence interval for relative risk (RR) is obtained.

Example 14 Find the confidence interval of relative risk for data in example 13.

Solution: In Example 13, the values were $a = 7$, $b = 944$, $c = 86$, $d = 31,444$ and RR = 2.70. To calculate the confidence interval, we must first calculate the standard error of the logarithm of relative risk ($\log_e$RR) as follows:

$$Se\left(\log_e RR\right) = \sqrt{\frac{1}{a} + \frac{1}{b} - \left(\frac{1}{a+c} + \frac{1}{b+d}\right)}$$

$$= \sqrt{\frac{1}{7} + \frac{1}{944} - \left(\frac{1}{93} + \frac{1}{32388}\right)} = 0.3649.$$

The corresponding value of the standard normal distribution table for 95% confidence level is 1.96 and $\log_e 2.70 = 0.993$.

$$\left(\log_e RR - Z_{1-\frac{\alpha}{2}} \times SE\left(\log_e RR\right) \text{ to } \log_e RR + Z_{1-\frac{\alpha}{2}} \times Se\left(\log_e RR\right)\right)$$
$$= (0.993 - 1.96 \times 0.3649 \text{ to } 0.993 + 1.96 \times 0.3649)$$
$$= (0.993 - 0.7152 \text{ to } 0.993 \text{ to } 0.7152)$$
$$= (0.2778 \text{ to } 1.7082).$$

Now, using back transformation (antilogarithm from the upper and lower boundaries), a 95% confidence interval is obtained for the relative risk.

$$(e^{0.2778} \text{ to } e^{1.7082}) = (1.320 \text{ to } 5.519).$$

The confidence interval is not wide because the sample size in both groups is large.

9.8   Odds Ratio (OR)

Before starting about odds ratio, we note that "odds" is a singular word, not the plural of "odd" and has advantages for some types of analysis, as it is not constrained to lie between 0 and 1 but can take any value from zero to infinity (Bland 2015).

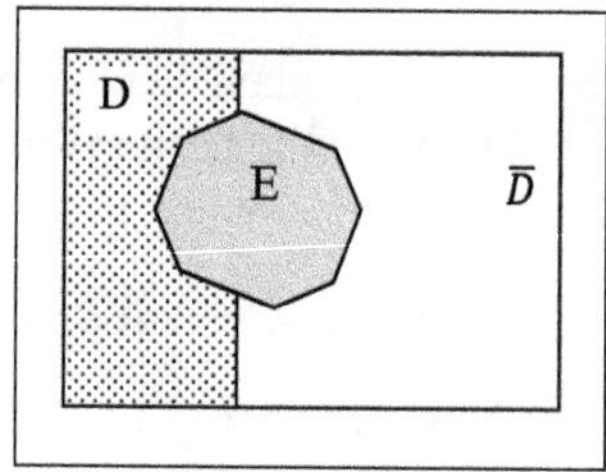

There is a main difference between the retrospective or case control studies, with prospective studies. In retrospective study, the selection of subjects is based on outcome (disease) not on exposure to a factor. Then it looks backwards to see/ examine if subjects have been exposed to a characteristic factor (such as a risk factor or a protection factor) in relation to the outcome that is established at the start of the study.

Tips

The odds of an event are a ratio of the frequency (or probability) of its occurrence to the frequency (or probability) of its non-occurrence.
The OR is a comparison of the odds of an event after exposure to a risk factor with the odds of that event in a control or reference situation. The OR is estimated as the odds of an event in the exposure group divided by the odds of that event in the control or reference group. An OR of 1.00 means that there is no increase or decrease in risk.

If the probability of occurrence of an event is p and the probability of its non-occurrence is $1 - p$, the odds of that event are defined as $\frac{p}{1-p}$. Now, if the probability of an event occurring in the two groups of patients and non-patients is p_1 and p_2, respectively, the odds of the event in the two groups will be $\frac{p_1}{1-p_1}$ and $\frac{p_2}{1-p_2}$ respectively, and the odds ratio in both groups is expressed as follows:

$$\text{OR} = \frac{\frac{p_1}{1-p_1}}{\frac{p_2}{1-p_2}}.$$

The odds ratio illustrates how strongly the presence or absence of a certain characteristic (disease) relates to the presence or absence of another characteristic (Exposure). We can use the odds ratio to observe if a certain outcome (e.g., developing the disease) is associated with exposure to a hypothesized risk factor (e.g., smoking). With an odds ratio, *the outcome can be the starting point* with which we can determine the relative odds of someone having been exposed to a risk factor. Alternatively, we can also use it to describe the ratio of disease odds given the exposure status. Once we know the exposure and disease status of a research population, we can fill in their corresponding numbers in Table 9.21.

In retrospective and case control studies, we look at the table from vertical direction (Table 9.22).

Thus, the odds of the exposed are:

$$\text{Odds for exposed} = \frac{\text{No. of diseased \& Exposed}}{\text{No. of Healthy \& Exposed}} = \frac{a}{b}.$$

Table 9.21 Disease status and exposure in retrospective (case control) studies

		Disease	
		Yes	No
Exposure	Yes	Diseased+Exposed	Healthy+Exposed
	No	Diseased+Unexposed	Healthy+Unexposed

Table 9.22 Disease status and exposure in case control studies

Case control retrospective		disease	
		D	$\overline{D}$
Exposure	E	a	b
	$\overline{E}$	c	d
total		$a+c$	$b+d$

And the odds of unexposed are:

$$\text{Odds for Unexposed} = \frac{\text{No. of Diseased \& Unexposed}}{\text{No. of Healthy \& Unexposed}} = \frac{c}{d}.$$

The odds ratio (OR) is obtained by dividing the probability of disease among the exposed by the probability of disease among unexposed.

$$OR = \frac{\frac{\text{Diseased \& exposed}}{\text{Healthy \& Exposed}}}{\frac{\text{Diseased \& Unexposed}}{\text{Healthy \& Unexposed}}} = = \frac{\frac{a}{c}}{\frac{b}{d}} = \frac{a \times d}{b \times c}.$$

We can define as:

$$OR = \frac{\frac{\text{Diseased \& exposed}}{\text{Diseased \& Unexposed}}}{\frac{\text{Healthy \& Exposed}}{\text{Healthy \& Unexposed}}} = \frac{\frac{a}{b}}{\frac{c}{d}} = \frac{a \times d}{b \times c}.$$

And then:

$$OR = \frac{\frac{p_1}{1-p_1}}{\frac{p_0}{1-p_0}}.$$

Example 15 In a case control study of people aged 45–54, they developed diabetes by residential areas (rural, urban) according to Table 9.23:

Solution: In this example, $a = 269$, $b = 504$, $c = 331$ and $d = 496$, and the odds ratio is equal to:

$$OR = \frac{a \times d}{b \times c} = \frac{269 \times 496}{504 \times 331} = 0.80.$$

Table 9.23 Diabetic patients by residential area

		Diabetes	
		Case	Control
Residential area	Rural	269	504
	Urban	331	496

Example 16 In a study of 96 low birth weight infants, mothers of 60 infants smoked during pregnancy, and out of 904 low birth weight infants, 440 mothers smoked. Estimate the odds ratio of low birth weight by smoking during pregnancy.

Solution: Since the study is through the mothers' history, the study can be considered retrospective. To calculate the odds ratio, we fill the following Table 9.24.

In the above table: $a = 60$, $b = 440$, $c = 360$, and $d = 464$.

The odds of low birth weight among smoking mothers are $\frac{60}{440}$ and the odds of low birth weight among non-smoking mothers are $\frac{36}{464}$. By dividing these two odds, we have:

$$\text{OR} = \frac{\frac{60}{440}}{\frac{36}{464}} = \frac{60 \times 464}{440 \times 36} = 1.76.$$

That is, the odds of low birth weight among mothers who smoked during pregnancy are 1.76 times the odds of low birth weight for mothers whose mothers did not smoke during pregnancy.

Confidence interval for relative risk.

Like relative risk (RR) we can calculate confidence interval for OR. Similar to RR, we first find the standard error of $\log_e \text{OR}$ using the following formula:

$$\text{Se}\left(\log_e \text{OR}\right) = \sqrt{\frac{1}{a} + \frac{1}{b} + \frac{1}{c} + \frac{1}{d}}.$$

So, the lower and upper boundaries of $(1 - \alpha)\%$ confidence intervals for $\log_e \text{OR}$ are:

$$\left(\log_e \text{OR} - Z_{1-\frac{\alpha}{2}} \times \text{Se}\left(\log_e \text{OR}\right) \text{ to } \log_e \text{OR} + Z_{1-\frac{\alpha}{2}} \times \text{Se}\left(\log_e \text{OR}\right)\right),$$

where $Z_{1-\frac{\alpha}{2}}$ is corresponding value obtained from standard normal distribution. Finally, using back transformation (antilogarithm) from these boundaries, the confidence interval for odds ratio (OR) is obtained.

Example 17 Calculate the 95% confidence interval for the odds ratio in the above example.

Table 9.24 Low birth weight infants in terms of smoking habit of mothers

		Low birth weight		
		Yes	No	Total
Smoking habit	Yes	60	440	500
	No	36	464	500
Total		96	904	1000

Solution: In this example, the values are $a = 60$, $b = 440$, $c = 36$, $d = 464$, OR $= 1.76$, and $Z_{0.957} = 1.96$.

$$\text{Se}\left(\log_e \text{OR}\right) = \sqrt{\frac{1}{a} + \frac{1}{b} + \frac{1}{c} + \frac{1}{d}} = \sqrt{\frac{1}{60} + \frac{1}{440} + \frac{1}{36} + \frac{1}{464}} = \sqrt{0.489} = 0.221.$$

By replacing the above formula, we have:

$$(1.76 - 1.96 \times 0.221 \ \text{ to } \ 1.76 + 1.96 \times 0.221)$$
$$= (1.76 - 0.433 \ \text{ to } \ 1.76 - 0.433)$$
$$= (1.33 \ \text{ to } \ 2.19).$$

Due to the large sample size, the calculated confidence interval of the odds ratio from 1.33 to 2.19 is narrow.

Tips

Risk difference provides a measure of the *impact* of the risk factor, and focuses on the number of cases that could potentially be prevented by eliminating the risk factor.

While Relative risk and odds ratios, provide a measure of the *strength* of the association between a factor and a disease or outcome.

If the probability of occurrence of the disease in one group is p_1 and in another group is p_2, it is obvious that the probability of non-occurrence of the disease in those two groups are $1 - p_1$ and $1 - p_2$, respectively. The odds ratio can then be estimated as follows:

$$\text{OR} = \frac{\frac{p_1}{1-p_1}}{\frac{p_2}{1-p_2}} = \frac{p_1(1 - p_2)}{p_2(1 - p_1)}.$$

Calculating the standard error requires knowing the sample size of each group.

Example 18 The probability of a disease in the two groups is $p_1 = 0.4$ and $p_2 = 0.3$, respectively. So $1 - p_1 = 1 - 0.4 = 0.6$ and $1 - p_2 = 1 - 0.3 = 0.$ and the odds ratio is:

$$\text{OR} = \frac{p_1(1 - p_2)}{p_2(1 - p_1)} = \frac{0.4 \times 0.7}{0.3 \times 0.6} = \frac{0.28}{0.18} = 1.56.$$

That is, this disease is 1.56 times more common in the first population than in the second population.

Note that RR and OR should not be interpreted equally. Because as it was said $OR = \frac{\frac{p_1}{1-p_1}}{\frac{p_2}{1-p_2}}$ and $RR = \frac{p_1}{p_2}$ where $p_1 = p(D|E)$ and $p_2 = p(D|\overline{E})$. If $p_1 = p(E|D)$ and $p_2 = p(E|\overline{D})$ are defined, the prevalence ratio for exposure is obtained.

9.9 Risk Difference or Excess Risk

The **risk difference** (RD), **excess risk**, or **attributable risk** is the difference between the risk of an outcome (disease) in the exposed group and the unexposed group. Suppose we fill the corresponding numbers of outcome and exposure situation in the table as following format then p_e is the incidence (proportion) of an outcome in the exposed group and p_u is the incidence (proportion) of an outcome in the unexposed group (Table 9.25).

The point estimation of the risk difference is:

$$RD = p_e - p_u,$$

where $p_e = \frac{a}{a+b}$, $p_u = \frac{c}{c+d}$ and $RD = p_e - p_u = \frac{a}{a+b} - \frac{c}{c+d}$.

If n_e and n_u are the sample sizes of exposed and unexposed groups, the sampling distribution of RD is approximately normal with standard error as follows:

$$Se(RD) = \sqrt{\frac{p_e(1-p_e)}{n_e} + \frac{p_u(1-p_u)}{n_u}} = \sqrt{\frac{a \times b}{(a+b)^3} + \frac{c \times d}{(c+d)^3}}.$$

Confidence interval $(1 - \alpha)\%$ is obtained as:

$$CI_{1-\alpha} = p_e - p_u + z_{1-\frac{\alpha}{2}} \times SE(RD).$$

Example 19 To illustrate the risk difference (RD), we use data from a prospective cohort study (from literature) in which was examined the association between

Table 9.25 Disease status and exposure

		disease		total
		D	$\overline{D}$	
Exposure	E	a	b	$a + b$
	$\overline{E}$	c	d	$c + d$
Total		$a + c$	$b + d$	

smoking and lung cancer. They classified anyone who smoked regularly as a "smoker" and was shown in Table 9.26.

Solution: In this example, $a = 40$, $b = 20$, $c = 10$, and $d = 130$. The proportion of lung cancer in the exposed and unexposed groups and risk difference (RD) is:

$$p_e = \frac{a}{a+b} = \frac{40}{40+20} = 0.667 \quad p_u = \frac{c}{c+d} = \frac{10}{10+130} = 0.071$$

$$RD = p_e - p_u = 0.667 - 0.071 = 0.596.$$

To estimate the 95% confidence interval of the risk difference, we must first calculate the RD standard error as follows.

$$\begin{aligned}
SE(RD) &= \sqrt{\frac{p_e(1-p_e)}{n_e} + \frac{p_u(1-p_u)}{n_u}} \\
&= \sqrt{\frac{0.667(1-0.667)}{60} + \frac{0.071(1-0.071)}{140}} \\
&= \sqrt{0.00418} = 0.0646
\end{aligned}$$

or

$$\begin{aligned}
SE(RD) &= \sqrt{\frac{a \times b}{(a+b)^3} + \frac{c \times d}{(c+d)^3}} \\
&= \sqrt{\frac{40 \times 20}{(40+20)^3} + \frac{10 \times 130}{(10+130)^3}} \\
&= \sqrt{0.00418} = 0.0646.
\end{aligned}$$

Assuming 95% confidence level, $z_{1-\frac{\alpha}{2}} = 1.96$ and confidence interval is estimated as:

$$\begin{aligned}
CI_{1-\alpha} &= p_e - p_u \pm z_{1-\frac{\alpha}{2}} \times SE(RD) \\
&= 0.667 - 0.071 \pm 1.96 \times 0.0646 \\
&= 0.596 \pm 0.127 = (0.469 \text{ to } 0.723).
\end{aligned}$$

Table 9.26 Incidence of lung cancer in smokers and non-smokers

	Lung cancer	No lung cancer	Total
Smokers	40	20	60
Non-smokers	10	130	140
Total	50	140	200

The risk difference means that in the smoking group, compared to the non-smoking group during the study period, there were 596 excess (additional) lung cancer cases per 1000 subjects. According to this study, out of 100 sampling times, 95 times the true risk difference lies between 469 and 723 (in 1000 cases) excess (additional) cases of lung cancer.

As a rule of thumb, to interpret the 95% confidence interval for the risk difference, we should use the following wording: "Based on this sample, out of 100 sampling times, 95 times the true risk difference lies between the estimated [upper limit] and the [lower limit] excess/fewer cases of [disease]".

The **attributable fraction among the exposed** (AF_e) is the proportion of incidents in the exposed group that are attributable to the risk factor and are usually expressed as percentage.

$$AF_e = \frac{p_e - p_u}{p_e} \times 100.$$

The proportion of the cases can be attributed to a particular exposure.

Example 20 For the example 19, AF_e is equal to:

$$AF_e = \frac{p_e - p_u}{p_e} = \frac{0.667 - 0.071}{0.667} \times 100 = 89.4\% \, .$$

And means that 89.4% of the adverse outcomes in the exposed group can be attributed to the exposure ($AF_e = 89.4\%$).

9.10 Exercises

1. Out of 1000 patients (538 males and 462 females) 15–50 years old referred to a general practitioner, 70 (42 males and 28 females) were diagnosed with asthma and 930 were diagnosed with other causes. Find the odds ratio of asthma among men and women.
2. Out of 200 breast cancer patients who have referred for chemotherapy, 100 received the new method and 100 received the conventional standard method. In the new method, 85 people and in the conventional method, 70 people have severe nausea and vomiting. Calculate the relative risk of severe nausea and vomiting with the new method.
3. Calculate the specificity, sensitivity, positive predictive value, and negative predictive value of screening tests conducted on 2150 pupils in a city to evaluate vision impairment as well as standard assessments. Of the 2107 patients whose standard examination was negative (70 positive and 2037 negative), and of the 43 patients whose standard examination was positive (11 positive and 32 negative).

| | | Screening | | |
		Positive	Negative	Total
Standard examination	Positive	11	32	43
	Negative	70	2037	2107
	Total	81	2069	2150

4. Determine the relative risk and risk difference between smokers and women who have never smoked, as well as the associated 95% confidence interval.

Smoking habit	Incidence rate per 100,000 person-years	Observed person-years	No. of strokes
Never smoked	17.7	395,594	70
Smoker	49.6	280,141	139

5. The table below shows the findings of a case control research. Subjects are placed into two groups based on their recent history of consuming meat (yes or no), and those who do not reach part of their intestinal tissue and have turned black (necrotic) are compared to those who do not have the sickness. Calculate the odds ratio for necrotizing illness and its confidence intervals for meat consumption exposure. 50 of the 66 participants who consumed meat had necrotizing illness, while 11 of the 52 who did not eat meat recently were unwell.

| | | Recent history of meat consumption | | |
		Yes	No	Total
Necrotizing disease	Yes	50	11	61
	No	16	41	57
	Total	66	52	118

6. The table below displays the fractured bone diagnosis provided by the surgeon and X-ray testing for 200 patients. Calculate their sensitivity, specificity, positive predictive value, negative predictive value, and confidence interval for each one.

| | | Diagnosis by the surgeon | | |
		Positive	Negative	Total
X-ray	Positive	54	36	90
	Negative	123	12	135
	Total	177	48	225

7. The results of the questionnaires and interviews conducted with faculty students regarding their smoking habits are compiled in the table below. Determine Cohen's Kappa's value and its 95% confidence interval.

Smoking habit		Interview		
		Yes	No	Total
Questionnaire	Yes	100	5	105
	No	4	41	45
	Total	104	46	150

Chapter 10
Foreground Toolbox for Hypothesis Testing

10.1 Introduction

Statistical inference (judgment, decision-making, and generalization of sample results to the population) includes two stages. The first step is to estimate the desired parameter from a random sample expressed in the previous chapters. The second step is to perform a significant test or hypothesis test to determine whether the observed difference between sample data and specified value on the null hypothesis is true or due to sampling random variations. Statistical tests depend on the subject, hypothesis, the nature, and structure of the population (distribution of the studied property), whether or not the variance is known, and.... is different.

> **Tip**
>
> The "statistical significance" and "clinical importance" are two completely separate issues. "Non-significant" is not equivalent with "no effect" or "no difference" between the two groups, because the researcher may not have been able to formulate a good null hypothesis or the sample size may not have been large enough.

Therefore, the requirement of these tests is knowledge of some statistical concepts such as null and alternative hypothesis, test statistic, critical value, significance level, p-value, one-sided or two-sided test and decision rule, which we will briefly describe first.

S. H. Saneii and H. Doosti, *Practical Biostatistics for Medical and Health Sciences*,
https://doi.org/10.1007/978-981-97-3083-4_10

10.2   Definitions

Statistical significance

Statistical significance refers to the likelihood that observed differences from an expected value or between studied groups are substantial, genuine, and not merely a result of random chance. It indicates that the disparity between groups is attributable to inherent differences rather than random fluctuations. Statistical hypothesis testing is employed to ascertain significance, yielding a p-value that quantifies the probability of obtaining results as extreme as those observed, assuming randomness as the sole cause.

Conversely, clinical significance assesses the extent to which an observed phenomenon manifests in real-world implications. While statistical significance aims to refute chance occurrences, clinical significance validates the extent to which an event or effect genuinely occurs in a measurable manner. In essence, statistical significance underscores the improbability of an event occurring by chance alone.

Null hypothesis

The null hypothesis, which is denoted by H_0, states that a population parameter (such as the mean, the standard deviation, the proportion and so on) is equal to a claimed value or hypothesized value. When two or more populations are compared, their parameters (such as mean and proportion) are equal. For this reason, the hypothesis is called null hypothesis, which considers the difference between the true value of the parameter with the specified value in the hypothesis as zero (for one population) or the difference of the parameter in two populations equal to zero. The null hypothesis is often an initial claim that is based on researchers' previous analyses or specialized knowledge. The null hypothesis may be written to one of the following forms.

$$\text{One Population:} \begin{cases} H_0: \mu = \mu_0 \rightarrow \mu - \mu_0 = 0 \\ H_0: p = p_0 \rightarrow p - p_0 = 0 \end{cases}$$

$$\text{Two populations:} \begin{cases} H_0: \mu_1 = \mu_2 \rightarrow \mu_1 - \mu_2 = 0 \\ H_0: p_1 = p_2 \rightarrow p_1 - p_2 = 0 \end{cases}$$

Alternative hypothesis

The alternative hypothesis, denoted as H_1 or H_a, posits that a population parameter differs from the hypothesized value stated in the null hypothesis. It suggests that the parameter may be smaller, greater, or simply different from the null hypothesis. The alternative hypothesis represents what the researcher either believes to be true or aims to demonstrate.

In contrast, the null and alternative hypotheses are mutually exclusive statements regarding population parameters. A hypothesis test utilizes sample data to determine whether to reject the null hypothesis.

It's important to note that hypothesis tests are not designed to determine whether to accept the null hypothesis but rather to decide whether to reject it. The alternative hypothesis can be formulated as either one-tailed or two-tailed.

Alpha and p-value are two critical terms frequently used in statistical analyses, particularly in hypothesis testing scenarios. They aid in determining whether a test statistic is considered "significant" or "non-significant" in a research study. These terms assist researchers in concluding whether data conforms to a specific distribution, assessing if two processes share the same average or variation, identifying variables with an impact on the response variable in experimental designs, and more.

Significance level (Alpha)

The significance level, also known as alpha (α), represents the probability that a researcher will erroneously reject the null hypothesis when it is actually true. This alpha value determines the likelihood of a type I error, where we reject a null hypothesis that is, in reality, true. It is termed a pre-defined threshold value because researchers establish its value before examining the data.

For instance, if a researcher sets a significance level of 0.05, it means they anticipate (or desire) to erroneously reject the true null hypothesis approximately once out of every 20 times. This is because 0.05 equals 1/20, and 20 multiplied by 0.05 equals 1.

Reality of H_0	True		Not true	
Test result H_0	Don't reject	Reject	Don't reject	Reject
	$1 - \alpha$	α	β	$1 - \beta$

The p-value

The p-value is another crucial component of significance tests, derived from statistical tests. Each test statistic is associated with a corresponding probability or p-value. Unlike alpha, the p-value arises from a distinct source. This probability represents the likelihood of observing the statistic solely by chance, under the assumption that the null hypothesis is true.

Tips

When the p-value is close to the significant level of α (e.g., between 0.045 and 0.055), caution should be exercised, and decision-making and judgment require more precision. Contradictory conclusions from these two values (0.045 and 0.055) are not logical. Because with a little change in the sample size, the value of p will increase and decrease and another result will be obtained.

It is recommended that in these cases, the estimated confidence interval be calculated and presented so that the value of P is interpreted according to the estimated confidence interval. However, the use of the words "difficult to reject" or "just reject" is recommended.

The *p*-value of a test statistic serves as an indicator of the extremity of that statistic given the sample data. It quantifies the probability of obtaining a more extreme value than the one observed in the experiment.

The *p*-value describes how likely we had observed a particular set of data if the null hypothesis were true. It also tells us, if the null hypothesis of the test was true, how often we would expect to see a test statistic as extreme or more extreme than the one calculated by the statistical test. The *p*-value can vary between 0 and 1.

A *p*-value lower than significant level (α) is often considered to be statistically significant. The smaller the *p*-value, the more unlikely the observed sample and is stronger evidence more likely to reject the null hypothesis. The greater the *p*-value (greater than alpha level), the more likely the observed sample, and there is no evidence to reject the null hypothesis.

Interpretation of *p*-value

Statistical significance is a method used to determine whether the *p*-value of a statistical test is sufficiently low to warrant rejecting the null hypothesis. But what constitutes a sufficiently low *p*-value? This is determined by the alpha level, which serves as the threshold criterion.

To assess the statistical significance of an observed outcome, it is essential to compare the *p*-value with the alpha level.

Tips

Never say "we accept the null hypothesis" instead of "fail to reject the null hypothesis", because the decision is based on sample size of n. If the sample size changes, the conclusion will be different.

The goal of significance tests is to provide evidence to reject the null hypothesis to demonstrate that it is not a defensible idea, not to accept or prove it.

Always we are allowed to say:

There is no evidence to reject H_0 or
We are not able to reject H_0 or
We fail to reject H_0.

There are two situations that emerge:

If the *p*-value is very small, less than or equal to pre-defined threshold value (α), the null hypothesis (H_0) is rejected. When this happens, we say that the result is statistically significant. In other words, we are reasonably sure that there is some evidence besides chance alone that gave us an observed sample.

If the *p*-value is greater than pre-defined threshold value alpha (α), we fail to reject the null hypothesis (don't say we accept or prove the null hypothesis). In this case, we say that the result is not statistically significant. In other words, we are reasonably sure that our observed data can be explained by chance alone.

> **Tip**
>
> Never use general terms such as: **non-significant, significant,** and **highly significant** in your research report or article. Specify the exact p-value on your report instead of its limits. For example, don't report $p < 0.05$ or $p > 0.05$ instead $p = 0.08$ or $p = 0.15$. Because statistical software calculates its exact value and the reader can better interpret it by seeing the exact value of p.

Basically, statistical significance tests are designed to refute hypotheses. They are never presented to prove anything. The aim is therefore to demonstrate that the idea is not defensible. In other words, the tests are designed to reject the issues raised. Although this point is important both in medical studies and in statistical judgments, it is unfortunately rarely observed.

Test statistic

A test statistic (without "s" at the end) is a random variable computed from sample data and serves as a tool for making quantitative decisions regarding hypothesis testing. Its purpose is to determine whether there is sufficient evidence to "reject" a hypothesis (null hypothesis) about a process. The test statistic compares the observed data with what is expected under the null hypothesis and is instrumental in calculating the p-value. Essentially, a test statistic quantifies the degree of agreement between a sample of data and the null hypothesis. Its observed value varies randomly across different random samples and encapsulates information critical for deciding whether to reject the null hypothesis.

Test statistic is defined as below:

$$\text{test statistic} = \frac{\text{Observed value} - \text{Hypothesized value}}{\text{Observed standard error}},$$

where the "hypothesized value" is the same value as claimed in the null hypothesis, and in many cases, it is considered zero.

If the above value is negative, its absolute value should be considered according to the symmetry property of the normal distribution.

Critical value

In hypothesis testing, a critical value represents a point on the scale of the test statistic (on the test distribution) that is compared to the observed test statistic to determine whether to reject the null hypothesis. This critical value serves as a threshold in the hypothesis test and is contingent upon the significance level (α). It delineates the boundary of the rejection region and establishes a cutoff point that delineates regions where the test statistic is improbable to reside. Consequently, any observed test statistic beyond this critical value prompts the rejection of the null hypothesis. While critical values can be computed, they are typically derived from statistical tables based on the chosen significance level and degrees of freedom for convenience.

Decision rule

An essential aspect of hypothesis testing is the decision rule, which dictates when there is adequate evidence to support the null hypothesis (H_0). Essentially, it establishes the criteria and guidelines under which the null hypothesis should be rejected. This rule is employed to determine whether the null hypothesis should be rejected by comparing the test statistic to the critical value. The decision rule for a specific test is contingent upon three key factors: the alternative hypothesis, the test statistic, and the significance level. This comparison yields two potential outcomes:

1. If the test statistic surpasses the critical value, the null hypothesis (H_0) is rejected. This indicates that there is ample evidence to suggest that the alternative hypothesis is valid.
2. Conversely, if the test statistic does not exceed the critical value, the null hypothesis is not rejected. This signifies that there is insufficient evidence to warrant the rejection of the null hypothesis. However, it is crucial to note that one should refrain from asserting acceptance of the null hypothesis.

When employing the p-value and significance level (α) instead of the test statistic and critical value, the decision rule is as follows:

- If the significance level exceeds the p-value ($\alpha > p$), the null hypothesis (H_0) is rejected.
- Otherwise, if the significance level is less than or equal to the p-value, the null hypothesis is not rejected.

One-tailed and two-tailed test

Statistical significance tests can be conducted as either one-tailed or two-tailed, representing alternative approaches. The selection between a one-tailed and a two-tailed test hinges on the formulation of the alternative hypothesis.

Two-tailed test: A two-tailed test can detect when the population parameter differs in either direction but has less power than a one-tailed test. When the estimated value is greater or less than a certain value of parameter, a two-tailed test, which is known as a non-directional hypothesis, is appropriate. In this case, the alternative hypothesis states that the population parameter may be either greater than or less than the hypothesized value. For example, assuming the following notation:

μ_0 the claimed or hypothesized mean of population,
μ_1 the mean of first population,
μ_2 the mean of second population,
p_0 the claimed or hypothesized proportion of population.
p_1 the proportion of first population,
p_2 the proportion of second population.

Different possibilities to write the alternative hypothesis (H_A) in two-tailed test are as follows:

$H_A: p_1 \neq p_2$ or $p_1 - p_2 \neq 0$	To compare proportion of two populations
$H_A: \mu_1 \neq \mu_2$ or $\mu_1 - \mu_2 \neq 0$	To compare mean of two populations
$H_A: p \neq p_0$ or $p - p_0 \neq 0$	To compare proportion of a population with a fixed p_0
$H_A: \mu \neq \mu_0$ or $\mu - \mu_0 \neq 0$	To compare mean of a population with a fixed μ_0

One-tailed test: A one-tailed test can detect when the population parameter differs only in one direction (left or right), but not both. It has greater power than a two-tailed test, but it cannot detect whether the population parameter differs in the opposite direction. When the estimated value departs from the reference value of parameter in only one direction, a one-tailed test, which is known as a directional hypothesis, is appropriate. In this case, the alternative hypothesis determines whether the population parameter differs from the hypothesized value in a specific direction. For instance, employing the above notations for two-tailed test, we can write different possibilities for the alternative hypothesis (H_A) in one-tailed test as follows:

$H_A: p_1 < p_2$ or $p_1 > p_2$	To compare proportion of two populations
$H_A : \mu_1 < \mu_2$ or $\mu_1 > \mu_2$	To compare mean of two populations
$H_A : p_1 < p_0$ or $p_1 > p_0$	To compare proportion of a population with a fixed p_0
$H_A : \mu_1 < \mu_0$ or $\mu_1 > \mu_0$	To compare mean of a population with a fixed μ_0

See Fig. 10.1.

One-tailed or two-tailed Tests

Because the true difference rarely occurs in only one side of the distribution, in the vast majority of tests, there is no reason to calculate the p-value with one-tailed test, and performing a two-tailed test is recommended. But if the alternative hypothesis (H_A) is limited to one side of the distribution, one-tailed test is preferred. The values corresponding to α level for the one-tailed test are slightly lower than the corresponding values for the two-tailed test. For example, the corresponding value for 0.05 in the one-tailed test is 1.64 (1.64 times of the standard deviation away from the mean) and in the two-tailed test is 1.96 (1.96 times of the standard deviation away from the mean), which can sometimes change the conclusion of the statistical test.

> **Tip**
> You must not perform first a two-tailed test, obtain non-significant result, and then try a one-tailed test to see if that is statistically significant.

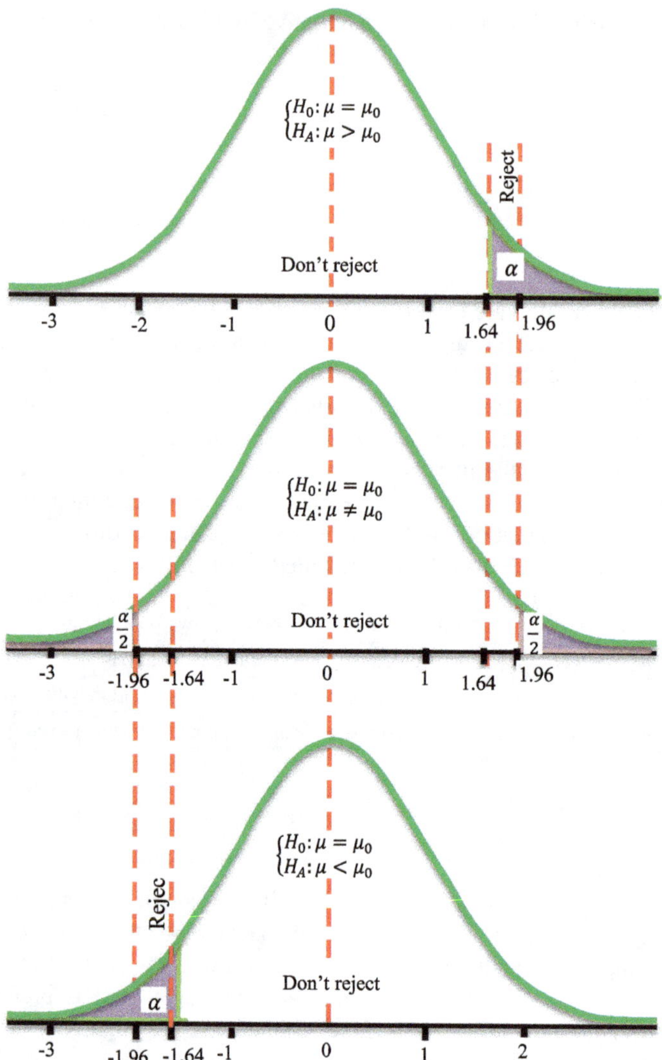

Fig. 10.1 One-tailed and two-tailed distributions

The two-tailed tests are the default choice, unless there is very good evidence to use one-tailed. Because in most studies, the researcher is interested to determine whether there is a positive effect or a negative effect and the results in either direction (positive or negative) are important for him.

One-tailed tests are rarely appropriate. The use of a one-tailed test is only permissible if there are many unexpected one-direction changes. However, in case of doubt, the use of one-tailed p is not recommended. However, if there is a

Fig. 10.2 Steps for testing
Hypothesis

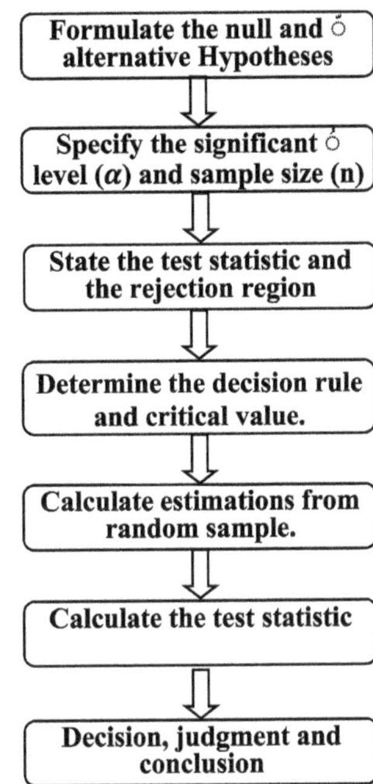

tendency to perform a one-tailed test, such a decision should be made at the beginning of the study before you look at the data.

There are some cases where one-tailed tests are not only a valid option, but truly are a requirement (Fig. 10.2).

10.3 The Steps for Testing Hypotheses

Various steps have been proposed for testing hypothesis. In this book, seven steps of performing the hypothesis test proposed by Niemann and Pearson will be explained.

- Formulate the null and alternative Hypotheses (H_0 and H_A).
- Specify the significant level (α) and sample size (n).
- State the test statistic and the rejection region.
- Determine the decision rule and critical value.
- Calculate estimations from random samples, such as means, proportions, and standard deviations.
- Calculate the test statistic.
- Decision, judgment, and conclusion.

10.4   Applications of Statistical Tests

The general applications of the common tests (Z, t, F and χ^2) will be briefly described here.

The Z Test

When the variance of population from which the sample is taken, is known, and the known value of variance is used in the test statistic, the Z test is applied for the following:

- Comparing the mean of a population with a fixed number (specified in null hypothesis).
- Comparing the mean of two populations.
- Comparison of the mean difference between two populations with a fixed number (specified in null hypothesis).
- This test should also be used when the goal is to compare proportions:
- Comparison of proportion in population with a fixed ratio (specified in null hypothesis).
- Comparing the proportion in two populations.
- Comparison of correlation coefficient with a fixed number (specified in null hypothesis).

The t-tests

This test will be useful if the variance of the reference population from which the sample was taken is unknown or the sample size is small (less than 30), and the estimation of sample variance is used in the calculation of test statistics:

- Comparing the mean of a population with a fixed number (specified in null hypothesis).
- Comparing the mean of two populations (independent, paired, matched).
- Comparison of the mean difference between two populations with a fixed number (specified in null hypothesis).
- Comparison of regression coefficient (β) with a fixed number (specified in null hypothesis).

The F tests

In this book, the F test is used only for the following two cases.

- Comparing the variance of two populations.
- Comparing the mean in more than two populations.

The χ^2 test

The Chi-squared (χ^2) test is used for the following two cases.

- Relationship or association between the two quantitative variables in contingency tables.
- Goodness of fit of an observed distribution to a theoretical one.

10.5 Common Statistical Tables

To conclude and judge about any statistical test and to find critical values, one should refer to the statistical distributions tables. Of course, in analyses performed by statistical software, test statistic and p-values are presented, but they still do not provide a critical value. In order to give readers easy access, the distribution tables of z, t, f, and χ^2 are given in the appendix only for some common confidence levels and limited degrees of freedom.

Chapter 11
Hypothesis Testing

11.1   Introduction

In the previous chapter, a foreground toolbox of statistical tests including the concepts, definitions, and stages of performing significant tests was explained in detail so that we can deal directly with statistical tests in this chapter. Thus, these two chapters are complementary to each other, and wherever you need to understand more, do not hesitate to refer to Chap. 10.

For each part of this section, an example is provided first to clarify what the researcher means by the test and what s/he wants to judge and decide. Then, we calculate 95% interval estimate (95% CI) and judge and conclude about it. We then execute the hypothesis test for the same example to reveal the tight relationship between the significant tests and the confidence interval.

Before explaining the hypothesis tests, the common symbols and Greek letters that are often used in this chapter are introduced.

μ	The actual proportion of the population	p	The actual proportion of the population
μ_0	Hypothesized mean of the population in the null hypothesis	p_0	Hypothesized proportion of the population in the null hypothesis
μ_1	The actual mean in the first population	p_1	The proportion of the property in the first population
μ_2	The actual mean in the second population	p_2	The proportion of the property in the second population
$\bar{x}$	Estimated sample mean	n	Total sample size
$\bar{x}_1$	The estimated mean in the first population	n_1	The first sample size
$\bar{x}_2$	The estimated mean in the second population	n_2	The second sample size
σ_1^2	The known variance in the first population	S_1^2	The estimated variance from the first sample
σ_2^2	The known variance in the second population	S_2^2	The estimated variance from the second sample
$\bar{p}$	The average of two proportions	S_p^2	The pooled variance of two samples

© The Author(s), under exclusive license to Springer Nature Singapore Pte Ltd. 2024

S. H. Saneii and H. Doosti, *Practical Biostatistics for Medical and Health Sciences*, https://doi.org/10.1007/978-981-97-3083-4_11

All tests in this book are done with two-tailed test. That is, the error will be on both sides $\frac{\alpha}{2}$ and $1 - \frac{\alpha}{2}$.

11.2 Comparing the Mean of a Population with a Fixed Value

The variance of population is Known

Example 11.1 Plasma volume in a sample of 9 from a population with a variance of 1.0 liter is: 3.11–3.25–3.49–3.26–2.76–3.37–2.86–2.75–3.05. Can the plasma volume of the population from which this sample is taken be equal to 3.5?

Solution: The available information show that $\mu_0 = 3.5$, $n = 9$, $\bar{x} = 3.10$, $\sigma^2 = 0.1$, $\sigma = 0.316$ and the estimated standard error of the mean can be calculated $Se = \frac{\sigma}{\sqrt{n}} = \frac{0.316}{\sqrt{9}} = 0.105$. Since the variancof the population is known, we use the standard normal distribution ($z_{0.975} = 1.96.$) and the 95% confidence interval is equal to:

$$95\% \, CI = \bar{x} \pm z_{0.975} \times Se = 3.10 \pm 1.96 \times 0.105$$
$$= 3.10 \pm 0.206 = (2.89 \text{ to } 3.31)$$

The 95% confidence interval does not contain the hypothesized and fixed value of 3.5, so there is no evidence that the plasma volume of the population from which this sample was taken should be equal to 3.5.

Using the Test:

The null and alternative hypotheses are: $\begin{cases} H_0 : \mu = 3.5 \\ H_A : \mu \neq 3.5 \end{cases}$

Sample size is 9 and significant level is 0.05.

Test statistic is $z = \frac{(\bar{x} - \mu_0)}{\frac{\sigma}{\sqrt{n}}}$.

Critical value is $z_{0.975} = 1.96$. If the test statistic is greater than the critical value of 1.96, then we reject the null hypothesis (H_0).

According to sample data $z = \frac{(\bar{x} - \mu_0)}{\frac{\sigma}{\sqrt{n}}} = \frac{(3.10 - 3.5)}{\frac{0.316}{\sqrt{9}}} = -3.80.$

Since $|-3.80| > 1.96$, then we reject the null hypothesis (H_0).

We cannot assume the mean plasma volume of population equal to 3.5. That is, the difference between mean of the plasma volume and the value and 3.5 is significant.

The Variance of Population is Unknown

Example 11.2 The recommended average level of energy intake for women aged 20–30 is 7725 kcal/day. In a random sample of 11 healthy women in the same age group, the average energy intake of 10 days was measured and the values of 5260–5470–5640–6180–6390–6515–6805–7515–8230–8770–7515 were recorded. Assuming that the distribution of the above population is normal, is their daily energy intake equal to the recommended level?

Solution: We can extract and calculate the following measures:

$$\mu_0 = 7725, \ n = 11, \ \bar{x} = 6753.6,$$

$$S = 1142.1 \text{ and } Se = \frac{S}{\sqrt{n}} = \frac{1142.1}{\sqrt{11}} = 344.4$$

Since the variance of the population is not given, instead of the z distribution, the t distribution with degree of freedom df $= n-1 = 11-1 = 10$ should be used. Its critical value from the t table is equal to $t_{(0.975,10)} = 2.228$ and the 95% confidence interval is obtained as follows:

$$95\% \text{ CI} = \bar{x} \pm t_{(0.975,10)} \times Se = 6753.6 \pm 2.228 \times 344.4$$
$$- (5986 \text{ to } 7521)$$

The 95% confidence interval does not include the recommended value. Therefore, the average energy intake of women in this population cannot be considered equal to the recommended level and is less than that.

Using the Test:

The null and alternative hypotheses are: $\begin{cases} H_0 : \mu = 7725 \\ H_A : \mu \neq 7725 \end{cases}$.

Sample size is 11 and significant level is 0.05.

Test statistic is $t = \frac{(\bar{x}-\mu_0)}{\frac{S}{\sqrt{n}}}$.

Critical value is $t_{(0.975,10)} = 2.228$. If the test statistic is greater than the critical value of 2.228, then we reject the null hypothesis (H_0).

According to sample data,

$$t = \frac{(\bar{x} - \mu_0)}{\frac{\sigma}{\sqrt{n}}} = \frac{(6753.6 - 7725)}{\frac{344.4}{\sqrt{11}}} = -2.82.$$

Since $|-2.82| > 2.228$, then we reject the null hypothesis (H_0).

Since the difference between the mean of energy intake level of women aged 20–30 and 7725 is significant, we cannot assume that the mean of energy intake level of women aged 20–30 is 7725 (Fig. 11.1).

To execute this test by SPSS, first we go to Analyze\Compare Means\One-Sample T-Test to open another window (Fig. 11.2).

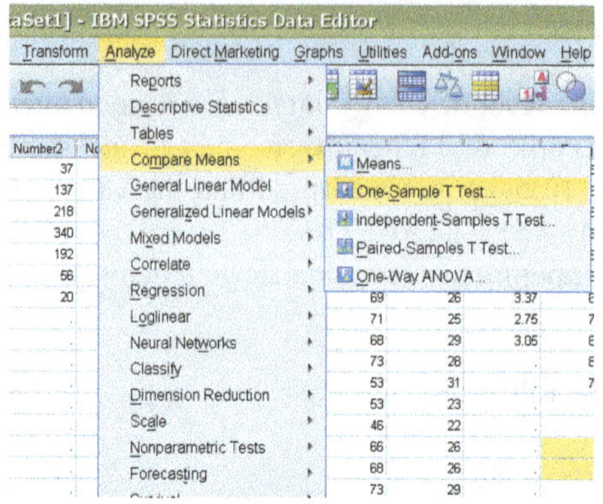
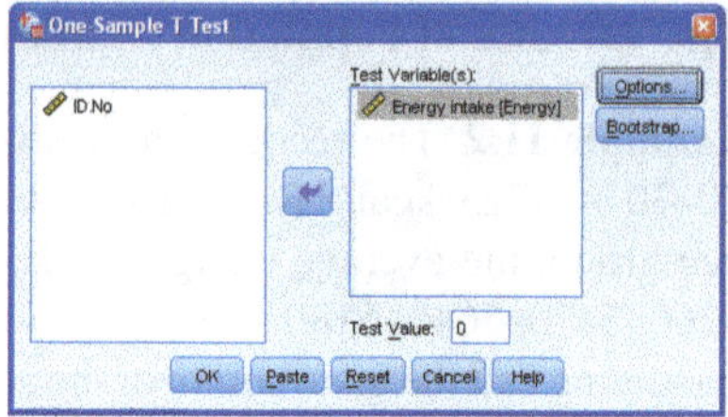

Fig. 11.1 Steps to run t-test in SPSS

Fig. 11.2 Data entry in SPSS

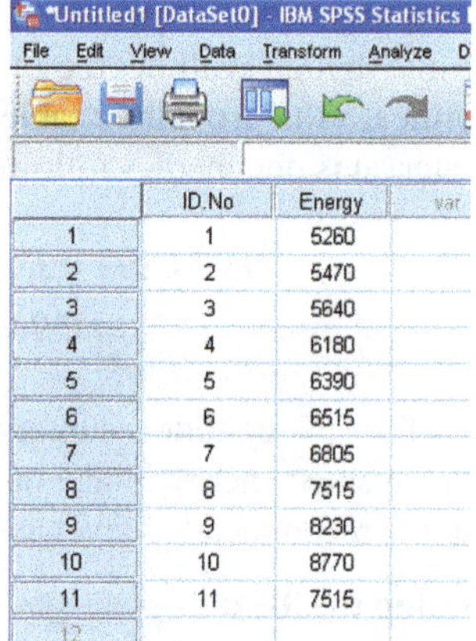

In this window, move the variable name to be tested to the "Test Variable (s)" box and type the hypothetical constant value in the "Test Value" box. After pressing the OK button, the test will run.

Example 11.3 Run the hypothesis test for the previous example (energy intake of women 20–30 years old) with SPSS.

Solution: The data should be entered in a single column in SPSS as follows.

Then, as shown in Fig. 11.2, select the SPSS path and enter the interested variable in the related box and press the OK button to have the following output.

One-sample statistics

	N	Mean	Std. deviation	Std. error mean
Energy intake	11	6753.64	1142.123	344.363

One-sample test

	Test value = 0					
	t	df	Sig. (two-tailed)	Mean difference	95% CI of the difference	
					Lower	Upper
Energy intake	19.612	10	0.000	6753.636	5986.35	7520.93

As can be seen, the answers are equal to manual calculations. Here the p-value is given in the fourth column. $p = 0.000$ is smaller than the significance level (α); therefore, the null hypothesis is rejected.

Example 11.4 A pharmaceutical company has stated that the percentage of alcohol in its cough syrup is 5%. The factory quality control unit, in order to investigate this claim, purchased 10 bottles of that syrup from different pharmacies, and after chemical analysis, the percentages of alcohol (4.81, 4.85, 4.70, 4.86, 4.86, 5.16, 4.80, 4.72, 5.04, 5.20) is obtained: Assuming the distribution of the percentage of alcohol in the bottles is distributed normally, do the sample observations confirm the factory claim?

Solution: Given information on the exercise is:

$$n = 10 \text{ and } \bar{x} = 4.90,$$

Note that in this example the average percentage of alcohol (not the proportion of alcoholic syrups) is considered. On the other hand, since the variance of the population is unknown, we have to find its variance from the sample data and use the t distribution with degree of freedom df $= n-1 = 10-1 = 9$ instead of z.

$$S^2 = \frac{(5.20 - 4.90)^2 + (5.04 - 4.90)^2 + \cdots + (4.81 - 4.90)^2+}{n - 1}$$

$$= \frac{0.2734}{9} = 0.0304$$

The standard deviation is $S = \sqrt{0.0304} = 0.174$ and the standard error of estimation is Se $= \frac{S}{\sqrt{n}} = \frac{0.174}{\sqrt{10}} = 0.055$.

Its critical value from the table is $t_{(0.975,10)} = 2.262$. Then, the 95% confidence interval is obtained as follows:

$$95\% \text{ CI} = \bar{x} \pm t_{(0.975,9)} \times \text{Se} = 4.90 \pm 2.262 \times 0.055 = 4.90 \pm 0.124$$
$$= (4.776 \text{ to } 5.024)$$

The 95% confidence interval includes the factory's claimed percentage. Therefore, the average alcohol of these syrups can be considered equal to the claimed amount of 5%.

Using the Test:

The null and alternative hypotheses are: $\begin{cases} H_0 : \mu = 5 \\ H_A : \mu \neq 5 \end{cases}$.

Sample size is 10 and significant level is 0.05.

Test statistic is $t = \dfrac{(\bar{x} - \mu_0)}{\frac{s}{\sqrt{n}}}$.

Critical value is $t_{(0.975,9)} = 2.262$. If the test statistic is greater than the critical value of 2.262, then we reject the null hypothesis (H_0).

According to sample data,

$$t = \frac{(\bar{x} - \mu_0)}{\frac{\sigma}{\sqrt{n}}} = \frac{(4.9 - 5)}{\frac{0.174}{\sqrt{10}}} = -1.818.$$

Since $|-1.818| < 2.262$, then we don't reject the null hypothesis (H_0).

The mean percentage of alcohol in syrups produced by this factory can be considered to be 5%. Because, the difference between the mean of the sample with the claimed percentage of 5 is not statistically significant.

To execute the significant test in SPSS, first go to Analyze\Compare Means\One-Sample T Test, to open the following window.

In this window, select the name of the variable (Syroup) from the list on the left and enter it in the right box titled Test Variable (s). In the Test Value box, enter the value 0.05, and then confirm the program by pressing OK to run and have the output as below.

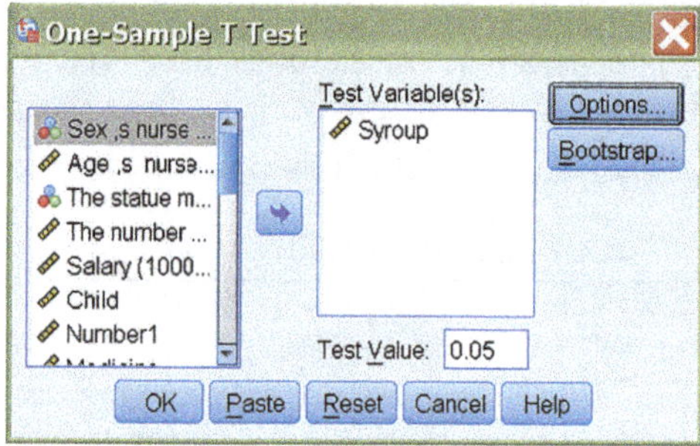

One-sample statistics

	N	Mean	Std. deviation	Std. error mean
Syroup	10	4.9000	0.17429	0.05512

One-sample test

	Test value $= 0$					
	t	df	Sig. (two-tailed)	Mean difference	95% CI of the difference	
					Lower	Upper
Syroup	88.903	9	0.000	4.90000	4.7753	5.0247

11.3 Comparing the Mean of Two Populations

The Variance of the Two Populations is Known.

Example 1.5 In a study, the age of 180 patients with coronavirus was recorded. The mean age of 49 deaths was 68.5 years with a standard deviation of 15.07 years and the mean age of 131 recovered patients was 55.69 years with a standard deviation of 18.08 years. Can the average age of the deaths be equal to the average age of the recovered ones? Similar studies show that the standard deviation of the age of the dead and the recovered is 19 and 14 years, respectively.

Solution: The given information in this example is:

$$n_1 = 49, \ \bar{x}_1 = 68.50, \ S_1 = 15.07, \ \sigma_1^2 = (19)^2 = 361$$

$$n_2 = 131, \ \bar{x}_2 = 55.69, \ S_2 = 18.08, \ \sigma_2^2 = (14)^2 = 196$$

In order to obtain the pooled standard error of the two groups, the known standard deviations of the groups, which are given in the example, must be used, and the samples standard deviation cannot be applied:

$$\text{Se} = \sqrt{\frac{\sigma_1^2}{n_1} + \frac{\sigma_2^2}{n_2}} = \sqrt{\frac{361}{49} + \frac{196}{131}} = \sqrt{8.863} = 2.98$$

The samples are large enough and the 95% confidence interval of the age mean difference between the two groups of recovers and deaths can be estimated using the normal approximation ($z_{0.975} = 1.96$).

$$\text{CI95\%} = (\bar{x}_1 - \bar{x}_2) \pm z_{0.975} \times \text{Se} = (68.50 - 55.69) \pm 1.96 \times 2.98$$
$$= 12.81 \pm 5.84 = (6.97 \text{ to } 18.64)$$

The estimated interval does not include the value zero. As a result, based on this example, the null hypothesis is rejected. That is, the mean age in the two groups of recovers and deaths cannot be equated.

Using the Test:

The null and alternative hypotheses are:
$$\begin{cases} H_0 : \mu_1 = \mu_2 \text{ or } \mu_1 - \mu_2 = 0 \\ H_A : \mu_1 \neq \mu_2 \text{ or } \mu_1 - \mu_2 \neq 0 \end{cases}.$$

Sample size in first and second groups is 49 and 131, respectively, and default significant level is 0.05.

Test statistic is

$$z = \frac{(\bar{x}_1 - \bar{x}_2)}{\sqrt{\frac{\sigma_1^2}{n_1} + \frac{\sigma_2^2}{n_2}}}$$

Critical value is $z_{0.975} = 1.96$. If the test statistic is greater than the critical value of 1.96, then we reject the null hypothesis (H_0).

The test statistic according to sample data in two groups is:

$$z = \frac{(68.5 - 55.69)}{\sqrt{\frac{361}{49} + \frac{196}{131}}} = \frac{12.81}{2.98} = 4.30$$

Since $4.30 > 1.96$, then we reject the null hypothesis (H_0).

The mean age of the two groups cannot be considered equal. That is, the age mean difference between the two groups recovered and the deaths was significant with zero.

Example 11.6 In a malaria-screening program, 150 children aged 1–4 years were tested for the presence of the P. falciparum parasite in their blood. The parasite was present in 70 children with a mean hemoglobin level of 10.6 per 100 ml. In 80 children without parasites, the mean hemoglobin level was 11.5 g per 100 ml. Can the mean hemoglobin level among children infected with the parasite be equal to that of children without the parasite? Similar studies show a standard deviation of hemoglobin levels among children in both groups which is 1.4 g.

Solution: The following information is provided in the example:

$$\sigma_1 = \sigma_2 = 1.4, \ \bar{x}_1 = 10.6, \ \bar{x}_2 = 11.5, \ n_1 = 70 \text{ and } n_2 = 80$$

Because the standard error of the two groups is equal, the pooled standard error can be obtained by the following formula:

$$Se = \sigma \times \sqrt{\frac{1}{n_1} + \frac{1}{n_2}} = 1.4 \times \sqrt{\frac{1}{70} + \frac{1}{80}} = 0.229$$

The samples are large enough and the 95% confidence interval of the difference in hemoglobin level between the two groups can be estimated using the normal approximation ($z_{0.975} = 1.96$).

$$CI95\% = (\bar{x}_1 - \bar{x}_2) \pm z_{0.975} \times Se = (11.50 - 10.6) \pm 1.96 \times 0.229$$
$$= 0.90 \pm 0.45 = (0.45 \text{ to } 1.35)$$

The estimated interval does not include the value zero. As a result, based on this example, the null hypothesis is rejected. That is, the mean hemoglobin level in the two groups cannot be equated.

Using the Test:

The null and alternative hypotheses are: $\begin{cases} H_0 : \mu_1 = \mu_2 \\ H_A : \mu_1 \neq \mu_2 \end{cases}$ or $\begin{cases} H_0 : \mu_1 - \mu_2 = 0 \\ H_A : \mu_1 - \mu_2 \neq 0 \end{cases}$

Sample size in first and second groups is 70 and 80, respectively, and default significant level is 0.05.

Test statistic is:

$$z = \frac{(\bar{x}_1 - \bar{x}_2)}{\sigma \times \sqrt{\frac{1}{n_1} + \frac{1}{n_2}}}$$

Critical value is $z_{0.975} = 1.96$. If the test statistic is greater than the critical value of 1.96, then we reject the null hypothesis (H_0).

The test statistic according to sample data in two groups is:

$$z = \frac{(11.5 - 10.6)}{1.4 \times \sqrt{\frac{1}{70} + \frac{1}{80}}} = \frac{0.90}{0.229} = 3.93$$

Since $3.93 > 1.96$, then we reject the null hypothesis (H_0).

The mean hemoglobin level of the two groups cannot be considered equal. That is, the mean difference of hemoglobin levels between two groups with zero was significant.

The Variance of the Two Populations is Unknown

Example 11.7 A study was performed to compare the birth weight of infants whose mothers smoked with those whose mothers did not smoke. Sample mean and standard deviation of weight in a sample of 14 smoking mothers were 3.20 and 0.492 and in a sample of 15 non-smoking mothers were 3.59 and 0.364. Can the birth weight of the two groups be equal?

Solution: The following information is provided in the example:

$$n_1 = 14, \ \bar{x}_1 = 3.20, \ S_1 = 0.492, \text{ and } n_2 = 15, \ \bar{x}_2 = 3.59, \ S_2 = 0.364$$

The pooled standard deviation and standard error of two groups can be obtained by the following formula:

$$S_p = \sqrt{\frac{(n_1 - 1) \times S_1^2 + (n_2 - 1) \times S_2^2}{n_1 + n_2 - 2}}$$

$$= \sqrt{\frac{(14 - 1) \times 0.492^2 + (15 - 1) \times 0.364^2}{14 + 15 - 2}} = 0.430$$

$$Se = S_p \times \sqrt{\frac{1}{n_1} + \frac{1}{n_2}} = 0.430 \times \sqrt{\frac{1}{14} + \frac{1}{15}} = 0.160$$

The variance of the population is unknown and the sample size is small, so instead of the z distribution, we should use the t distribution with degree of freedom $df = n_1 + n_2 - 2 = 14 + 15 - 2 = 27$. Its critical value from the table is $t_{(0.975,27)} = 2.05$. The 95% confidence interval for mean difference in the newborn weights between two groups is:

$$CI95\% = (\bar{x}_1 - \bar{x}_2) \pm t_{(0.975,27)} \times Se = (3.59 - 3.2) \pm 2.05 \times 0.16 = 0.39$$
$$\pm 0.328 = (0.062 \; to \; 0.717)$$

The estimated interval does not include the zero value. As a result, based on this example, the null hypothesis is rejected. That is, the mean weight newborns in the two groups (smoker mothers and non-smoker mothers) are not equal.

Using the Test:

The null and alternative hypotheses are $\begin{cases} H_0 : \mu_1 = \mu_2 \\ H_A : \mu_1 \neq \mu_2 \end{cases}$ or $\begin{cases} H_0 : \mu_1 - \mu_2 = 0 \\ H_A : \mu_1 - \mu_2 \neq 0 \end{cases}$:

Sample size in first and second groups is 14 and 15, respectively, and default significant level is 0.05.

Test statistic is:

$$t = \frac{(\bar{x}_1 - \bar{x}_2)}{S_p \times \sqrt{\frac{1}{n_1} + \frac{1}{n_2}}}$$

Critical value is $t_{(0.975, 27)} = 2.05$. If the test statistic is greater than the critical value of 2.05, then we reject the null hypothesis (H_0).

The test statistic according to sample data in two groups is:

$$t = \frac{(3.59 - 3.20)}{0.43 \times \sqrt{\frac{1}{14} + \frac{1}{15}}} = \frac{0.39}{0.16} = 2.44.$$

	InfantWeight	Smoking	
1	2.75	0	
2	3.68	Yes	
3	3.97	Yes	
4	3.23	Yes	
5	2.84	Yes	
6	3.52	Yes	
7	2.98	Yes	
8	2.29	Yes	
9	3.38	Yes	
10	2.44	Yes	
11	3.49	Yes	
12	3.29	Yes	
13	3.76	Yes	
14	3.18	Yes	
15	4.11	No	
16	3.57	No	
17	2.74	No	
18	3.50	No	
19	4.11	No	
20	3.51	No	
21	3.58	No	
22	3.76	No	
23	3.82	No	
24	3.51	No	

Since $2.44 > 2.05$, then we reject the null hypothesis (H_0).

This interval tells us that the weight difference between the infants of the two groups is between 62 and 717 g. The 95% confidence interval does not contain zero, so the mean weight of infants in these two groups is not equal, and women smokers give birth to infants with less weight.

To run this significant test in SPSS, it should be noted that the weight data of both groups should be entered in only one column and the information about the group of smoking and non-smoking mothers should be entered in a separate column. Then select Analyze\Compare Means\ Independent Samples T Test to open following window.

In this window, from the list on the left, we first select the name of the variable whose mean in the two groups should be compared and shift it to the Test Variable (s) box. This variable must be on a quantitative scale. In the box below, titled Grouping Variable, enter the name of the variable that separates the data into two categories (such as gender or smoker and non-smoker). This variable must be on categorical scale. By entering the name of the grouping variable, the "Define Groups" button is activated. Pressing it will open another sub-window into which to enter the codes of the two groups being compared. Then, by pressing the "Continue" button and confirming "OK", the program will run.

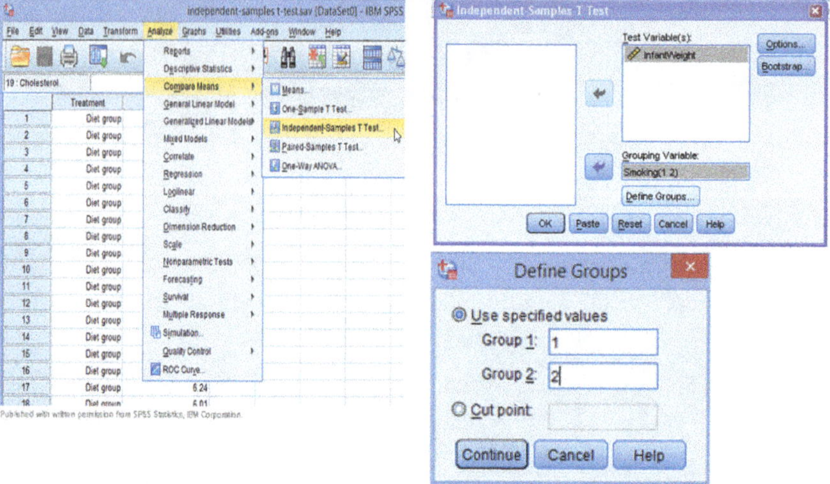

Published with written permission from SPSS Statistics, IBM Corporation.

The output of the SPSS will be as follows:

Group statistics

	Smoking	N	Mean	Std. deviation	Std. error mean
Infant weight	Yes	14	3.2000	0.49236	0.13159
	No	15	3.5900	0.36401	0.09399

Levene's test for equality of variances		t-test for equality of means							
f	Sig	t	df	Sig. (two-tailed)	Mean difference	Std. error difference	95% CI of the difference		
							Lower	Upper
2.127	0.156	-2.437	27	0.022	-0.39000	0.16002	-0.71833	-0.06167

11.4   Comparison of the Mean of the Two Paired Samples

Paired samples (dependent samples or related sample) are two samples (measurements) drawn from the same coupled, matched, or paired individuals. In this type of sample, we have a dataset in which each data point in one sample is uniquely paired to a data point in another sample. Paired samples may include: 1—pre-test/post-test samples, 2—duplicate measurements on the same samples, and 3—any situation in which each data point in one sample is uniquely matched to a data point in the second sample. With paired samples, we need an approach, which looks at the differences between pairs. Because differences between two samples are calculated from the matched or paired samples and we can use paired t-test (equivalently, a one-sample t-test of the differences).

Example 11.8 To determine the levels of blood cholesterol, two blood samples were taken from 11 people and sent to two different laboratories. The results of these two laboratories (x and y) are given in Table 11.1. Can the results of these two laboratories be considered equal?

Solution: The data are not independent of each other and are paired data. Therefore, we are dealing with the difference in the values of the two laboratories for each one ($d_i = x_i - y_i$). The difference between the two measurements and its square is given in the fourth and fifth columns of Table 2.7.

$$\overline{d} = \frac{55}{11} = 5, \quad S = \sqrt{\frac{\sum d_i^2 - \frac{(\sum d_i)^2}{n}}{n-1}} = \sqrt{\frac{2661 - \frac{(55)^2}{11}}{11-1}} = 15.45,$$

$$Se\left(\overline{d}\right) = \frac{15.45}{\sqrt{11}} = 4.66$$

In this example, df $= n - 1 = 11 - 1 = 10$ and the critical value is $t_{(0.975,10)} = 2.228$, and the 95% confidence interval is calculated as follows:

$$CI95\% = \overline{d} \pm t_{(0.975,10)} \times Se = 5 \pm 2.228 \times 4.66 = 5 \pm 10.38$$
$$= (-5.38 \text{ to } 15.38)$$

Table 11.1 Results of two laboratories (x and y)

Sample	Lab 1 (x)	Lab 2 (y)	$d_i = x_i - y_i$	d_i^2
1	310	285	+25	623
2	228	234	−6	36
3	265	279	−14	196
4	247	231	+16	256
5	243	212	+31	961
6	238	250	−12	144
7	285	285	0	0
8	263	260	+3	9
9	269	256	+13	169
10	221	210	+11	121
11	208	220	−12	144
Total	–	–	+55	2661

This interval means that the mean differences in blood cholesterol levels in the two laboratories are between -5.38 and 15.38. Since the 95% confidence interval includes zero, the difference between the two laboratories can be considered zero. That is, the values of these two laboratories are equal.

To run the significance test in SPSS, it must be noted that the data of each laboratory must be entered in a separate column (e.g., lab 1 and lab 2). Then select path Analyze\Compare Means\Paired-Samples T Test to open the following window.

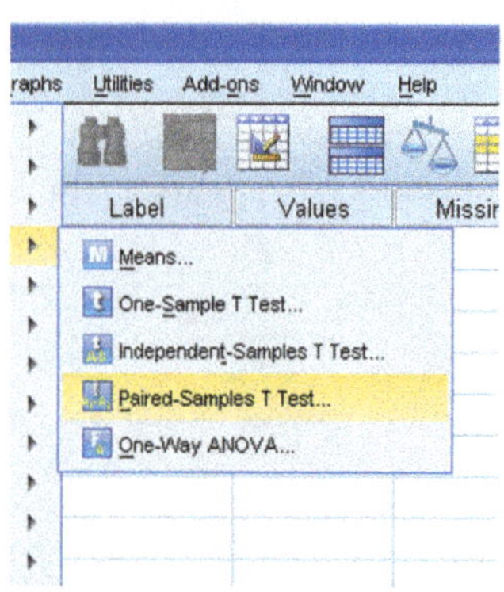
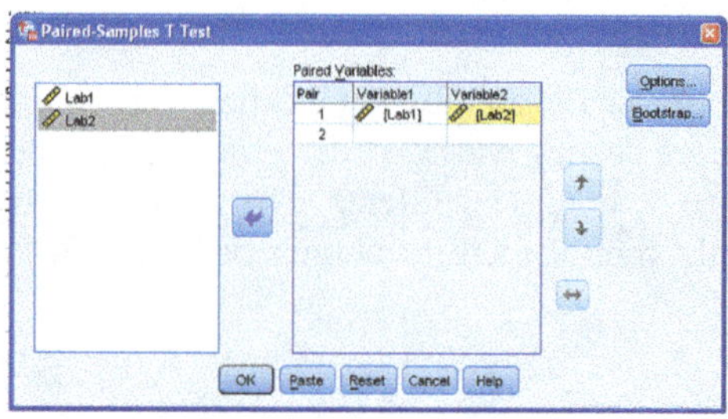

In this window, first select the paired variables from the left box and enter in the right box named Paired Variables. Then, the test is executed by confirming the OK button.

The SPSS output for the paired t-test is given below.

Paired samples statistics

		Mean	N	Std. deviation	Std. error mean
Pair 1	Lab 1	252.45	11	29.673	8.947
	Lab 2	247.4545	11	28.13668	8.48353

Paired samples' correlations

		N	Correlation	Sig
Pair 1	Lab 1 and lab 2	11	0.859	0.001

Paired samples' test

	Paired differences					t	df	Sig (two-tailed)
	Mean	Std. deviation	Std. error mean	95% CI of the difference				
				Lower	Upper			
Pair 1: Lab 1–Lab 2	5.00	15.44668	4.65735	−5.37722	15.37722	1.074	10	0.308

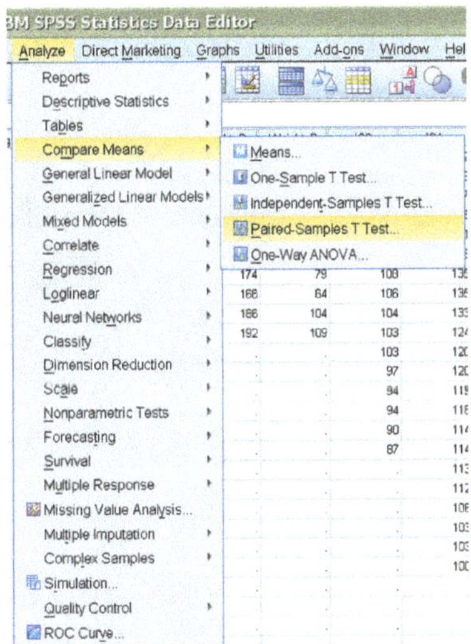

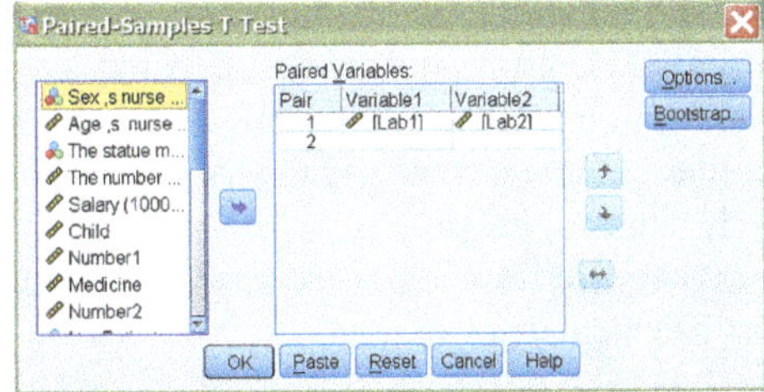

Using the Test:

The null and alternative hypotheses are:

$$\begin{cases} H_0 : \overline{d} = 0 \\ H_A : \overline{d} \neq 0 \end{cases}$$

where the $\overline{d}$ is the mean difference of two paired samples.

Sample size is 11 and default significant level is 0.05.

Test statistic is:

$$t = \frac{\overline{d}}{\frac{S}{\sqrt{n}}}, \quad \text{where the } S = \sqrt{\frac{\sum d_i^2 - \frac{(\sum d_i)^2}{n}}{n-1}}.$$

Critical value is $t_{(0.975, 10)} = 2.228$. If the test statistic is greater than the critical value of 2.28, then we reject the null hypothesis (H_0).

The test statistic according to sample data is $t = \frac{5}{\frac{15.45}{\sqrt{11}}} = \frac{5}{4.66} = 1.07.4$.

Since $1.07 < 2.228$, then we don't reject the null hypothesis (H_0).

The mean blood cholesterol levels measured by the two laboratories can be considered equal.

This result was obtained exactly from the estimation of the confidence interval.

11.5 Comparing Population Proportion with a Fixed Proportion

When the variable is binomial and the answers are in the form of failure (0) and success (1), proportion is equal to the number of successes divided by the total number of trials. Thus, the proportion is a kind of mean and their significant tests are similar to the mean tests (z test).

If the sample size n is large enough (i.e., both r and $n-r$ should exceed 5), it is reasonable to use the normal approximation. So, in practice the standard error of p can be calculated as:

$$\text{Se} = \sqrt{\frac{\hat{p}(1-\hat{p})}{n}}.$$

The appropriate margin of error is:

$$\text{Margine of error} = z_{1-\frac{\alpha}{2}}\sqrt{\frac{\hat{p}(1-\hat{p})}{n}}$$

The confidence interval is:

$$\textbf{Point estimate} \pm \textbf{Margin of error},$$

$$\text{CI}(p) = \hat{p} \pm z_{1-\frac{\alpha}{2}}\sqrt{\frac{\hat{p}(1-\hat{p})}{n}}$$

Since we are testing the null hypothesis, we can use the standard error of the expected proportion if the null hypothesis is true, i.e.,

$$\text{Se}(p_0) = \sqrt{\frac{p_0(1-p_0)}{n}}$$

which is slightly different with $\text{Se}(p)$ used for confidence interval (Altman, 1996, p231).

We will have an approximately normal distribution under the null hypothesis. Thus, in practice the test statistics is:

$$z = \frac{(\hat{p} - p_0)}{\text{Se}(p)}$$

where p_0 is the "pre-specified" or "expected" proportion.

The above test statistic uses the continuous normal distribution as an approximation to the discrete binomial distribution. If the variable only takes integer values, we are allowed applying the above method with a small correction of $\frac{1}{2}$ to the observed frequency, which is called continuity correction.

$$z_c = \frac{|\hat{p} - p_0| - \frac{1}{2n}}{\text{Se}(p)}$$

The continuity correction reduces the difference between the observed and expected proportions. But if the sample size increases, the effect of the correction weakens.

Example 11.9 The Health Ministry has announced that the vaccination coverage rate of Covid-19 is 22%. In a random sample of 800 residents of a city, 144 people were vaccinated. Judge about the Ministry of Health's claim based on this random sample.

Solution: The following information is given in this exercise: $n = 800$, $r = 144$, $p_0 = 0.22$, $\hat{q} = 1 - 0.18 = 0.82$, the default significant level is 0.05, and $z_{1-\frac{\alpha}{2}} = z_{0.975} = 1.96$.

As mentioned, rate is a kind of proportion and has a binomial distribution that may accept two answers: "vaccinated and unvaccinated". On the other hand, since both $r = 144$ and $n - r = 800 - 144 = 656$ are greater than 5, the standard error of the observed ratio is appropriate for calculating the confidence interval using the normal approximation as follows:

$$\text{Se}(p) = \sqrt{\frac{p.q}{n}} = \sqrt{\frac{0.18 \times 0.82}{800}} = 0.0136.$$

The 95% confidence interval for proportion is:

$$\text{CI95\%} = \bar{p} \pm z_{1-\frac{\alpha}{2}} \times \text{Se}(p) = 0.18 \pm 1.96 \times 0.0136 = 0.18 \pm 0.027$$
$$= (0.153 \text{ to } 0.207)$$

According to this sample, the vaccination coverage of 22% is rejected for this population, because the 95% confidence interval does not include the claim of the Health Ministry (0.22).

Using the Test:

The null and alternative hypotheses are:

$$\begin{cases} H_0 : p = 0.22 \\ H_A : p \neq 0.22 \end{cases}$$

Sample size is 800 and default significant level is 0.05.
Test statistic is:

$$z = \frac{(\hat{p} - p_0)}{\mathrm{Se}(p)}$$

Critical value is $z_{0.975} = 1.96$. If the test statistic is greater than the critical value of 1.96, then we reject the null hypothesis (H_0).

The test statistic according to sample data is:

$$z = \frac{(0.18 - 0.22)}{0.0136} = -2.94$$

Since $|-2.94| > 1.96$, then we reject the null hypothesis (H_0).

The claimed coverage of vaccination in the population cannot be 22%.

This result was obtained exactly from the estimation of the confidence interval.

The test statistic with the continuity correction for the Covid-19 vaccination data is:

$$z_c = \frac{|0.18 - 0.22| - \frac{1}{2 \times 800}}{0.0136} = 2.895,$$

which is slightly smaller than previous test statistic, because the sample size is quite large.

To perform a significant test to compare the population proportion with a fixed number in SPSS, first choose the path Analyze\Nonparametric tests\Legacy Dialogs\Binomial to open a new window.

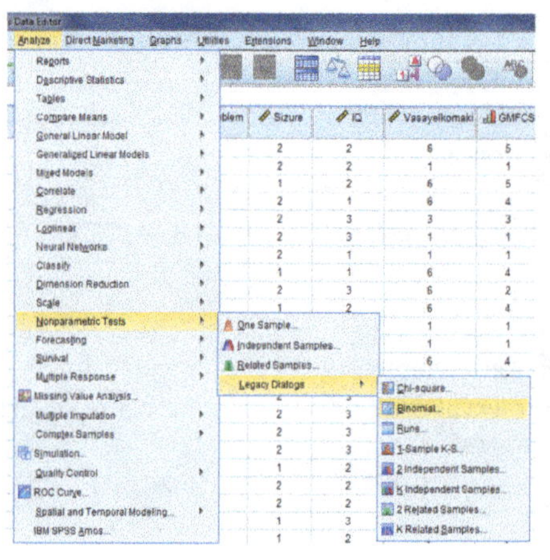
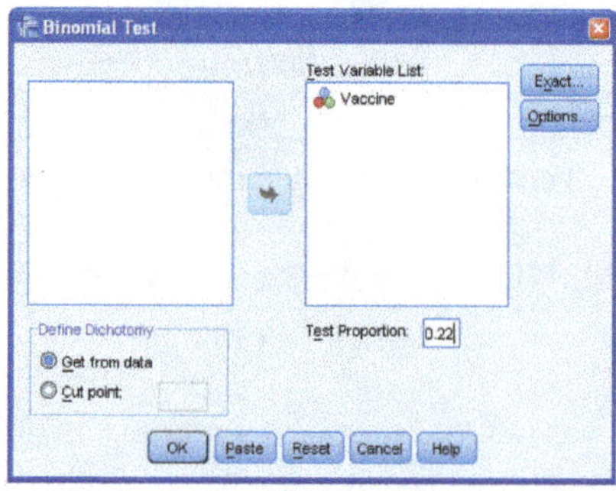

In this window, we first select the name of the binomial variable which the sample proportion should be estimated, from the list on the left, and move it to the box on the right, called the "Test variable list". If the data is entered as binary (1 and 0), mark in the front of the "Get from Data" option, but if the data is entered in quantities or polynomials, we specify a value as the cut point in the "Cut point" box. Then, write the number to which the population proportion should be compared, in the "Test Proportion" box. This number must be less than 1. Then, the program is run with OK confirmation.

Descriptive statistics

	N	Mean	Std. deviation	Minimum	Maximum
Vaccine	800	0.18	0.384	0	1

Binomial test

		Category	N	Observed prop.	Test prop.	Exact sig. (one-tailed)
Vaccine	Group 1	1	144	0.18	0.22	0.003[a]
	Group 2	0	656	0.82		
	Total		800	1.00		

[a] Alternative hypothesis states that the proportion of cases in the first group < 0.22

Example 11.10 The prevalence of asthma among women in a city was 15%. In a random sample of 215 women who referred to the health center of that city, 39 of them had a history of asthma. Based on this random sample, can it be said that the prevalence of asthma in women in this city has not changed?

Solution: The following information is provided in this example:

$$p_0 = 0.15, \; n = 215, \; r = 39, \; \hat{p} = \frac{39}{215} = 0.18,$$

$$\hat{q} = 1 - 0.18 = 0.82 \text{ and } z_{0.975} = 1.96.$$

As mentioned, the prevalence rate itself is a kind of proportion with the binomial distribution that accepts both "asthmatic and healthy" responses. On the other hand, since both $n = 215$ and $r = 39$ are greater than 5, we use the normal approximation:

$$Se(p) = \sqrt{\frac{p.q}{n}} = \sqrt{\frac{0.18 \times 0.82}{215}} = 0.026$$

And its 95% confidence interval is:

$$95\%\text{CI} = \hat{p} \pm z_{0.975} \times \text{Se}(p) = 018 \pm 1.96 \times 0.026 = (0.129 \text{ to } 0.231)$$

Based on this sample, the prevalence rate of the population can be assumed to be the same as 15%, because the above 95% confidence interval includes the previous prevalence rate of 0.15.

Using the Test:

The null and alternative hypotheses are:

$$\begin{cases} H_0 : p = 0.15 \\ H_A : p \neq 0.15 \end{cases}$$

Sample size is 215 and default significant level is 0.05.
Test statistic is:

$$z = \frac{\frac{r}{n} - p_o}{\sqrt{\frac{p_o(1-p_o)}{n}}} :$$

> **Tip**
>
> When we are testing the null hypothesis and that hypothesis is true, the same standard error of the expected proportion (p_0) should be used. In fact, it is like when the variance of population is known $(\sigma^2 = p_0(1 - p_0)$.

It should be noted that when we are testing the null hypothesis, if that hypothesis is true, the same standard error of the expected proportion (p_0) should be used. In fact, it is like when the variance of population is known $(\sigma^2 = p_0(1 - p_0))$.

Critical value is $z_{0.975} = 1.96$. If the test statistic is greater than the critical value of 1.96, then we reject the null hypothesis (H_0).

The test statistic according to sample data is:

$$z = \frac{\frac{r}{n} - p_o}{\sqrt{\frac{p_o(1-p_o)}{n}}} = \frac{\frac{39}{215} - 0.15}{\sqrt{\frac{0.15 \times 0.85}{215}}} = \frac{0.031}{0.024} = 1.29$$

Since $1.29 > 1.96$, then we reject the null hypothesis (H_0).

Based on this sampling, the prevalence rate of asthma among women in this population has changed and it cannot be considered as 15%.

This result was obtained exactly from the estimation of the confidence interval.

11.6 Comparison the Proportions in Two Independent Groups

Comparison of observed proportions in two independent groups can arise in all types of studies (observational or experimental) and is the most interested question in clinical research.

Example 11.11 In a clinical trial, 20 out of 240 people who received the flu vaccine and 80 out of 220 others who received the placebo became infected with the flu. Is the proportion of infected people with the flu equal in both the vaccinated and the unvaccinated groups?

Solution: The provided information in this exercise is:

$$n_1 = 240, \ r_1 = 20, \ p_1 = \frac{20}{240} = 0.083, \ q_1 = 1 - 0.083 = 0.917,$$

$$n_2 = 220, \ r_2 = 80, \ p_1 = \frac{20}{240} = 0.083, \ q_1 = 1 - 0.083 = 0.917,$$

To estimate 95% confidence interval for difference between proportions in the two groups, the standard error of the difference is calculated as follows:

$$\mathrm{Se}(p_1 - p_2) = \sqrt{\frac{p_1 \cdot q_1}{n_1} + \frac{p_2 \cdot q_2}{n_2}} = \sqrt{\frac{0.083 \times 0.917}{240} + \frac{0.364 \times 0.636}{220}} = 0.037$$

When the sample size and proportions are not small, i.e., r_1, r_2, $n_1 - r_1$ and $n_2 - r_2$ are all greater than 5, the sample distribution of $(p_1 - p_2)$ has an approximately normal distribution and the confidence interval can be obtained by the following formula:

$$\mathrm{CI}(1 - \alpha)\% = (p_1 - p_2) \pm z_{1-\frac{\alpha}{2}} \times \mathrm{Se}(p_1 - p_2)$$

In this case, 95% confidence interval is:

$$\mathrm{CI}95\% = (0.364 - 0.083) \pm 1.96 \times 0.037 = 0.281 \pm 0.073$$
$$= (0.208 \ to \ 0.354)$$

This confidence interval does not include zero, so the proportion of people with the flu in the two groups cannot be equated.

Using the Test:

Testing the hypothesis of equality of proportions in two groups means that under the true null hypothesis, both groups are taken from the same population with a proportion of p. Although we do not know the true value of p, then p_1 and p_2 will each be estimates of p, but our best estimate of p is when the two groups merge. That is, it will be $n = n_1 + n_2$ and $r = r_1 + r_2$.

In this case:

$$n_1 + n_2 = 240 + 220 = 460, \quad r_1 + r_2 = 20 + 80 = 100$$

$$\overline{p} = \frac{r_1 + r_2}{n_1 + n_2}$$

We calculate the grand proportion as: $\overline{p} = \frac{100}{460} = 0.217$ and $\overline{q} = 1 - 0.217 = 0.783$

Assuming that the proportion in each group is equal to $\overline{p}$, the standard error of $(p_1 - p_2)$ is calculated under the null hypothesis with the following formula:

$$Se(p_1 - p_2) = \sqrt{\frac{\overline{p}.\overline{q}}{n_1} + \frac{\overline{p}.\overline{q}}{n_2}} = \sqrt{\overline{p}.\overline{q} \times \left(\frac{1}{n_1} + \frac{1}{n_2}\right)}$$

In this example, we have:

$$Se(p_1 - p_2) = \sqrt{\overline{p}.\overline{q} \times \left(\frac{1}{n_1} + \frac{1}{n_2}\right)} = \sqrt{0.217 \times 0.783 \times \left(\frac{1}{220} + \frac{1}{240}\right)} = 0.0385$$

As noted in previous section, this standard error is slightly different from that calculated for confidence interval.

The sampling distribution of $(p_1 - p_2)$ is approximately normal, and the test statistic is:

$$z = \frac{(p_1 - p_2)}{Se(p_1 - p_2)}$$

The null and alternative hypotheses are:

$$\begin{cases} H_0 : p_1 = p_2 \\ H_A : p_1 \neq p_2 \end{cases}$$

Sample size is 800 and default significant level is 0.05.

Test statistic is:

$$z = \frac{(p_1 - p_2)}{Se(p_1 - p_2)}$$

Critical value is $z_{0.975} = 1.96$. If the test statistic is greater than the critical value of 1.96, then we reject the null hypothesis (H_0).

The test statistic according to sample data is:

$$z = \frac{(0.364 - 0.083)}{0.0385} = 7.30$$

Since $|7.30| > 1.96$, then we reject the null hypothesis (H_0).

The incidence rate of flu is not equal in the vaccinated and unvaccinated groups.

This result was obtained exactly from the estimation of the confidence interval.

To perform a significant test in order to compare the proportions in two independent groups in SPSS, first choose the path Analyze\Descriptive statistics\Crosstab... to open a new window.

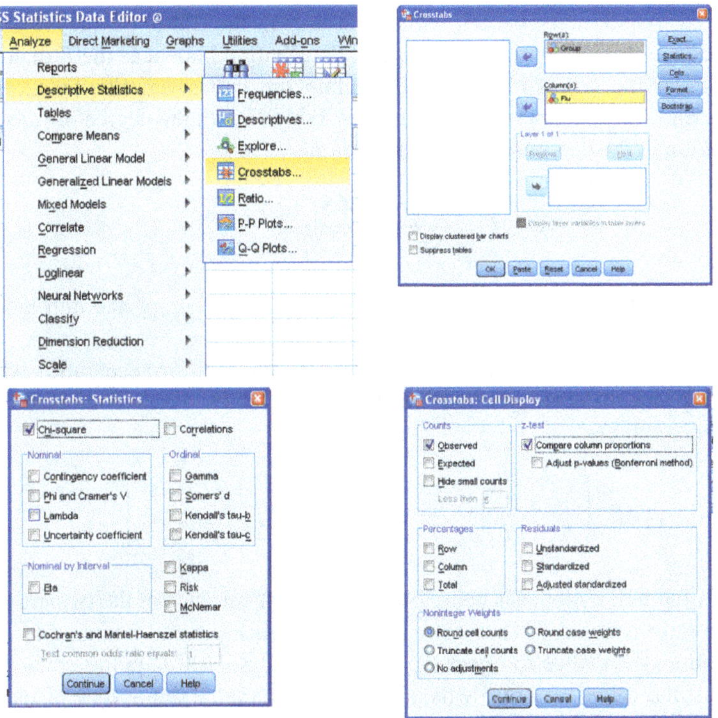

In this window, we first select the name of the binomial variables from the list on the left and move them to the boxes on the right, called the "Row(s)" and "Column(s)".

Then, we choose the "Cells" option to open another sub-window. In the right box titled "z-test", mark "Compare Column proportions" option and close the window by pressing "Continue".

In the main windows, again choose the "Statistics" option to open another sub-window. In the top left, mark "Chi-square" option and close the window by pressing "Continue". Then press "OK" to run z test and have the following output.

11.7   Testing the Equality of Variance of Two Groups

Sometimes researchers are keen on assessing the equality of variance (standard deviation) between two distinct groups. This inquiry arises from the common prerequisite of numerous statistical models, which necessitate equal variance (homogeneity). For instance, one might investigate whether the variability of body temperature is consistent across different blood types, or compare the variance of cholesterol levels between males and females. To address this, researchers employ the f test or variance ratio test. Under the null hypothesis positing that two normally distributed populations exhibit equal variances, the ratio of the two sample variances is expected to follow a sampling distribution known as the F distribution. The test statistic (F) is obtained by dividing the larger variance by the smaller variance and comparing the critical value in Table F with the degree of freedoms of numerator and the denominator of the fraction.

$$F = \frac{\text{Larger of } S_1^2 \text{ and } S_2^2}{\text{Smaller of } S_1^2 \text{ and } S_2^2}$$

If test statistic $F > F_{(1-\frac{\alpha}{2}, n_1-1, n_2-1)}$ or $F < F_{(\frac{\alpha}{2}, n_1-1, n_2-1)}$, the null hypothesis ($H_0$), equality of two variances is rejected.

Also, we can calculate confidence interval for the variance ratio, when the underlying data distribution is normal. Two-tailed $100(1-\alpha)\%$ confidence interval for variance ratio is:

$$\left[\frac{S_1^2}{S_2^2} \times \frac{1}{F_{(\frac{\alpha}{2}, n_1-1, n_2-1)}} \text{ to } \frac{S_1^2}{S_2^2} \times F_{(\frac{\alpha}{2}, n_1-1, n_2-1)} \right]$$

Under the null hypothesis being true (equality of variance of the two groups), the variance ratio is equal 1. Therefore, if the confidence interval contains 1, there are no reason and evidence to reject the null hypothesis. Null hypothesis can only be rejected if the value 1 is outside the confidence interval $100(1-\alpha)\%$. That is, both the upper and lower limits are either greater than one or both are less than one.

However, the two-tailed $100(1-\alpha)\%$ confidence interval for a single variance can be estimated by the following equation:

$$\left[\frac{(n-1)S^2}{\chi^2_{\frac{\alpha}{2}}} < \sigma^2 < \frac{(n-1)S^2}{\chi^2_{1-\frac{\alpha}{2}}} \right]$$

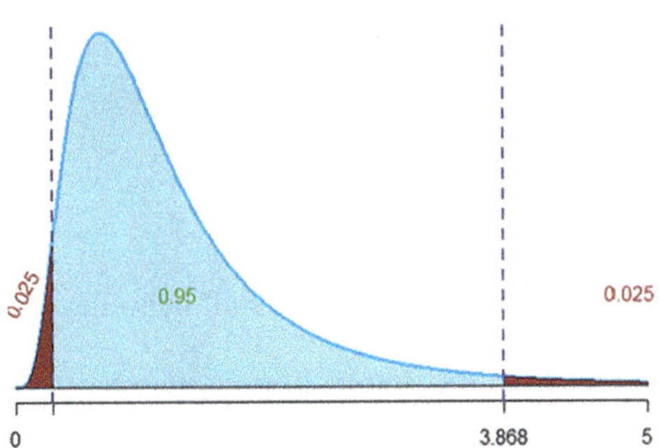

Example 11.12 A researcher aims to ascertain whether there is a difference in the variance of heart rates (measured in beats per minute, bpm) between smokers and non-smokers. He selects a random sample of 8 smokers and 9 non-smokers and the data is shown in Table 11.2. Using a $\alpha = 0.050$, is there enough evidence to decide about hypothesis?

Solution: First we calculate mean and variance of each group.

$$\bar{x}_1 = \frac{\sum x_i}{n} = \frac{640}{8} = 80, \ \bar{y}_1 = \frac{\sum y_i}{n} = \frac{666}{9} = 74$$

$$S_x^2 = \frac{\sum (x_i - \bar{x})^2}{n-1} = \frac{256}{8-1} = 36.57, \quad S_y^2 = \frac{\sum (y - \bar{y})^2}{n-1} = \frac{252}{9-1} = 31.5$$

Null and alternative hypotheses are:

$$\begin{cases} H_0 : \sigma_x^2 = \sigma_y^2 \\ H_A : \sigma_x^2 \neq \sigma_y^2 \end{cases}$$

Sample sizes are 8 and 9, degrees of freedom are $df_1 = n_1 - 1 = 8 - 1 = 7, \ df_2 = n_2 - 1 = 9 - 1 = 8$, and default significant level is 0.05.

Test statistic is the ratio of larger variance to the smaller variance:

$$F = \frac{S_{larger}^2}{S_{smaller}^2}$$

Table 11.2 Heart rates (bpm) of samples by smoking habit

Smoker	83	78	83	78	80	70	77	91	–
Non-smoker	67	76	78	76	79	78	79	66	67

Critical value is $F_{(\%2.5,7,8)} = 4.23$. If the test statistic is greater than the critical value of 4.23, then we reject the null hypothesis (H_0.).

The test statistic according to sample data is:

$$F = \frac{S^2_{\text{larger}}}{S^2_{\text{smaller}}} = \frac{36.57}{31.5} = 1.16$$

Since $1.16 < 4.23$, then we don't reject the null hypothesis (H_0).

The variance of 2 groups is equal.

To calculate confidence interval for variance ratio, critical value of F from the F distribution table is $F_{(\frac{\alpha}{2},n_1-1,n_2-1)} = 4.23$. Therefore, 95% confidence interval is calculated as below:

$$\left[\frac{S_1^2}{S_2^2} \times \frac{1}{F_{(\frac{\alpha}{2},n_1-1,n_2-1)}} \text{ to } \frac{S_1^2}{S_2^2} \times F_{(\frac{\alpha}{2},n_1-1,n_2-1)} \right]$$

$$= \left(\frac{36.57}{31.5} \times \frac{1}{4.23} \text{ to } \frac{36.57}{31.5} \times 4.23 \right)$$

$$= 0.27 \text{ to } 4.91$$

The estimated confidence interval includes 1, then we can assume that the variance of two groups is equal.

11.8   Goodness-Of-Fit Tests

The notion of "goodness-of-fit" entails examining whether sample data, organized in a one-way frequency table, aligns with a specific distribution under consideration, such as normal or uniform. To assess this alignment, we employ a hypothesis test to determine whether the observed frequency counts correspond with the expected or theoretical frequency distribution. This test, applicable to qualitative and categorical data, is referred to as the Chi-square test.

Goodness-of-fit test Is used to consider the hypothesis that an observed frequency distribution fits (or conforms to) some theoretical or expected distribution.

$$\chi^2 = \sum_{i=1}^{k} \frac{(O_i - E_i)^2}{E_i}$$

where O_i is the observed frequency in a cell and E_i is the expected frequency in each cell that is found by the following equation:

$$E = \frac{\text{Row total} \times \text{Column total}}{\text{Grand total}}$$

If data is provided in a column, E_i is the product of the probability of ith row by the grand total.

Example 11.13 To check the accuracy of the recorded weight of individuals, a sample of 96 adults was selected and last digit of their weight is given in Table 11.3. Judge the accuracy of this scale.

Solution: If the weights of individuals are accurately measured and recorded without rounding to the nearest 0 or 0.5, it is expected that the last digits of these weights will follow a uniform distribution. In such a distribution, each digit from 0 to 9 should occur with approximately the same frequency, as the probability of observing each digit is equal to 0.1 ($p\_0 = p\_1 = p\_2 = \ldots = p\_8 = p\_9$). Despite a subjective observation suggesting that the digits 0 and 5 occur more frequently than others, we will conduct a formal hypothesis test to corroborate this subjective assessment.

To do this, we employ the χ^2 test statistic, which is based on differences between the observed and expected values. If the observed and expected values are *close,* the χ^2 test statistic will be small, the P-value will be large, and we don't reject the null hypothesis. If the observed and expected frequencies are *far apart,* the χ^2 test statistic will be large, the P-value will be small, and we reject the null hypothesis.

Table 11.3 Last digits of individuals' weights

Last digit	Observed frequency
0	13
1	8
2	10
3	9
4	10
5	14
6	5
7	12
8	11
9	4
Total	96

The hypothesis tests of this section are always right-tailed, because the critical value and critical region are located at the extreme right of the distribution.

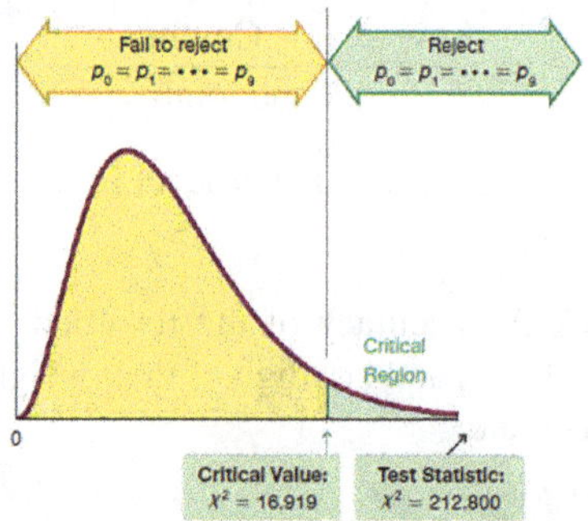

To find each corresponding expected frequency (E) in this exercise, we multiply the probability of each digit by the grand total ($96 \times 0.6 = 9.6$). Because the 96 digits would be uniformly distributed among the 10 categories (see Table 11.3a). This set of columns is called theoretical distribution.

$$\chi^2 = \sum_{i=1}^{k} \frac{(O_i - E_i)^2}{E_i} = 9.833$$

The critical value with 9 degree of freedom is $\chi^2_{(0.95,9)} = 16.92$. Since the test statistic is less than the critical value ($9.833 < 16.92$), then it can be concluded that the observed frequency distribution is consistent with the theoretical frequencies.

Table 11.3a Uniform distribution of Table 11.3

Last digit	Observed frequency (O)	Expected frequency (E)	($O-E$)	$(O-E)^2$	$\frac{(O-E)^2}{E}$
0	13	9.6	3.4	11.56	1.204
1	8	9.6	−1.6	2.56	0.266
2	10	9.6	0.4	0. 16	0.017
3	9	9.6	−0.6	0.36	0.037
4	10	9.6	0.4	0.16	0.017
5	14	9.6	4.4	19.36	2.017
6	5	9.6	−4.6	21.16	2.204
7	12	9.6	2.4	5.76	0.60
8	11	9.6	1.4	1.96	0.204
9	4	9.6	−5.6	31.36	3.267
Total	96	96	$\chi^2 = 9.883$		

The sum of last column is the test statistic of Chi-square which should be compared with the critical value of the table with degree of freedom df $= n-1 = 10-1 = 9$

> **Tip**
>
> For the goodness-of-fit test to be valid, the **expected frequency** of **each** cell must be **at least 5**.

11.9 Association and Dependence of Variables in Contingency Tables

Contingency **table** or **cross-tabulation** is two dimensions' table and a type of table in a matrix format that displays the frequency distribution of the two variables in terms of each other. It can provide a basic picture of the interrelation between two variables and can help find interactions between them. The term *contingency table* refers to a dependence of a variable on another variable. We use the term contingency table in this book because we test for independence between the row and column variables.

To test independence of two qualitative or categorical variables, we use test statistic of Chi-square, which is an one-tailed test. The formula and degree of freedom of Chi-square test statistic are as below, which is explained by an example below.

$$\chi^2 \sum \frac{(O - E)^2}{E}$$

$$df = (c - 1)(r - 1)$$

where O is observed value and E is expected value for each cell.

$$E = \frac{\text{Row total} \times \text{Column total}}{\text{Grand total}}$$

The closer the observed frequencies are to the expected frequencies (i.e., the row variable and the column variable are independent), the calculated test statistic is closer to zero; it becomes. Conversely, the farther away they are (i.e., the more dependent the row variable and the column variable), the larger the test statistic.

The calculated test statistic is compared with the critical value obtained from the Chi-square distribution table with degree of freedom $df = (c-1)(r-1)$. If the test statistic is greater than the critical value, the null hypothesis (independence of two variables) is rejected, i.e., the two variables are dependent. Otherwise, there is no reason to reject it.

The Chi-square test is valid when the sample size is large enough. That is, when 80% of the expected frequencies are greater than 5 and none are less than 1. Otherwise, some categories must be merged to meet this condition or use Fisher's exact test.

Example 11.14 In a random sample, 200 university students were asked about their preference to use e-books and hard copies, and the results are shown in Table 11.4. Can the priority of students be considered independent of the level of education?

Solution: The null and alternative hypotheses are:

$$\begin{cases} H_0 : & \text{The row and column variables are independent. or } p_1 = p_2 = p_3 \\ H_A : & \text{The row and column variables are dependent. or } p_1 \neq p_2 \neq p_3 \end{cases}$$

In order to be independent, the students' education level and their preference of the book type, the proportion of students using e-books must be approximately equal in all three levels of education. That is, we expect the proportion of e-book preference among BSc, MSc, and Ph.D. students to be equal to the proportion of e-book preferences for all students. It is also expected that the proportion of hard copy priorities in all three levels is equal to the proportion of total preferences of hard copy. For this reason, these values are called expected frequencies.

Thus, the expected frequencies for each cell can be obtained by multiplying the total proportion of e-book preference by the number of students in each level of education, that is, by multiplying the row sum by the sum of the column divided by the total number. For example, the expected frequency for BSc students who prefer e-books and also for MSc students who prefer hard copy is calculated as follows:

$$E = \frac{140 \times 120}{200} = 84 \text{ and } E = \frac{40 \times 80}{200} = 16$$

The rest of the table cells can be obtained by subtracting from the marginal sums.

We know that in this study, we have 140 BSc students, and the expected number of BSc students with an e-book priority is 84. Therefore, the number of BSc students with the priority of hard copy is equal to $140\text{–}84 = 56$.

There are 40 MSc students, 16 of whom prefer the hard copy. Therefore, the number of MSc students with e-book priority is equal to $40\text{–}16 = 24$.

On the other hand, 120 students were interested in e-books (first row), of which 24 were MSc students and 84 were BSc students. Therefore, the number of PhD students who want to prefer e-books is equal to $120 - (24 + 84) = 12$.

Thus, when the type of book is independent of the level of education (null hypothesis is true), we should have the expected frequencies according to Table 11.4a.

Table 11.4 Preference of students to use e-books and hard copies

Education level	BSc	MSc	Ph.D.	Sum
e-book	90	22	8	120
Hard copy	50	18	12	80
Total	140	40	20	200

Table 11.4a Expected frequencies of preference for e-books and hard copy in terms of educational level

Education level	BSc	MSc	Ph.D.	Sum
e-book	84	24	12	120
Hard copy	56	16	8	80
Sum	140	40	20	200

It can be seen that by calculating the expected frequency of two of the table cells, we were able to complete all the table cells. For this reason, the degree of freedom of this table is 2.

$$df = (c - 1)(r - 1) = (3 - 1)(2 - 1) = 2$$

Now we have to examine how big is the difference between the observed frequencies and the expected frequencies? Are they so big that they cannot be ignored? Obviously, the magnitude of these differences must be weighed against the expected frequencies. Because the difference of the cell with a frequency of 8 is not equal to the difference of the cell with a frequency of 84. So, we have to calculate the "relative difference". On the other hand, some frequencies are positive and some other are negative, and the sum of these differences $\left(\sum (O_{ij} - E\_{ij}) \right)$ is always zero, because the observed and expected frequencies both add up to the sample size. In order not to neutralize each other, we square them and divide them by the expected frequency of each cell. Then their sum will be the test criterion. A summary of these calculations is given in Table 11.4b.

Note: In the special case, the contingency tables are based on two dichotomous characteristics and with 2×2 format such as Table 11.5.

Table 11.4b Calculations for Chi-square test statistic

Education level	BSc	MSc	Ph.D.	Sum
e-book	$\frac{(90 - 84)^2}{84} = 0.43$	$\frac{(22 - 24)^2}{24} = 0.17$	$\frac{(12 - 8)^2}{12} = 1.33$	–
Hard copy	$\frac{(50 - 56)^2}{50} = 0.72$	$\frac{(18 - 16)^2}{16} = 0.25$	$\frac{(12 - 8)^2}{8} = 2.00$	–
Sum	–	–	–	4.90

Thus, the test statistic of Chi-square in this example is equal to $\chi^2 = 4.90$. The critical value with degree of freedom 2 is equal to $\chi^2 = 5.991$. Because the test statistic is less than the critical value, there is no reason to reject the null hypothesis. That is, the priority to e-books and hard copy has no significant relationship with students' educational level

Table 11.5 Sample distribution in terms of two characteristics

	First column (yes)	Second column (No)	Sum
First row (yes)	a	b	$E = a + b$
Second row (no)	c	D	$F = c + d$
Sum	$G = a + c$	$H = b + d$	N

In this case, the test statistic can be calculated with the following simplified formula:

$$\chi^2 = \frac{N(a.d - b.c)^2}{E.F.G.H}$$

This simplified form of the Chi-square formula can only be used for 2×2 contingency tables.

Example 11.15 In a random sample, 40 athletes were asked about their favourite exercise between Pilates and swimming, and the results are shown in Table 11.6. Is there a difference in the choice of sports between men and women?

Solution: This table is 2×2 with 1 degree of freedom. First, we calculate the Chi-square test statistics by obtaining expected frequencies. Just finding one of the table cells with this method is enough (Table 11.7).

$$E = \frac{16 \times 18}{40} = 7.2$$

$$\chi^2 = \frac{(10 - 7.2)^2}{7.2} + \frac{(8 - 10.8)^2}{10.8} + \frac{(6 - 8.8)^2}{8.8} + \frac{(16 - 13.2)^2}{13.2} = 3.2997$$

To calculate it using simplified formula, the following information is provided:

Table 11.6 Type of exercise by the gender

	Swimming	Pilates	Sum
Men	10	8	18
Women	6	16	22
Sum	16	24	40

Table 11.7 Expected frequencies for Table 10.7

	Swimming	Pilates	Sum
Men	7.2	10.8	18
Women	8.8	13.2	22
Sum	16	24	40

$$a = 10,\ b = 8,\ c = 6,\ d = 16,$$
$$E = 18,\ F = 22,\ G = 16,\ H = 24 \text{ and } N = 40$$

By replacing them in the above-simplified formula, the result is exactly the same:

$$\chi^2 = \frac{N(a.d - b.c)^2}{E.F.G.H} = \frac{40 \times (10 \times 16 - 8 \times 6)^2}{16 \times 24 \times 18 \times 22} = \frac{501760}{152064} = 3.2997$$

It should be remembered that if the sample size is small, we will have some bias in the Chi-square calculation and calculate it larger than its actual value. For this reason, Yates correction is used to eliminate the bias value. The Chi-square test statistics in tables 2×2 after applying Yates correction will be as follows:

$$\chi_Y^2 = \frac{N\left(|a.d - b.c| - \frac{N}{2}\right)^2}{E.F.G.H}$$

Altman (p. 252), however, recommends the use of Yates correction even for large examples.

11.10 Fisher's Exact Test for Small Samples

The general guideline for using the Chi-square test is that at least 80% of cells should have expected values greater than or equal to 5. Specifically, for a 2×2 table, all cells should meet this criterion. Therefore, even with Yates correction, the expectation for the size of expected frequencies remains. However, in practice, this rule can be slightly relaxed to permit one cell to have an expected value slightly lower than 5.

In cases where tables have very small expected frequencies, the renowned statistician R. A. Fisher proposed an alternative approach known as Fisher's exact test.

This method is also based on the observed row and column totals, but has a little difference in principle from previous ones. The Fisher's exact test is purely a hypothesis test, which provides the exact p-value instead of test statistic and should be compared with significant level of α. The null hypothesis for it, like the Chi-squared test, is that the row and column variables are unrelated. Considering Table 7. 6, Fisher's exact test formula will be as follows:

$$p = \frac{\binom{a+b}{a}\binom{c+d}{c}}{\binom{n}{a+c}} = \frac{\binom{a+b}{b}\binom{c+d}{d}}{\binom{n}{b+d}} = \frac{E! \times F! \times G! \times H!}{N! \times a! \times b! \times c! \times d!},$$

Table 11.8 Type of exercise by the gender

	Swimming	Pilates	Sum
Men	4	3	7
Women	2	6	8
Sum	6	9	15

where the symbol "$x!$" is "x factorial", which means that we multiply together all the integers from 1 up to x. For instance, $5! = 1 \times 2 \times 3 \times 4 \times 5 = 120$.

The above formula is obviously based on the exact hypergeometric probability distribution.

Example 11.16 Consider the previous example with a sample size of 15 athletes as in Table 11.8. Is the type of exercise independent of the athlete's gender?

Solution: To calculate Fisher's exact probability, we use the given information:

$$a = 4, \ b = 3, \ c = 2, \ d = 6, \ E = 7, \ F = 8, \ G = 6, \ H = 9 \text{ and } N = 15$$

By placing them in the above-simplified formula, the result is exactly the same:

$$p = \frac{E! \times F! \times G! \times H!}{N! \times a! \times b! \times c! \times d!} = \frac{7! \times 8! \times 6! \times 9!}{15! \times 4! \times 3! \times 2! \times 6!} = 0.196$$

The obtained p-value is greater than the significance level of 0.05, so the null hypothesis is not rejected, that is the type of sport is independent of athlete's gender.

11.11 Tests of Homogeneity

In contingency tables, we typically focus on testing the independence between the row and column variables, where the sample data is drawn from a single population. In such cases, both the row and column variables are considered random.

However, there are situations where samples are randomly selected from different populations, or where values of one variable (e.g., the row variable) are assigned by the investigator and are not random. In these scenarios, the assigned values can be treated as separate populations. The objective then becomes testing whether the distribution of the randomly selected variable (typically the column variable) is the same across each category of the other variable (the row variable). This type of analysis is referred to as a "test of homogeneity" or a contingency table with one margin fixed. "Homogeneity" here refers to the idea of having the same quality or proportion.

For example, consider a study comparing two types of diet, A and B, involving 100 adults. Sixty adults are randomly assigned to diet A, while the remaining 40 are assigned to diet B. After four months, each participant's satisfaction level is recorded as "Satisfied", "Moderate", or "Unsatisfied". The frequencies recorded in Table 11.9 are used to test the null hypothesis that there is no difference in satisfaction levels between the two diets.

In Table 11.9a, the samples were assigned to row categories (Diet A or Diet B), while only the column variable (satisfaction level) is random. Our objective is to conduct a test of homogeneity to determine whether the distribution of outcomes (satisfaction levels) is the same for each row (diet type). This test follows the same methodology as a test of independence.

Since the two rows of Table 11.10 represent independent samples, it would be appropriate to compute the probabilities (relative frequencies) for each row to provide a better description of the data.

These are given in Table 11.9a.

Now, this table allows us to describe the null hypothesis more clearly. The null hypothesis of "no difference" is equivalent to the statement that, for each response category, the probability is the same for diet A and diet B, i.e.:

$$\begin{cases} H_0 : p(A1) = p(B1) \quad \text{and} \quad p(A2) = p(B2) \quad \text{and} \quad p(A3) = p(B3) \\ H_A : p(A1) \neq p(B1) \quad \text{or} \quad p(A2) \neq p(B2) \quad \text{or} \quad p(A3) \neq p(B3) \end{cases}$$

Solution: As previous example, we calculate the expected frequencies for all cells. For example, expected frequency for satisfied by and Moderate cells of Diet A are:

Table 11.9 Point of view under two different diets

	Satisfied	Moderate	Unsatisfied	Sum
Diet A	24	22	14	60
Diet B	11	23	6	40
Total	35	45	20	100

Table 11.9a Point of view under two different diets

	Satisfied	Moderate	Unsatisfied	Sum
Diet A	$p(A1)$	$p(A2)$	$p(A3)$	1
Diet B	$p(B1)$	$p(B2)$	$p(B3)$	1

Table 11.9b Expected frequencies of Table 11.10

	Satisfied	Moderate	Unsatisfied	Sum
Diet A	21	27	12	60
Diet B	14	18	8	40
Total	35	45	20	100

Table 11.9c Calculations for Chi-square test statistic

	Satisfied	Moderate	Unsatisfied	Sum
Diet A	$\dfrac{(24-21)^2}{21}$ $=0.43$	$\dfrac{(22-27)^2}{27}$ $=0.93$	$\dfrac{(14-12)^2}{12}$ $=0.33$	–
Diet B	$\dfrac{(11-14)^2}{14}$ $=0.64$	$\dfrac{(23-18)^2}{18}$ $=1.39$	$\dfrac{(6-8)^2}{8}$ $=0.50$	–
Total	–	–	–	4.22

$$E = \frac{35 \times 60}{100} = 21 \text{ and } E = \frac{45 \times 60}{100} = 27$$

Then, the expected frequencies are summarized in Table 11.9b:

Calculation for Chi-square test statistic is shown in Table 11.9c:

Thus, the test statistic of Chi-square in this example is equal to $\chi^2 = 4.22$. The critical value with degree of freedom 2 is equal to $\chi^2 = 5.991$. Because the test statistic is less than the critical value, there is no reason to reject the null hypothesis. That is, the probability of adults point view level in two types of diet is the same.

11.12 Chi-Squared Test for Trend

When there is a binary variable with two categories (say, yes and no) and an ordinal variable with k categories (e.g., non-smoker, light-smoker, moderate-smoker, and heavy-smoker), William Cochran (https://en.wikipedia.org/wiki/William_Gemm ell_Cochran) and Peter Armitage (https://en.wikipedia.org/wiki/Peter_Armitage) modified the Chi-square test to assess the presence of an association (https://en.wik ipedia.org/wiki/Association_%28statistics%29) between a dichotomous variable and an ordinal variable and named "Chi squared test for trend" or **Cochran–Armitage test for trend**. The Chi-square test for trend provides a more powerful test than the usual Chi-squared test, because it is a Chi-squared distribution with one degree of freedom instead k–1 degrees of freedom for the usual Chi-squared test. Although the value of χ^2_{trend} for trend will always is less than χ^2 for the overall comparison. For example, for different doses of a treatment as "low", "medium", and "high", we may consider that the treatment benefit cannot become smaller as the dose increases or we are interested to study wheezing in different groups of smokers (non-smoker, light-smoker, moderate-smoker, and heavy-smoker).

To conduct this test, we need to assign ranks to each group. If the variable has a clear quantitative score, we can use these scores as ranks. For instance, in the case of shoe size data, we can assign ranks such as 3.5, 4.0, 4.5, 5.0, 5.5, and 6.0. Alternatively, integer scores like 1, 2, 3, 4, 5, and 6 can also be used. Equally spaced scores are typically recommended when the dose-dependent shape of the binomial proportions is unknown beforehand. The test statistic is obtained as:

$$\chi^2_{\text{trend}} = \frac{N\left[N\sum_{i=1}^{k} r_i.x_i - R.\sum n_i.x_i\right]^2}{R(N-R)\left[N\sum_{i=1}^{k} n_i.x_i^2 - \left(\sum n_i.x_i\right)^2\right]}$$

Or

$$\chi^2_{\text{trend}} = \frac{\left[\sum_{i=1}^{k} r_i.x_i - R.\bar{x}\right]^2}{p(1-p)\left[\sum_{i=1}^{k} n_i.x_i^2 - N.\bar{x}^2\right]}$$

where

$$N = \sum_{i=1}^{k} n_i, \quad R = \sum_{i=1}^{k} r_i, \quad p = \frac{R}{N}, \quad \text{and } \bar{x} = \frac{\sum_{i=1}^{k} n_i.x_i}{N}$$

Example 11.18 In a sample of 200, people are categorized as Table 11.10 based on the severity of smoking and wheezing. Researcher is interested to investigate whether wheezing increases with increasing smoking levels.

Solution: To obtain the test statistic, first add a column for the smoking level to the table above and then complete the calculations in Table 11.11.

By replacing the calculated values in the above formula, the Chi-square statistic for trend is obtained with both formulas:

$$\chi^2_{\text{trend}} = \frac{\left[\sum_{i=1}^{k} r_i.x_i - R.\bar{x}\right]^2}{p(1-p)\left[\sum_{i=1}^{k} n_i.x_i^2 - N.\bar{x}^2\right]}$$

Table 11.10 Distribution of people with wheezing in terms of smoking level

Smoker	Wheezing		
	Yes	No	Total
Non	8	60	68
Light	12	36	48
Moderate	15	28	44
Heavy	25	15	40
	70	130	200

Table 11.11 Required calculation for Chi-square test for trend for table above

Smoker	Score (x)	Wheezing			$r_i.x_i$	$n_i.x_i$	$n_i.x_i^2$
		Yes = r	No	Total = n			
Non	1	8	60	68	8	68	68
Light	2	12	36	48	24	96	192
Moderate	3	15	28	44	45	132	396
Heavy	4	25	15	40	100	160	640
Sum		$R = 70$	$N-R = 130$	$N = 200$	$\sum r.x = 177$	$\sum n.x = 456$	$\sum n.x^2 = 1296$

$$p = \frac{R}{N} = \frac{70}{200} = 0.35 \quad \text{and} \quad \bar{x} = \frac{\sum n.x}{N} = \frac{456}{200} = 2.28$$

$$= \frac{[177 - 70 \times 2.28]^2}{0.35(1 - 0.35)\left[1296 - 200 \times 2.28^2\right]} = 5.192.$$

$$\chi^2_{\text{trend}} = \frac{N\left[N \sum_{i=1}^{k} r_i.x_i - R. \sum n_i.x_i\right]^2}{R(N - R)\left[N \sum_{i=1}^{k} n_i.x_i^2 - \left(\sum n_i.x_i\right)^2\right]}$$

$$= \frac{200[200 \times 177 - 70 \times 456]^2}{70(200 - 70)\left[200 \times 1296 - (456)^2\right]} = 5.192.$$

The Chi-square test statistic (5.192) is larger than the critical value in the Chi-square distribution with one degree of freedom (3.841) and the null hypothesis (H_0) is rejected.

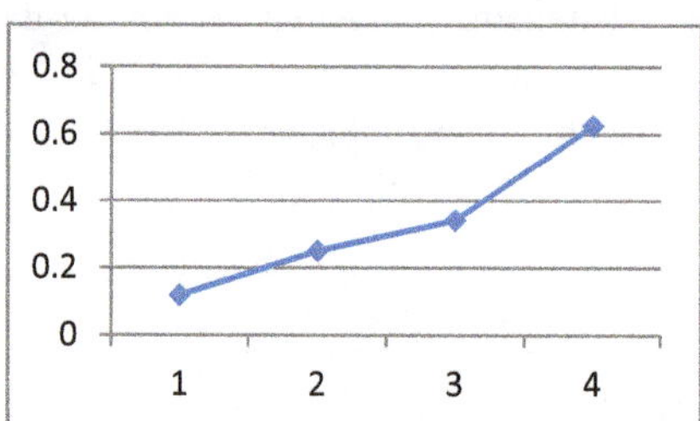

To run the Chi-square test in SPSS, select analyze\Descriptive Statistics\Crosstab or Analyze\Nonparametric Tests\Legacy Dialogs\Chi-square to open a window like the one below.

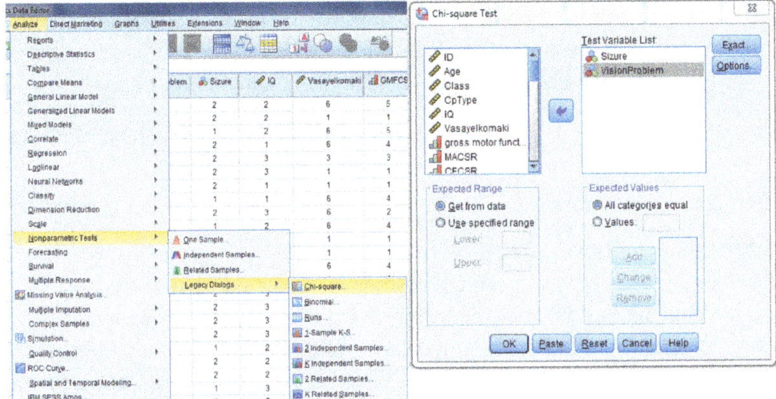

In this window, enter categorical variable into right box and press OK.

11.13   McNemar's Test for Paired Proportions

In scenarios where measurements are taken before and after an intervention on the same group of individuals, or when two experts evaluate the same group of patients based on certain characteristics, or when paired observations are made, such as in age and gender-matched individuals with similar clinical conditions randomly assigned to different treatment methods (*A* and *B*), the measurements, opinions, or observations within each pair are not independent. This is because they pertain to the same individuals.

> **Tips**
>
> A **discordant** pair is a matched pair in which the outcome differs for each member of the pair with the other.
>
> A **concordant** pair is a matched pair in which the outcome is the same for each member of the pair.

In such cases, traditional tests like the Chi-square test or Fisher's exact test are not suitable. Instead, we can employ the McNemar test, which is specifically designed for analyzing paired categorical data or repeated measures data.

> Paired samples mean that the two "groups" consist of data from the same group observed at multiple points in time, or measured by two raters.

In this test, the variables must have nominal or ordinal scales, or if they are quantitative, they can be transformed into binary variables. In this test, the values of two proportions, which are in the form of pairs of binomial samples (p_A, p_B), are compared. In this test, the null hypothesis $\left(\begin{cases} H_0 : p_A = p_B \\ orb = c \end{cases} \right)$ is tested against the alternative hypothesis $\left(\begin{cases} H_A : p_A \neq p_B \\ orb \neq c \end{cases} \right)$. Paired data can be summarized as Table 2 $\times$ 2 below (Table 11.12).

In this table, cells a and d are concordant observations with both methods and cannot provide useful information. However, cells b and c present discordant observations and are of particular importance in this test. If the sample size is large $(b + c \geq 20)$, then both z (normal) and Chi-square statistics can be used for this test.

$$z = \frac{(|b - c| - 1)}{\sqrt{b + c}} \quad \text{and} \quad \chi^2_{1,\,(1-\alpha)} = \frac{(|b - c| - 1)^2}{b + c}$$

These two tests are exactly equivalent. If the calculated test statistic at the significant level of $(\alpha)\%$ is greater than the critical value in the corresponding tables, then the null hypothesis is rejected.

When, say, $(b + c) < 20$, then χ^2 is not well-approximated by the Chi-squared distribution and a two-tailed exact test, based on the cumulative binomial distribution with $p = 0.5$, can be used instead.

$$p = 2 \times \sum_{i=\max b \text{ or } c}^{n} \binom{n}{i} \times 0.5^n$$

where "$n = b + c$" and the factor 2 is used to compute the two-tailed p-value (be careful, here "n" is not the total sample size).

Example 11.20 In order to compare survival rate with breast cancer treatments, A and B, a study of 207 pairs of cancer patients matched in terms of age, sex, and clinical conditions was investigated. Each member of these pairs was assigned to one of the two treatments A and B randomly. After 5 years, 75 patients in group A and 72 patients in group B survived which their information has been presented in Table 11.13.

Table 11.12 How the paired observations are located

Rater A		Rater B		
		Positive	Negative	
Rater A	Positive	a	b	$d + a$
	Negative	c	d	$d + c$
		$c + a$	$d + b$	N

Table 11.13 Outcomes of the disease treatment in terms of two methods A and B

		Treatment B		
		Alive	Died	Sum
Treatment A	Alive	70	5	75
	Died	2	30	32
	Sum	72	35	207

At a significant level of 5% and after 5 years, is the survival rate differed with these two treatments?

Solution: In this example, $a = 70$, $b = 5$, $c = 2$, and $d = 30$. Since $b + c = 5 + 2 = 7$ is less than 20, z and χ^2 test statistics are not appropriate and we will apply cumulative binomial distribution with $p = 0.5$.

The maximum value of b and c is the max $(5, 2) = 5$, then $i = b = 5$ and $n = b + c = 5 + 2 = 7$. Replacing above elements we have:

$$\sum_{i=\max b\ or\ c}^{n} \binom{n}{i} \times 0.5^n = \sum_{i=5}^{7} \binom{7}{5} \times 0.5^7$$

$$= \binom{7}{5} \times 0.5^7 + \binom{7}{6} \times 0.5^7 + \binom{7}{7} \times 0.5^7$$

$$= \left[\binom{7}{5} + \binom{7}{6} + \binom{7}{7}\right] \times 0.5^7$$

$$= [21 + 7 + 1] \times 0.0078125 = 0.226565$$

and

$$p = 2 \times \sum_{i=\max borc}^{n} \binom{n}{i} \times 0.5^n = 2 \times 0.226565 = 0.453$$

which is greater 0.05 and we cannot reject the null hypothesis.

However, if we ignore the assumption $b + c < 20$, the χ^2 statistic will be as:

$$\chi^2 = \frac{(|b - c| - 1)^2}{b + c} = \frac{(|5 - 2| - 1)^2}{5 + 2} = 0.571$$

To execute McNemar test in SPSS, first select the path to open a new window.

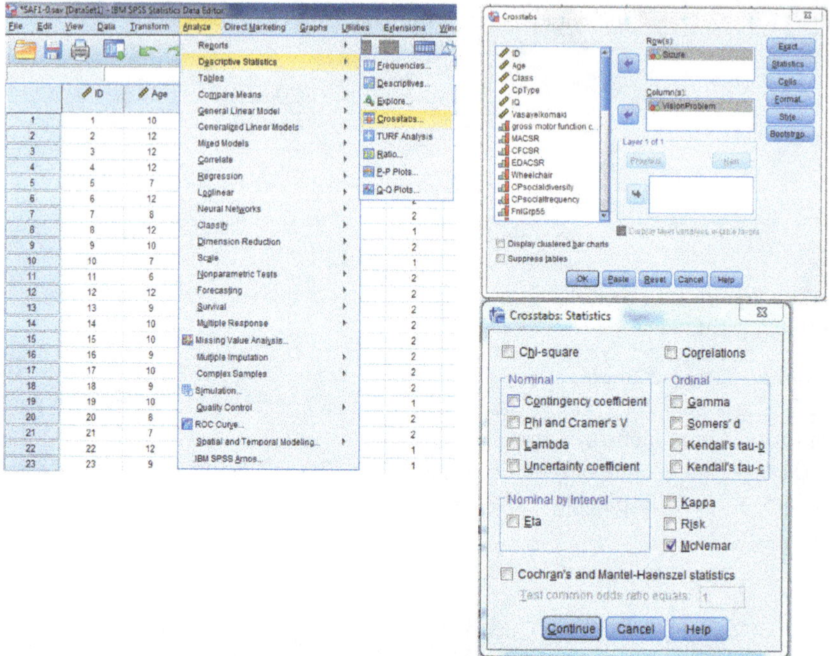

In this window, we first move the dichotomous variables to the right boxes called Row(s) and Column(s). Then, by selecting the Statistics option, another window opens. On the bottom right of this sub-window, mark the McNemar option and close it by pressing Continue. Then it runs McNemar test with confirming OK.

It should be noted that the McNemar test can be run through the path Analyze\Nonparametric tests\Related samples. After selecting this path, a window opens.

In this window, click "customize analysis" in the "What is your objective" box? Then go to the "Fields" section at the top of the window and enter your variables into the "Test Fields" box. Then, from the top of the Settings window, select "customize tests" option to enable its options. You can now mark "McNemar's test (2 samples)" to run the test.

11.14   Confidence Intervals and Hypothesis Tests?

Looking at the formulas of confidence interval and two-tailed statistical tests, it can be found that there is a logical and close link between hypothesis test and confidence intervals. Because four elements: confidence coefficient, standard error, sample size and point estimation of parameter have a direct role in the formula of both. Therefore, there is a clear and strong relationship between confidence interval and p-value which is the outcome of the hypothesis test and the conclusions are consistent with both methods, because both methods (confidence interval and p-value) are based on the same theoretical distribution of the test statistic.

Thus, when $p < \alpha$ shows that the data is inconsistent with the null hypothesis and the null hypothesis should be rejected, the confidence interval of $\% \ 100(1 - \alpha)$ will not include the value specified in the null hypothesis and indicates rejecting the null hypothesis. When $p > \alpha$, shows the compatibility of the data with the null hypothesis that there is no reason to reject the null hypothesis, the confidence interval of $\% \ 100(1 - \alpha)$ will also include the value specified in the null hypothesis and the null hypothesis cannot be rejected. For example, if the significance level is considered at 0.05 and the value is $p < 0.05$, the null hypothesis is rejected at the confidence level of 95%, and the 95% confidence interval will not include the specified value in the null hypothesis, which is an evidence to reject the null hypothesis.

The confidence interval shows uncertainty and inaccuracy in estimating the studied variable. Consequently, it has more and more useful information than p. However, if the confidence interval is presented alone, only the experienced statistician can estimate the approximate value of p. Therefore, it is desirable to present both p and confidence intervals in scientific articles and reports.

11.15 How to Present the Results of a Statistical Test

The results of the research in scientific papers and reports should provide the exact p-value in addition to the confidence interval for the means and percentages and avoid stating the limits for p-value as a range ($p < 0.05$). In cases where the test is meaningless, avoid using phrases such as "$p = NS$". Because by providing the exact p-value (e.g., $p = 0.092$), the reader will have a clearer interpretation of the results of your research. Only in cases where the p-value is obtained by the researcher himself from the statistical tables (not by software), he can express the p-value between the two given values in the table (e.g., $0.05 < p < 0.01$). Usually, it is enough to provide p-value up to two decimal places after zeros (e.g., $p = 0.14$ or $p = 0.0034$). Only in cases where the p-value is too small, its presentation is reported as $p < 0.001$.

11.16 Exercises

1. The mean and standard deviation of pain relief time of a type of reliever were announced 25 min and 18 min after consumption. A researcher bought a sample of 10 of those painkillers randomly from the city and gave them to patients. The mean pain relief time in this sample was 32 min. Does the researcher's sample support the factory claim based on the time of pain relief of these pills?

Table 11.14 Shoe size of women by hospital wards

Shoe size	36	37	37.5	38	38.5	39	40+	Total
Cesarean ward	5	7	6	7	8	10	43	86
Other wards	17	28	36	41	46	140	308	616

2. An entomological researcher has applied a statistical model to investigate the effect of aphids on cotton bolls (buds). He counted the number of aphids on the 20 cotton blossoms on a particular day, which are: 42–26–15–19–27–32–26–24–17–19–22–18–24–16–14–30–22–25–27–29. Can he assume the mean of aphids on each boll at 20 per day at a significance level of 5%?

3. In the previous exercise and for a t-test, are 20 aphids enough for this researcher's sample size?

4. Based on a sample, the shoe size of some women in the cesarean ward and other hospital wards is according to Table 11.14.

   A. Can the standard deviation of mother's shoe size in the two groups be considered equal?
   B. Can the mean size of mothers' shoes be equal in both groups?

5. If in a sample of 20 female students 8 without eating breakfast and in a sample of 15 boys 5 without eating breakfast have come to the university, do the following three methods and compare and interpret the results.

   A. Find the confidence interval for the difference between the two proportions of "without breakfast".
   B. Test the equality of the proportion of students without breakfast among boys and girls.
   C. By running the Chi-squared test, determine if not eating breakfast is related to gender

6. The number of births by months of the year is as follows:

January	February	March	April	May	June	July	August	September	October	November	December
528	503	496	46	518	411	45	44	37	41	331	416

Assuming that the birth rate is the same in all months of the year, check the observed frequencies with the theoretical distribution.

7. Systolic blood pressure of 16 patients was measured with two manometers A and B and is shown in Table 11.15. Can the difference between the two methods be ignored at a significance level of 5%?

8. The mean and standard deviation of adult height in a population by gender are estimated and are given in Table 11.16. Can height be equal in men and women?

9. According to Table 3.2, are albumin levels equal in two groups of women?

10. According to Table 3.3, are the levels of urea acid equal in two groups of women?

11. According to Table 2.11 and assuming that the first and second groups are the same subjects whose calcium levels were measured before and after an experiment, is it permissible to consider the difference in calcium levels before and after this experiment as non-significant?

12. Do the observations of Table 2.10 confirm the claim that the average prescribed painkillers for each version are 2 items?

13. One cigarette factory claims that its cigarettes contain a maximum of 40 mg of nicotine. The mean and standard deviation of 5 random samples of the products of this factory are 42.6 mg and 3.7 mg. If we assume that the nicotine content of cigarettes has a normal distribution. Test the claim of this factory at the significance level of 0.01.

Table 11.15 Systolic blood pressure of 16 patients with two manometers A and B

Patient	A	B	Patient	A	B
1	220	190	9	190	174
2	221	202	10	182	195
3	210	180	11	162	160
4	216	208	12	150	168
5	204	188	13	160	172
6	194	182	14	158	150
7	191	175	15	188	204
8	182	168	16	183	200

Table 11.16 Mean and standard deviation of adult height (Cm) in a population by gender

Gender	Mean	Sd	Sample n
Women	171.5	5.72	81
Men	179.1	5.85	64

Table 11.17 Daily protein intake in grams over a 24-hour period for two income groups: below the poverty line (Group A) and above the poverty line (Group B)

Group A:	87.7-69.7-56.4-77.2-98.6-78.1-68.6-80.2-59.7-69.0-73.6
Group B:	73.7-55.0-65.8-76.7-72.0-49.7-51.4-86.0-73.3-62.0-65.4-65.5-75.8-62.1

14. A study was conducted to evaluate the effect of diet with exercise (group A) versus diet without exercise (group B) on obese people. In this study, 37 people from group A and 36 people from group B were selected and the results of weight loss in these two groups are given below:

Group	Mean	Sd	Sample n
A	7.63	1.59	37
B	7.79	2.36	36

   Compare the effect between these two interventions at the significance level of 0.05.

15. A study was carried out to compare daily protein uptake with income level. In this study, the individuals were divided into two groups below the poverty line (group A) and above the poverty line (group B). The protein uptake in grams in a 24-h period is shown in Table 11.17:

   A. Exam the equality of the variances of the two groups.
   B. Are the mean of protein uptake equal in two groups at 5% significant level?

Chapter 12
Correlation and Regression

12.1 Introduction

In previous chapters, we've covered two main areas of inferential statistics: confidence intervals and hypothesis tests. Another important aspect of inferential statistics involves examining whether there's a relationship between two or more variables. For instance, we might want to know if there's a relationship between a person's age and their blood pressure, or if the number of hours a student studies is linked to their exam scores. When we collect observations on two variables for a group of individuals (such as their height and weight), we turn to bivariate analysis to explore how one variable is influenced by another.

Correlation and regression are two fundamental techniques used to assess the strength of association between two variables. Correlation analysis helps determine if there's a linear relationship between variables, offering a unit-less measure of association, typically on a linear scale. On the other hand, regression analysis delves deeper into describing the nature of the relationship between variables, whether it's positive or negative, linear or nonlinear. Regression analysis also provides a framework for predicting one variable (the dependent variable) based on another (the predictor variable).

For these methods, one of the following goals can be enumerated:

- Initial knowledge and understanding of the scope of the subject.
- Finding the reasons and justifications for the phenomenon studied.
- Trying to predict the future status of the studied variable.

In this chapter, we will explain about the relationship between two variables.

S. H. Saneii and H. Doosti, *Practical Biostatistics for Medical and Health Sciences*,
https://doi.org/10.1007/978-981-97-3083-4_12

> **Tip**
>
> Correlation and causality are two different concepts. Correlation is the mutual relationship between two things, while causality is the reason for causing something.

12.2   Correlation

Correlation, as discussed earlier, describes the statistical relationship between two variables, indicating how they change together. It's important to note that correlation and causality are distinct concepts. While correlation reveals a mutual relationship between variables, causality delves into the reasons behind causing something.

> **Tip**
>
> The closer the correlation coefficient is to 1, the more perfect and stronger it is. On the contrary, the closer it is to zero, the weaker it is and the more indication of absence of linear correlation.

Correlations possess two key properties: strength and direction. The strength of a correlation is quantified by the correlation coefficient, which ranges between 0 and 1. A correlation coefficient of 1 signifies perfect correlation, while 0 denotes no linear correlation at all. The direction of the correlation is determined by whether it's positive or negative.

Positive correlation occurs when both variables move in the same direction. This means that as one variable increases, the other also increases, and vice versa. Graphically, positive correlation is depicted by points rising from left to right on a scatter plot.

Conversely, **negative correlation** arises when the variables move in opposite directions. Here, as one variable increases, the other decreases, and vice versa. Visually, negative correlation is represented by points falling while moving from left to right on a scatter plot.

> **Tip**
>
> There is no rule for determining what size of correlation is considered strong, moderate, or weak.

No Correlation. When data points are dispersed throughout a scatter graph without forming any clear pattern, it signifies the absence of correlation between the variables. In this scenario, it becomes difficult to discern any consistent trend or directionality in the relationship between the variables. The randomness in the distribution of points indicates that there is no meaningful relationship or association between the variables being studied.

Regardless of being the positive and negative correlations, the strength of correlation can be defined as follows in terms of perfect correlation, strong correlation, moderate correlation, and weak correlation. There is no rule for determining what size of correlation is considered strong, moderate, or weak. Therefore, this classification is based on the researcher's personal experiences and others may use other categories.

Perfect correlation shows convergence to complete and matched variations of two variables together. So that, both variables change with the same exact proportion. The value of perfect correlation is $|1.0|$ and indicates that every change in the x variable is accompanied by a proportional and corresponding change in the y variable and vice versa. Perfect correlation could be negative (in different direction) or positive (the same direction). Some examples of perfect positive linear correlation are: (1) relation between Fahrenheit temperature and Celsius temperature. (2) The number of patients who go to visit a clinician and the total income of clinician, (3) number of kilometers traveled and gasoline consumption for your car. (4) Number of Covid-19 patients and the number of deaths from Covid-19, because number of death is a function of mortality rate. When plotted on a graph, correlation coefficient equals to 1. Some examples of perfect negative linear correlation are: (1) The number of daily consumed packs of cigarettes and people's life expectancy has perfect negative correlation. (2) The slower you drive (speed) in your car, the longer it will take you (time) to reach your destination. (3) Two variables that have perfect negative linear correlation are the distance from a door and the height of a wheelchair ramp. Because every inch of ramp height requires 12 inches of distance to the door.

Packs of cigarette	Life expectancy
0	85
0.5	81
1.0	77
1.5	73
2.0	69
2.5	65

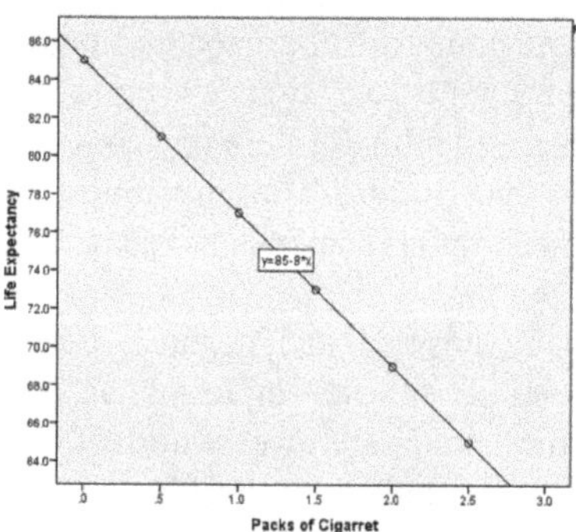

Strong correlation shows convergence to similar changes between the two variables. Both variables change with the same fashion and trend. Strong correlation value varies between |0.70| and larger thought to perfect correlation and shows that when the variable X changes one unit, the Y variable changes not equally, but by a similar value, and vice versa. The same value may be less than one unit or more than one (for example, greater than 0.70 and less than 1.30 units).

Moderate correlation between the two variables is somewhat diluted for "aligned" changes with each other, so that the two variables change in the same direction but not much. The moderate correlation value between |35.0| 0.70 and larger thought to strong correlation and shows that when the variable X changes with one unit, the variable Y varies by almost half.

Weak correlation is a dilute convergence and less clear for changes between two variables, so that the two variables change but slightly and minimal. Weak correlation value is less than |0.35| and shows that when the variable X changes with one unit, the Y variable changes by a smaller amount of variation (Fig. 12.1).

The scatter diagram is a very good visual tool for having a general glance of how the relationship between the two variables is.

Correlation and correlation coefficient indicate the relationship between the two variables and its numerical value, which varies between -1 and $+1$. Correlation coefficient is a measure used to express similarities between two distributions (two sets of data). The correlation coefficient in the population is represented by the ρ (Rho) and its sample estimation with r. The sign $(+)$ or $(-)$ in the correlation coefficient expresses the direction of changes.

Positive correlation coefficient indicates direct correlation between the two variables. That is, the increase and decrease of both variables are in the same direction. If one variable increases, the other variable increases. If one of the variables decreases, the other variable decreases.

Negative correlation coefficient indicates the inverse correlation between the two variables. That is, two variables increase or decrease in the opposite direction. When one of the variables increases, the other variable decreases. If one of the variables decreases, the other variable increases.

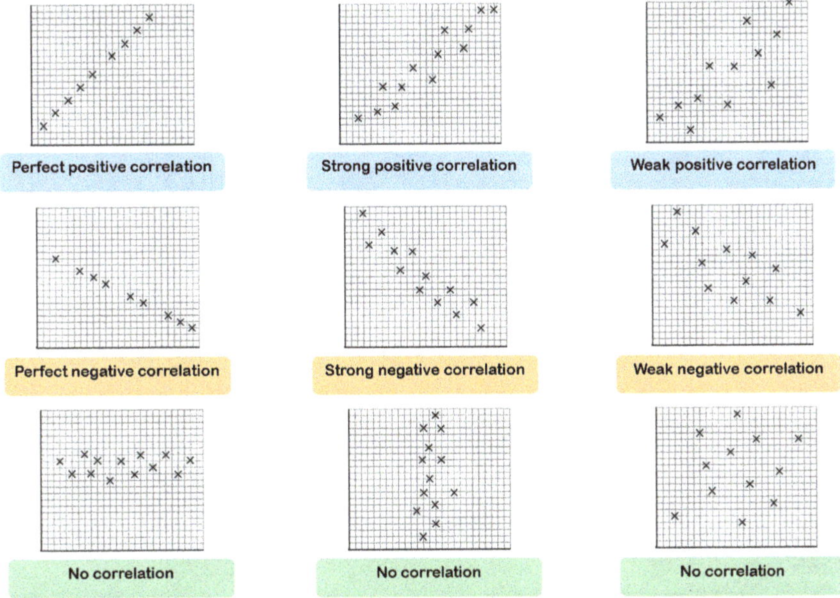

Fig. 12.1 Different patterns of correlation between two variables with scatter gram

When the correlation coefficient is zero or close to zero, it means that the changes of one variable do not provide any information about the changes of the other variable. The correlation coefficient of zero should not be considered as independency of two variables. Because correlation only examines the linear relationship between the two variables and does not address the nonlinear relationship of variables. However, the correlation coefficient of two independent variables is always zero. But the conclusions on the contrary are not true.

12.2.1 *Pearson Correlation Coefficient*

The most common coefficient of correlation is Pearson's correlation coefficient and is used for observations on a quantitative scale (interval, ratio). Pearson correlation coefficient for variables X and Y is obtained using the following formula:

$$\rho = \frac{\text{Cov}(x, y)}{\sqrt{\text{var}(x).\text{var}(y)}} = \frac{\sigma_{xy}}{\sigma_x.\sigma_y} = \frac{\sum_{i=1}^{N} (x_i - \mu_x)(y_i - \mu_y)}{\sqrt{\left[\sum_{i=1}^{N} (x_i - \mu_x)^2\right]\left[\sum_{i=1}^{N} (y_i - \mu_y)^2\right]}}$$

The above formula looks somewhat complicated, but using the practical formula below makes correlation coefficient in sample somewhat easier to determine the value of r.

$$\hat{\rho} = r_{xy} = \frac{1}{n-1}\left(\frac{x-\bar{x}}{S_x}\right)\left(\frac{y-\bar{y}}{S_y}\right) = \frac{n\sum x.y - (\sum x)(\sum y)}{\sqrt{\left[n\sum x^2 - (\sum x)^2\right]\left[n\sum y^2 - (\sum y)^2\right]}}$$

This formula may be seen as follows:

$$r_{xy} = \frac{\sum x_i.y_i - \bar{x}.\bar{y}}{(n-1).S_x.S_y}$$

Even to simplify calculations, it is recommended to make a table such as Table 12.2 in which five columns are assigned for x, y, and. Then replace the sum of each column directly in the above formula.

> **Tip**
>
> The correlation coefficient is very sensitive to outliers in the sense that a single outlier could dramatically affect its value. It may be misleading when outliers are present.

Assumptions for the Pearson Correlation Coefficient

Before applying the Pearson correlation to a dataset, it's essential to examine the following assumptions.

- The data is selected through a random sampling process.
- The relationship between the two datasets is linear and measured on an interval or ratio scale.
- The variables exhibit a joint normal distribution. This implies that given any specific value of one variable, the values of the other variable are normally distributed, and vice versa.
- There are no outliers present in the data, which could significantly skew the results of the correlation analysis.

Example 12.1 The education office aims to investigate the potential relationship between the number of study hours and the final test scores on an exam. To do so, they randomly select a sample of 13 faculty students, as shown in Table 12.1. The objective is to estimate the correlation coefficient between the study hours and the corresponding final grades achieved in the exam.

> **Tip**
>
> Round the value of correlation coefficient r to three decimal places.

Table 12.1 Number of absences, number of hours of study, and final test scores on an exam of a random sample

Hours of study	Final grade	Absence
6	82	6
5	86	2
1	50	15
3	74	9
1	58	12
4	90	5
4	78	8
2	70	10
2	65	12
7	94	2
8	96	0
3	75	8
5	82	5

Table 12.2 Necessary calculations for estimating correlation coefficient for data on Table 12.1

Student	Hours (x)	Grade (y)	x^2	y^2	$x.y$
1	6	82	36	6724	492
2	5	86	25	7396	430
3	1	50	1	2500	50
4	3	74	9	5476	222
5	1	58	1	3364	58
6	4	90	16	8100	360
7	4	78	16	6084	312
8	2	70	4	4900	140
9	2	65	4	4225	130
10	7	94	49	8836	658
11	8	96	64	9216	768
12	3	75	9	5625	225
13	5	82	25	6724	410
Sum	$\sum x = 51$	$\sum y = 1000$	$\sum x^2 = 259$	$\sum y^2 = 79170$	$\sum x.y = 4255$

Solution: Let's to denote number of hours of study as x, final grade as y, no. of absences as z, and obviously $n = 13$. Then a Table 12.2 can be made for the data calculations, as shown here.

Now we substitute values from the last row in the formula and estimate the coefficient correlation of r.

$$\hat{\rho} = r_{xy} = \frac{n \sum x.y - (\sum x)(\sum y)}{\sqrt{\left[n \sum x^2 - (\sum x)^2\right]\left[n \sum y^2 - (\sum y)^2\right]}}$$

$$= \frac{13 \times 4255 - 51 \times 1000}{\sqrt{\left[13 \times 259 - (51)^2\right]\left[13 \times 79170 - (1000)^2\right]}}$$

$$= \frac{4315}{4730.21} = 0.912$$

The coefficient correlation of r has no unit of measurement and the value of r will remain unchanged, if the x and y-values are switched.

12.2.2   Confidence Interval for Pearson Correlation Coefficient

Like other estimators, it's possible to compute a confidence interval for the correlation coefficient in the population. However, because the sampling distribution of Pearson's r is not normal, we need to transform r to z using the formula below to achieve a normal sampling distribution. Subsequently, we can reverse this transformation to obtain a confidence interval for the population correlation coefficient.

$$z = \frac{1}{2} \ln\left(\frac{1+r}{1-r}\right)$$

The standard error of z is approximately:

$$Se_z = \frac{1}{\sqrt{n-3}}$$

Hence, the 95% confidence interval for z is:

$$z_1 = z - 1.96 \times Se_z = z - \frac{1.96}{\sqrt{n-3}}$$

$$z_2 = z + 1.96 \times Se_z = z + \frac{1.96}{\sqrt{n-3}}$$

The 95% confidence interval for correlation coefficient of r is reachable by back transformation of the above value as below:

$$\frac{e^{2z_1} - 1}{e^{2z_1} + 1} \quad \text{to} \quad \frac{e^{2z_2} - 1}{e^{2z_2} + 1}$$

Example 12.2 The correlation coefficient of number of hours study and final grade in Example 12.1 with 13 samples was 0.912. Find the 95% confidence interval for r.

Solution: We first transform $r = 0.912$ as below:

$$z = \frac{1}{2} \ln\left(\frac{1+r}{1-r}\right) = \frac{1}{2} \ln\left(\frac{1+0.912}{1-0.912}\right) = \frac{1}{2} \ln 21.7272 = 1.539$$

The standard error of z is:

$$Se_z = \frac{1}{\sqrt{n-3}} = \frac{1}{\sqrt{13-3}} = 0.316$$

The 95% confidence interval for z is:

$$z_1 = z - \frac{1.96}{\sqrt{n-3}} = 1.539 - 6.198 = 0.9196$$

$$z_2 = z + \frac{1.96}{\sqrt{n-3}} = 1.539 + 6.198 = 2.1584$$

After back transformation of the above values, 95% confidence interval for correlation coefficient of r is:

$$\text{Lower limit} = \frac{e^{2z_1} - 1}{e^{2z_1} + 1} = \frac{e^{2(0.9196)} - 1}{e^{2(0.9196)} + 1} = \frac{5.2915}{7.2915} = 0.726$$

$$\text{Upper limit} = \frac{e^{2z_2} - 1}{e^{2z_2} + 1} = \frac{e^{2(2.1584)} - 1}{e^{2(2.1584)} + 1} = \frac{73.948}{75.948} = 0.974$$

The both confidence limits suggest that there really is quite a strong association between the two variables in the population.

12.2.3 The Significant Test of the Pearson Correlation Coefficient

Tips

- Two variables have a **linear relationship** if the data tends to cluster around a straight line when plotted on a scatterplot.
- Two variables have a **positive correlation** if large values of one variable are associated with large values of the other and small values of one variable are associated with small values of the other.
- Two variables have a **negative correlation** if large values of one variable are associated with small values of the other and if small values of one variable are associated with large values of the other.

As previously noted, the correlation coefficient ranges from -1 to $+1$. Values close to -1 or $+1$ indicate a strong linear relationship between two variables, while values close to 0 suggest a weak or non-existent linear relationship. Since r is estimated from sample data and influenced by sample size, when r is not zero, two scenarios arise: Either r is sufficiently high to infer a significant linear relationship between the variables, or r is attributable to chance and sampling fluctuations.

The hypothesis test makes this decision (there is no association in the population: $\rho = 0$) easy. The null and alternative hypotheses are:

$$\begin{cases} H_0 : \rho = 0 \\ H_A : \rho \neq 0 \end{cases}$$

When there is a significant difference between the estimated values of r and 0, then the null hypothesis must be rejected at a significant level α. If the value of r is not significantly different from 0 (zero) and is probably due to chance, then the null hypothesis must not be rejected. The test statistics is:

$$t = \frac{r}{Se(r)} = r\sqrt{\frac{n-2}{1-r^2}} \quad \text{with } df = n - 2$$

The test statistic should compare with two-tailed critical value from t distribution table with degree of freedom $(n-2)$.

Example 12.3 Test the significance of the correlation coefficient ($r = 0.912$) found in Example 12.1 at significant level of 0.05.

Solution: The null and alternative hypotheses are: $\begin{cases} H_0 : \rho = 0 \\ H_A : \rho \neq 0 \end{cases}$. Degree of freedom is $df = n - 2 = 13 - 2 = 11$ and critical value at 0.05 significant level is $t_{(0.975,11)} = 1.796$.

The test statistic is calculated as below:

$$t = r\sqrt{\frac{n-2}{1-R^2}} = 0.912\sqrt{\frac{13-2}{1-0.912^2}} = 7.374$$

Since test statistics is greater than the critical value ($7.374 > 1.796$) and fall in reject region, we reject the null hypothesis and mean that there is a significant difference with zero.

Example 12.4 Hemoglobin level (y) and age of women (x) in a random sample of 9 are as follows. In order to simplify the calculations, hemoglobin level values have been rounded.

Obtain the correlation coefficient of hemoglobin level and women's age and its confidence interval.

Solution: Let's to denote age of women as x, hemoglobin level as y, and obviously $n = 9$. Then a Table 12.3 can be made for the data calculations, as shown here (Table 12.4).

Now we substitute values from the last row in the formula and estimate the coefficient correlation of r.

$$\hat{\rho} = r_{xy} = \frac{n \sum x.y - (\sum x)(\sum y)}{\sqrt{\left[n \sum x^2 - (\sum x)^2\right]\left[n \sum y^2 - (\sum y)^2\right]}}$$

$$= \frac{9 \times 4805 - 340 \times 123}{\sqrt{\left[9 \times 14044 - (340)^2\right]\left[9 \times 1707 - (123)^2\right]}}$$

$$= \frac{1425}{1589.42} = 0.897$$

To carry out the significant test, the null and alternative hypotheses are:

$$\begin{cases} H_0 : \rho = 0 \\ H_A : \rho \neq 0 \end{cases}$$

Degree of freedom is $df = n - 2 = 9 - 2 = 7$ and critical value at 0.05 significant level is $t_{(0.975,7)} = 1.895$.

Table 12.3 Hemoglobin level (y) and age of women (x) in a random sample

Sample	1	2	3	4	5	6	7	8	9
Age	20	25	28	35	38	40	45	54	55
Hemoglobin	11	12	14	12	13	14	15	16	16

Table 12.4 Necessary calculations for estimating correlation coefficient for data on Table 12.3

Sample	Age (x)	Hemoglobin (y)	x^2	y^2	$x.y$
1	20	11	400	121	220
2	25	12	625	144	300
3	28	14	784	196	392
4	35	12	1225	144	420
5	38	13	1444	169	494
6	40	14	1600	196	560
7	45	15	2025	225	675
8	54	16	2916	256	864
9	55	16	3025	256	880
Sum	$\sum x = 340$	$\sum y = 123$	$\sum x^2 = 14044$	$\sum y^2 = 1707$	$\sum x.y = 4805$

The test statistic is calculated as below:

$$t = r\sqrt{\frac{n-2}{1-r^2}} = 0.897\sqrt{\frac{9-2}{1-0.897^2}} = 5.369$$

Since test statistics is greater than the critical value ($5.369 > 1.895$) and fall in reject region, we reject the null hypothesis and mean that there is a significant difference with zero.

To construct confidence interval, we first transform $r = 0.897$ as below:

$$z = \frac{1}{2}\ln\left(\frac{1+r}{1-r}\right) = \frac{1}{2}ln\left(\frac{1+0.897}{1-0.897}\right) = \frac{1}{2}\ln 18.4175 = 1.457$$

The standard error of z is:

$$S_z = \frac{1}{\sqrt{n-3}} = \frac{1}{\sqrt{9-3}} = 0.408$$

The 95% confidence interval for z is:

$$z_1 = z - \frac{1.96}{\sqrt{n-3}} = 1.457 - 0.800 = 0.657$$

$$z_2 = z + \frac{1.96}{\sqrt{n-3}} = 1.457 + 0.800 = 2.2572$$

After by back transformation of the above values, 95% confidence interval for correlation coefficient of r is:

$$\text{Lower limit} = \frac{e^{2z_1}-1}{e^{2z_1}+1} = \frac{e^{2(0.657)}-1}{e^{2(0.657)}+1} = \frac{2.721}{4.721} = 0.5764$$

$$\text{Upper limit} = \frac{e^{2z_2}-1}{e^{2z_2}+1} = \frac{e^{2(2.2572)}-1}{e^{2(2,2572)}+1} = \frac{90.3228}{92.3228} = 0.978$$

Tips

It should be noted that when the sample size is too large, a weak correlation may also be statistically significant. On the contrary, if the sample size is too small, a strong and intense correlation may not be statistically significant.

However, the sample size is small, but its confidence interval suggests that there really is quite a strong association between the two variables in the population.

12.2.4 Running Correlation Coefficient on SPSS

To calculate the Pearson correlation coefficient in SPSS, first select the Analyze\Correlate\Bivariate path to open the following window.

In this window, we select the two variables we want to calculate their correlation coefficient from the list on the left side and move to the Variables box. In the same box, we mark Pearson. Then, the program runs with OK confirmation and the value r can be seen in the correlations table.

For example, SPSS output for data of Example 12.4 is as:

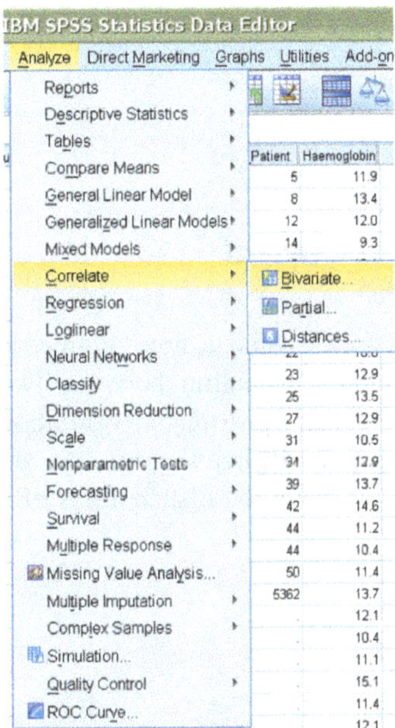

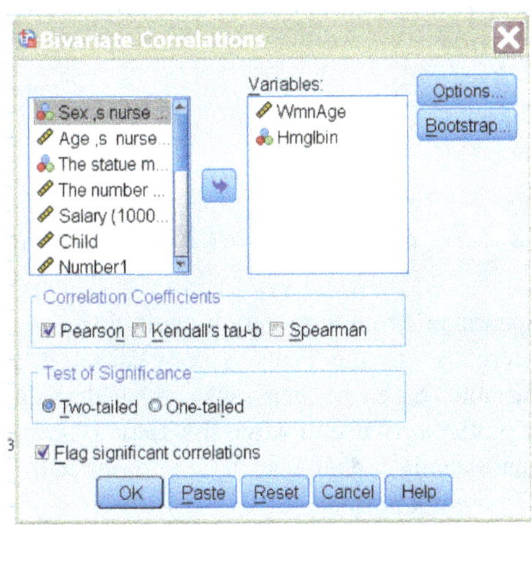

Correlations

		Hemoglobin	Age
Hemoglobin	Pearson correlation	1	0.897**
	Sig. (2-tailed)		0.001
	N	9	9
Age	Pearson correlation	0.897**	1
	Sig. (2-tailed)	0.001	
	N	9	9

**. Correlation is significant at the 0.01 level (two-tailed)

If one wishes to classify the strength of the relationship between two variables, for absolute values of r, 0–0.19 is regarded as very weak, 0.2–0.39 as weak, 0.40–0.59 as moderate, 0.6–0.79 as strong and as very strong correlation, but these are rather arbitrary limits, and the context of the results should be considered.

Value of r	Interpretation n
0–0.19	Very weak
0.2–0.39	Weak
0.40–0.59	Moderate
0.6–0.79	Strong
0.8–0.99	Very strong
1.0	Perfect

12.2.5   The Spearman Rank Correlation Coefficient

Spearman's rank correlation coefficient, denoted as r_s, provides a non-parametric alternative to the Pearson correlation. It assesses the correlation between two variables based on their ranks rather than their raw values. Spearman's correlation is particularly useful when the assumptions underlying the Pearson correlation are significantly violated, such as in cases with outliers or when the distributions of x or y are not normal.

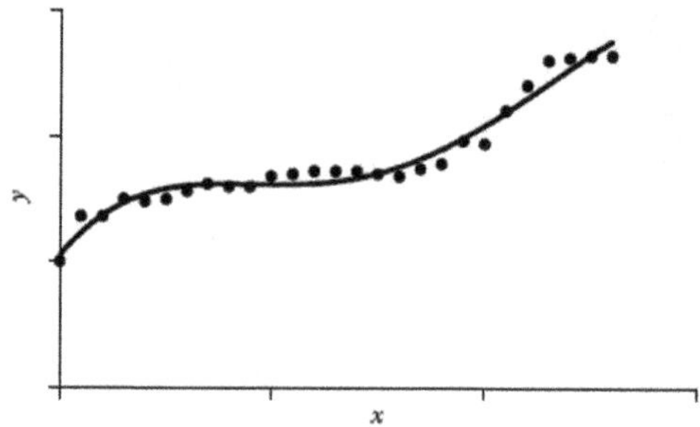

When the two variables (x and y) clearly increase or decrease together, but the relationship is not necessarily linear, this is called a **monotonic** relationship (Fig. 12.3). The dependence of height on age of children is an example of such a relationship. This is not linear but definitely exists. Monotonicity is "less restrictive" than that of a linear relationship. A monotonic relationship is not strictly an assumption of Spearman's correlation. But, if a scatterplot shows that the relationship between your two variables looks monotonic, you would run a Spearman's correlation to measure the strength and direction of this monotonic relationship.

The Spearman correlation coefficient has simpler computations than those for the Pearson correlation coefficient and involves ranking each set of data rather than numeric values. The difference in ranks is obtained for any pairs of data ($d_i = x_i - y_i$) and then r_s is estimated by equivalent formula for correlation which is given, except using these differences instead of raw data.

$$r_S = 1 - \frac{6 \sum_{i=1}^{n} d_i^2}{n(n^2 - 1)}$$

If both datasets have the same ranks, r_s will be $+1$, and if two datasets are ranked in exactly the opposite way, r_s will be -1 and $r_s = 0$ indicates no relationship between ranks of two datasets.

If two or more observations have the same value, it means that all of them should be given the same rank. They are called "tied" and cause problems in calculations. For this purpose, tied observations should be given average ranks as if there are no tied observations in the dataset. For example, suppose the dataset of 7 figures (1, 1, 2, 4, 5, 5, 5). Usually they are expected to be given ordinal ranks 1, 2, 3, 4, 5, 6, 7. However, since the value 1 is repeated twice and both of them must have the identical rank, we give the average of the two ordinal ranks to both equal values of 1. Therefore, the fractional rank for both values of 1 is $\frac{1+2}{2} = 1.5$. The rank of values 2 and 4 is unchanged equal to 3 and 4. The rank of the last three values (5, 5, 5) will be equal to the average ranks of 5, 6, and 7, i.e., $\left(\frac{5+6+7}{3} = 6\right)$. Thus the fractional ranks are: 1.5, 1.5, 3, 4, 6, 6, 6.

Example 12.5 Table 12.5 shows the age, accommodation (ACC), and intraocular pressure (IOP) and gender in a sample of 20 clients to an optometry clinic. Find the Spearman correlation coefficient for age and right intraocular (R.IOP) pressure in women.

Solution: First we create two more columns (x_i and y_i) for the ranks of each variable and create another column for their difference ($d_i = x_i - y_i$). Write the square of differences (d_i^2) in the last column on the right and get the sum of this column (Table 12.6)

Now by substituting the sum of the last column in the corresponding formula, the Spearman correlation coefficient is obtained.

$$r_S = 1 - \frac{6 \sum d_i^2}{n(n^2 - 1)} = 1 - \frac{6 \times 101}{10(10^2 - 1)} = 1 - 0.612 = 0.388$$

Table 12.5 Age, accommodation (ACC), intraocular pressure (IOP), and gender in a sample of 20 clients to an optometry clinic

Age	Gender	R. Acc	L. Acc	R.IOP	L.IOP
6	M	15	16	12	10
8	M	14	14	15	14
29	M	11.5	12	8	9
30	M	9	10.5	18	17
34	M	7.5	8	14	14
37	M	5	6	15	16
38	M	6	6	18	20
38	M	5.5	6	9	9
43	M	2.5	2.5	17	20
43	M	3.5	4	16	15
16	F	10	10.5	15	12
20	F	9.5	10	10	11
22	F	12	11	16	16
22	F	11	9	13	13
26	F	10	10	22	20
28	F	10.5	11.5	12	14
34	F	6.5	6.5	12	13
35	F	8	8.5	13	13
40	F	5	6.5	18	19
58	F	0	0	25	23

Table 12.6 Calculations for Spearman correlation coefficient in Table 12.3

Age	R.IOP	Age rank	R.IOP rank	$d_i = x_i - y_i$	d_I^2
16	15	1	6	−5	25
20	10	2	1	1	1
22	16	3.5	7	−3.5	12.25
22	13	3.5	4.5	−1	1
26	22	5	9	−4	16
28	12	6	2.5	3.5	12.25
34	12	7	2.5	4.5	20.25
35	13	8	4.5	3.5	12.25
40	18	9	8	1	1
58	25	10	10	0	0
Sum					101

Example 12.6 Two inspectors visited separately nine hospitals in terms of compliance of health protocols and ranked them according to Table 12.6. Obtain correlation coefficient between the ranks of the two inspectors.

Solution: In this example, because we have ordinal data, like the previous example, we find the difference in ranks ($d_i = x_i - y_i$) and their squares (d_i^2) and substitute them in the corresponding formula (Table 12.7).

$$r_s = 1 - \frac{6 \sum d_i^2}{n(n^2 - 1)} = 1 - \frac{6 \times 21.5}{9(9^2 - 1)} = 1 - 0.179 = 0.821$$

There is a strong positive correlation between the rankings of these two health inspectors.

Example 12.7 Find Spearman correlation coefficient between number of absence and final grade for data in Table 12.1.

Solution: In this example, the number of absences more than 10 sessions seems to be outliers and affects Pearson correlation coefficient. Therefore, using Spearman correlation coefficient would be more appropriate. We sort the rankings of both variables in columns 3 and 4 and record their difference in column 5. Then, the differences are squared and the sum is placed in the corresponding formula to obtain Spearman correlation coefficient (Table 12.8).

$$r_S = 1 - \frac{6 \sum d_i^2}{n(n^2 - 1)} = 1 - \frac{6 \times 716}{13(13^2 - 1)} = 1 - 1.967 = -0.967$$

Table 12.7 Ranking of 9 visited hospitals

Hospital	First inspector	Second inspector	d_i	d_i^2
A	9	8	1	1
B	6	2.5	3.5	12.25
C	7	7	0	1
D	5	6	−1	1
E	8	10	−2	4
F	1	1	0	0
G	3	4	−1	1
H	2	2.5	−0.5	0.25
I	4	5	−1	1
			Sum	21.5

Table 12.8 Ranking and calculation Spearman correlation coefficient for data in Table 12.1

Final grade	Absence	Rank of grade (x_i)	Rank of absence (y_i)	$d_i = x_i - y_i$	d_i^2
82	6	8.5	6	2.5	6.25
86	2	10	2.5	7.5	56.25
50	15	1	13	−12	144
74	9	5	9	−4	16
58	12	2	11.5	−9.5	90.25
90	5	11	4.5	6.5	42.25
78	8	7	7.5	−0.5	0.25
70	10	4	10	−6	36
65	12	3	11.5	−8.5	72.25
94	2	12	2.5	9.5	90.25
96	0	13	1	12	144
75	8	6	7.5	−1.5	2.25
82	5	8.5	4.5	4	16
Sum					716

12.2.6 *Confidence Interval for Spearman Correlation Coefficient*

As with the Pearson correlation coefficient, we use the Fisher transformation for the Spearman correlation coefficient to normalize its distribution. Then, by back transformation of the calculated values, we obtain the confidence interval for the Spearman correlation coefficient of the population.

$$z = \frac{1}{2}\ln\left(\frac{1+r}{1-r}\right)$$

The standard error of z is approximately:

$$Se_z = \sqrt{\frac{1.060}{n-3}}$$

Hence the 95% confidence interval for z is:

$$z_1 = z - 1.96 \times Se_z = z - 1.96 \times \sqrt{\frac{1.060}{n-3}}$$

$$z_2 = z + 1.96 \times S_z = z + 1.96 \times \sqrt{\frac{1.060}{n-3}}$$

The 95% confidence interval for correlation coefficient of r is reachable by back transformation of the above value as below:

$$\frac{e^{2z_1} - 1}{e^{2z_1} + 1} \quad \text{to} \quad \frac{e^{2z_2} - 1}{e^{2z_2} + 1}$$

The null and alternative hypotheses for hypothesis test are:

$$\begin{cases} H_0 : \rho = 0 \\ H_A : \rho \neq 0 \end{cases}$$

The test statistics is:

$$t = \frac{r}{\text{Se}\,(r)} = r\sqrt{\frac{n-2}{1-r^2}} \quad \text{with} \quad df = n - 2$$

The test statistic should compare with two-tailed critical value from t distribution table with degree of freedom $(n-2)$.

Example 12.8 The correlation coefficient of the number of absences and the final grade of a sample of 13 in Example 12.6 is equal to -0.967. Find the 95% confidence interval and perform a significant test.

Solution: We first transform $r = -0.967$ to z as below. Negative sign in calculated r is very important.

$$z = \frac{1}{2}\ln\left(\frac{1+r}{1-r}\right) = \frac{1}{2}\ln\left(\frac{1+(-0.967)}{1-(-0.967)}\right) = \frac{1}{2}\ln 0.0168 = -2.044$$

The standard error of z is:

$$\text{Se}_z = \sqrt{\frac{1.060}{n-3}} = \sqrt{\frac{1.060}{13-3}} = 0.3256$$

The 95% confidence interval for z is:

$$z_1 = z - 1.96 \times \text{Se}_z = -2.044 - 1.96 \times 0.3256 = -2.682$$

$$z_2 = z + 1.96 \times S_z = -2.044 + 1.96 \times 0.3256 = -1.406$$

After back transformation of the above values, 95% confidence interval for correlation coefficient of r is:

$$\text{Lower limit} = \frac{e^{2z_1} - 1}{e^{2z_1} + 1} = \frac{e^{2(-2.682)} - 1}{e^{2(-2.682)} + 1} = \frac{-0.9953}{1.0047} = -0.991$$

$$\text{Upper limit} = \frac{e^{2z_2} - 1}{e^{2z_2} + 1} = \frac{e^{2(-1.406)} - 1}{e^{2(-1.406)} + 1} = \frac{-0.9399}{1.0601} = -0.887$$

The both confidence limits suggest that there really is quite a strong negative association between the two variables in the population.

For significant test we need to the degree of freedom ($df = n - 2 = 13 - 2 = 11$) and critical value at 0.05 significant level is $t_{(0.975,11)} = 1.796$.

$$t = \frac{r}{\text{Se}(r)} = r\sqrt{\frac{n-2}{1-r^2}} = (-0.967)\sqrt{\frac{13-2}{1-(-0.967)^2}} = -12.588$$

Since test statistics is greater than the critical value ($|-12.588| > 1.796$) and fall in reject region, we reject the null hypothesis and mean that there is a significant difference with zero.

> **Tip**
>
> When the data is not in the ordinal scale, they should be assigned a ranking and then the Spearman correlation coefficient should be calculated for the assigned ranks (instead of the original data).

To calculate the Spearman correlation coefficient in SPSS, similar to run Pearson correlation coefficient, first select the Analyze\Correlate\Bivariate path to open a window. Then mark Spearman option to run program.

Significance test for correlation coefficient is also performed, and test statistic and p-value are presented below the output table.

Example 12.9 If the rankings of hospitals by two health officers for example 12.5 (Table 12.5) are exactly the same, the Spearman correlation coefficient should be equal to $+ 1$. But if it is exactly in opposite order such as Table 12.7, the Spearman correlation coefficient should be equal to -1. Do the calculations yourself to make sure.

12.2.7 Kendall's Rank Correlation Coefficient

Tips

Concordant: A pair of observations is concordant if the subject who is higher on one variable is also higher on the other variable or $(x_i - x_j)$ and $(y_i - y_j)$ have the same sign.

Discordant: A pair of observations is discordant if the subject who is higher on one variable is lower on the other variable or $(x_i - x_j)$ and $(y_i - y_j)$ have opposite signs.

Kendall's tau coefficient and Spearman's rank correlation coefficient both evaluate statistical relationships based on the ranks of the data. Kendall rank correlation, a non-parametric method, serves as an alternative to the parametric Pearson's correlation when one or more assumptions of the test are violated. It also offers a viable alternative to the non-parametric Spearman correlation, particularly when dealing with small sample sizes and a significant number of tied ranks.

Kendall rank correlation is particularly suited for assessing similarities in the ordering of data when ranked by quantities. It relies on pairs of observations to determine the strength of association, focusing on the pattern of concordance and discordance between these pairs. Typically, Kendall's tau coefficient yields smaller values compared to Spearman's rho correlation.

The calculation of Kendall's tau involves considering concordant and discordant pairs within the dataset. This method tends to provide more accurate p-values, especially in scenarios involving smaller sample sizes or the presence of outliers.

Kendall's tau coefficient correlation can be obtained by the following formula:

$$\tau = \frac{2(n_C - n_D)}{n(n-1)}$$

where n_C is the number of concordant pairs and n_D is the number of disconcordant pairs.

Solution: To estimate Kendall's tau, we first order one of the variables in ascending ranks as in Table 12.9, which is ordered by age, and placed the ranks of the second variable alongside (natural order); the body fat percentages now shift entirely to the second variable. Then we allocate two columns for concordant pairs (C) and discordant pairs (D).

Now we compare each rank of the second variable (body fat percentage) with its subsequent ranks to see how many ranks are larger and how many ranks are below it. We count the number of ranks larger than it (right order, concordant) and record in column (C) and the number of ranks less than it (wrong order, disconcordant) and write in column (D). We do this for all second variable ratings. The last rank or last row remains empty because then there are no rankings to be compared. Then we sum two columns n_C and n_D and replace them in the formula (Table 12.10).

Table 12.9 Ranking of 9 hospitals in opposite order

Subject	Age	Body fat%	Rank Age	Rank Body fat%	C	D
E	23	9.5	1	2	12	1
B	25	10.1	2	3	11	1
A	27	7.8	3	1	11	0
D	41	25.9	4	5	9	1
H	45	27.5	5	6	8	1
F	49	25.4	6	4	8	0
M	50	31.0	7	8	6	1
G	53	54.7	8	14	0	6
K	55	42.0	9	13	0	5
N	56	32.5	10	9	3	1
G	57	30.3	11	7	3	0
I	58	33.8	12	10	2	0
O	60	41.1	13	12	0	1
C	61	34.5	14	11	-	-
				Sum	73	18

Table 12.10 Ascending order of Table 12.8

Subject	Age	Body fat%	Rank age	Rank body fat%	C	D
E	23	9.5	1	2	12	1
B	25	10.1	2	3	11	1
A	27	7.8	3	1	11	0
D	41	25.9	4	5	9	1
H	45	27.5	5	6	8	1
F	49	25.4	6	4	8	0
M	50	31.0	7	8	6	1
G	53	54.7	8	14	0	6
K	55	42.0	9	13	0	5
N	56	32.5	10	9	3	1
G	57	30.3	11	7	3	0
I	58	33.8	12	10	2	0
O	60	41.1	13	12	0	1
C	61	34.5	14	11	–	–
	Sum				73	18

So looking at the first rank, 2, it is followed by a 3, which is in the right order, and therefore contributes concordant. It is then followed by 1, which is in the wrong order, and therefore contributes concordant. It is then followed by 5, 6, 4, 8, 14, 13, 9, 7, 10, 12, and 11, all of which are greater than 2 and in the right order, so all contribute concordant (concordant $= 12$ and disconcordant $= 1$). We then go to the third row, rank in the body fat %, 1, and find (naturally) that all subsequent ranks 5, 6, 4, 8, 14, 13, 9, 7, 10, 12, and 11 are in the right order, so this column contributes 11 concordant and none disconcordant. And so it goes, and eventually we end up with a total of $n_C = 72$ concordant and $n_D = 19$ disconcordant. Now substitute $n = 14$, $n_D = 19$, and $n_C = 72$ in the related formula to estimate Kendall's tau.

$$\tau = \frac{2(n_C - n_D)}{n(n-1)} = \frac{2(73 - 18)}{14(14-1)} = 0.604$$

12.2.8 Test Statistic and Confidence Interval for Kendall's Tau Correlation Coefficient

To calculate the confidence interval for Kendall's tau, the same method as the Spearman correlation coefficient must be performed. First, we use the z Fisher transformation to normalize its sample distribution.

$$\tau_z = 0.5 \ln\left(\frac{1 + \tau}{1 - \tau}\right)$$

The standard error of τ_z is

$$\mathrm{Sc}(\tau_z) = \sqrt{\frac{0.437}{n - 4}}$$

The lower and upper confidence limits for τ_z are calculated by:

$$\tau_{z,L} = \tau_z - z_{1-\frac{\alpha}{2}} \times \mathrm{Se}(\tau_z)$$

$$\tau_{z,U} = \tau_z + z_{1-\frac{\alpha}{2}} \times \mathrm{Se}(\tau_z)$$

where $\tau_{z,L}$ and $\tau_{z,U}$ are the lower and upper bounds, respectively, and $z_{1-\frac{\alpha}{2}}$ is the unit normal critical z value for specified level of confidence $1 - \alpha$.

Using back transformation of the above-mentioned lower and upper limits, the true confidence intervals of τ for the population correlation are found.

$$\tau_L = \frac{\exp(2 \times \tau_{z,L}) - 1}{\exp(2 \times \tau_{z,L}) + 1} \quad \text{and} \quad \tau_U = \frac{\exp(2 \times \tau_{z,U}) - 1}{\exp(2 \times \tau_{z,U}) + 1}$$

The test statistic using the standard normal distribution is as follows:

$$z = \frac{3 \times \tau \times \sqrt{n(n-1)}}{\sqrt{2(2n+5)}}$$

This statistic is compared with its corresponding value in the standard normal distribution table. If it is greater than the value of the table, the null hypothesis is rejected, meaning that the Kendall's tau correlation coefficient cannot be considered equal to zero.

Example 12.10 Kendall's tau correlation coefficient in Example 12.9 with a sample size of 13 is 0.604. Find the test statistic and 95% confidence interval.

Solution: First, we transform the Kendall's tau using the Fisher transformation as follows and obtain its standard error.

$$\tau_z = 0.5 \ln\left(\frac{1+\tau}{1-\tau}\right) = 0.5 \ln\left(\frac{1+0.604}{1-0.604}\right) = 0.6994$$

Then, we get the upper and lower limits for the transformed values:

$$\tau_{z,L} = \tau_z - z_{1-\frac{\alpha}{2}} \times \text{Se}(\tau_z) = 0.6994 - 1.96 \times 0.22035 = 0.2675$$

$$\tau_{z,U} = \tau_z + z_{1-\frac{\alpha}{2}} \times \text{Se}(\tau_z) = 0.6994 + 1.96 \times 0.22035 = 1.1313$$

Then, by using back transformation of the upper and lower limits obtained above ($\tau_{z,L}$ and $\tau_{z,U}$), the confidence intervals for the population τ are obtained.

$$\tau_L = \frac{\exp(2 \times \tau_{z,L}) - 1}{\exp(2 \times \tau_{z,L}) + 1} = \frac{\exp(2 \times 0.2675) - 1}{\exp(2 \times 0.2675) + 1} = \frac{0.7074}{2.7074} = 0.261$$

$$\tau_U = \frac{\exp(2 \times \tau_{z,U}) - 1}{\exp(2 \times \tau_{z,U}) + 1} = \frac{\exp(2 \times 1.1313) - 1}{\exp(2 \times 1.1313) + 1} = \frac{8.608}{10.608} = 0.811$$

Thus, 95% confidence interval of Kendall's tau correlation coefficient in the population will be from 0.261 to 0.811.

The test statistic for this example is:

$$z = \frac{3 \times \tau \times \sqrt{n(n-1)}}{\sqrt{2(2n+5)}} = \frac{3 \times 0.604 \times \sqrt{13(13-1)}}{\sqrt{2(2 \times 13 + 5)}} = \frac{22.632}{7.874} = 2.874$$

which is greater than corresponding value of standard normal distribution (1.96) and the null hypothesis is rejected.

12.2.9 Point-Biserial Correlation

The biserial correlation and point-biserial correlation are specialized forms of the Pearson correlation coefficient, specifically used when one variable is binary (having only two levels) and the other variable is quantitative (ratio or interval data). Binary variables are also referred to as dichotomous or dummy variables, particularly in regression analysis. The binary variable can represent naturally dichotomous traits, such as gender, or artificially dichotomized variables, like categorizing blood pressure as normal or abnormal. However, it's generally not advisable to artificially dichotomize ordinal or quantitative data because they contain more variance information than nominal data, thus potentially providing more reliable correlation analysis.

Both the point-biserial correlation and biserial correlation aim to gauge the strength and direction of the association between a quantitative variable and a binary variable, which takes two distinct values (such as 0 and 1, or 2 and 3).

In medical contexts, questions that can be addressed using a point-biserial correlation include inquiries into the effectiveness of a cancer drug in extending life, assessing the association between drug administration (placebo versus drug) and survival duration post-treatment, examining the relationship between gender (female versus male) and cigarette consumption, or exploring the correlation between hospitalization duration and gender among individuals with Covid-19.

Biserial Correlation: When the two responses of binomial variable are superior to the other, the biserial correlation is a better estimate of the original Pearson correlation. For example, when we want to calculate the correlation between IQ and exam score, but the only available measurement about the exam scores is passed or failed, employing biserial correlation has a more conceptual estimate. Because we know that success is better than failure.

Point-Biserial Correlation: When dealing with binary variables that lack a natural ordering, such as gender (where coding males as zero or one is arbitrary), they are termed nominal binary variables. In such cases, the point-biserial correlation coefficient is employed. This book focuses solely on the point-biserial correlation coefficient due to its prevalence.

Mathematically, the point-biserial correlation coefficient (denoted as r_{pb}) is computed similarly to the Pearson's bivariate correlation coefficient. However, one variable is dichotomous. As with other correlation coefficients, assumptions such as normal distribution and homogeneity of variance for the quantitative variable are necessary.

Like all correlation coefficients, the point-biserial correlation coefficient ranges from -1 to $+1$, with similar interpretations regarding strength and direction of the relationship.

Suppose a quantitative random variable Y and a binary random variable X which takes the values zero and one are available, the point-biserial correlation (r_{pb}) between X and Y can be calculated by the formula below:

$$r_{\text{pb}} = \left(\frac{\overline{Y}_1 - \overline{Y}_2}{S_Y}\right)\sqrt{\frac{n \times p \times (1-p)}{(n-1)}}$$

Or equivalently:

$$r_{pb} = \left(\frac{\overline{Y}_1 - \overline{Y}_2}{S_Y}\right)\sqrt{\frac{n_1 \times n_2}{n(n-1)}}$$

where S_Y is the standard deviation of Y for all observations, which is calculated by:

$$S_Y = \sqrt{\frac{\sum(Y_i - \overline{Y})^2}{n-1}}$$

$\overline{Y}_1$ is the mean value on the quantitative variable Y for all data points in group 1 and $\overline{Y}_2$ is the mean value on the quantitative variable Y for all data points in group 2, p is the proportion of number subjects in one of the groups to the total numbers, and n_1 is the sample size in in first group, n_2 is the sample size in second group, and n is the total sample size in both groups.

Estimation of r_{pb} is:

$$\widehat{\text{Var}}(r_{pb}) = \frac{\left(1 - r_{\text{pb}}^2\right)^2}{n}\left\{1 - 1.5 \times r_{\text{pb}}^2 + \frac{r_{pb}^2}{4p(1-p)}\right\}$$

An approximate confidence interval based on the normal distribution can be calculated from these quantities using:

$$r_{\text{pb}} \pm Z_{1-\frac{\alpha}{2}} \times \sqrt{\frac{\left(1 - r_{\text{pb}}^2\right)^2}{n}\left[1 + r_{\text{pb}}^2\left(\frac{1 - 6p(1-p)}{4p(1-p)}\right)\right]}$$

The null hypothesis that $H_0 : \rho = 0$ can be tested using the two-sample t-test.

$$t_{\text{pb}} = r_{\text{pb}} \times \frac{\sqrt{n-2}}{\sqrt{1 - r_{\text{pb}}^2}}$$

This test statistic follows Student's t distribution with $n-2$ degrees of freedom.

Example 12.11 As part of Covid-19 project, the number of hospitalization days for 28 patients by gender is listed in Table 12.8. Find a point-biserial correlation coefficient between sex and duration of hospitalization, %95 confidence interval, and test statistic.

Solution: Standard deviation of the number of hospitalization days for 28 patients was $S_Y = 6.130$, the first sample size $n_1 = 13$, the second sample size $n_2 = 15$, the total sample size $n = 28$, the proportion of the first sample size to the total sample size was $p = \frac{13}{28} = 0.4643$ and $(1 - p) = 1 - 0.4643 = 0.5357$, the total mean is $\overline{Y} = 11.2143$, the mean of the first group $\overline{Y}_1 = 7.846$, and the mean of the second group is $\overline{Y}_2 = 14.133$. By substituting these values in the formula, the point-biserial correlation coefficient is obtained (Table 12.11).

$$r_{\text{pb}} = \left(\frac{\overline{Y}_1 - \overline{Y}_2}{S_Y}\right)\sqrt{\frac{n_1 \times n_2}{n(n - 1)}} = \left(\frac{14.133 - 7.846}{6.130}\right)\sqrt{\frac{13 \times 15}{28(28 - 1)}} = 0.5209$$

The 95% confidence interval is:

$$r_{\text{pb}} \pm Z_{1-\frac{\alpha}{2}} \times \sqrt{\frac{\left(1 - r_{pb}^2\right)^2}{n}\left[1 + r_{pb}^2\left(\frac{1 - 6p(1 - p)}{4p(1 - p)}\right)\right]}$$

$$= 0.5209 \pm 1.96 \times \sqrt{\frac{\left(1 - 0.5209^2\right)^2}{28}\left[1 + 0.5209^2\left(\frac{1 - 6 \times 0.4643(1 - 0.4643)}{4 \times 0.4643(1 - 0.4643)}\right)\right]}$$

$$= 0.5209 \pm 1.96 \times \sqrt{0.01896\left[1 + 0.271\left(\frac{-0.4924}{0.9949}\right)\right]}$$

$$= 0.5209 \pm 1.96 \times 0.1281$$

$$= (0.2698 \text{ to } 0.7720)(0.183 \text{ to } 0.748)$$

And test statistic is:

Table 12.11 Number of hospitalization days for Covid-19 patients according to gender

Male	17	17	8	20	22	8	6	13	11	8	16	13	18	28	7
Female	10	12	7	9	10	5	7	4	15	4	12	3	4	–	–

$$t_{\mathrm{pb}} = r_{\mathrm{pb}} \times \frac{\sqrt{n-2}}{\sqrt{1-r_{\mathrm{pb}}^2}} = \frac{0.521\sqrt{28-2}}{\sqrt{1-0.52^2}} = \frac{2.6515}{0.8542} = 3.104$$

which is greater than the critical value with a degree of freedom of 28–2 = 28 (3.104 > $t_{(0.975,26)}$ = 2.056) and the null hypothesis is rejected. That is, the point-biserial correlation coefficient cannot be considered equal to zero.

If we plot the scatter diagram of two variables with SPSS and then fit the regression line, it can be observed that the mean difference between the two groups is equal to the slope of the regression line and the p-value obtained for the t-test and for the point-biserial correlation coefficient is exactly identical.

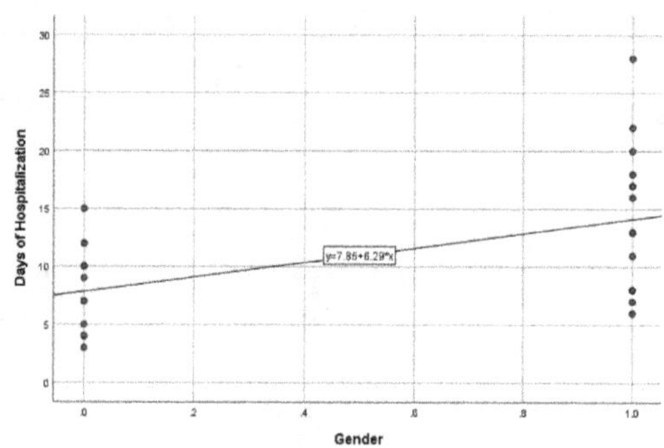

Correlations

		Gender Covid-19	Days hospitalization
Gender Covid-19	Pearson correlation	1	0.521**
	Sig. (two-tailed)		0.004
	N	28	28
Days hospitalization	Pearson correlation	0.521**	1
	Sig. (two-tailed)	0.004	
	N	28	28

**. Correlation is significant at the 0.01 level (two-tailed)

12.2.10 Running Point-Biserial Correlation on SPSS

To calculate the point-biserial correlation coefficient in SPSS, first select the Analyze\Correlate\Bivariate path like the Pearson correlation coefficient to open a window.

Select the Dichotomous and the quantitative variable from the box on the left and move it to the box on the right. Then mark the Pearson option to calculate the point-biserial correlation coefficient.

As mentioned, the point-biserial correlation coefficient is a special case of Pearson correlation coefficient, one of the variables of which is dichotomous. Therefore, we need to record the quantities of both groups together in one column and assign another column to the dichotomous variable to determine which value belongs to group one and which value belongs to the other group (figure on the top left). Then, by confirming OK, the program is executed and the value of the point-biserial correlation coefficient is obtained.

Note that the dichotomous variable in SPSS must be entered by nominal scale so that the digits entered in this column are considered code. Therefore, code (0 and 1) or code (1, 2) or code (4 and 5) will not change the calculations.

12.2.11 *Partial Correlation Coefficient*

Partial correlation is a linear correlation between two quantitative variables (x and y) while controls or eliminates the effect of another quantitative confounding variable (z). Therefore, when the correlation coefficient between the two variables (x and y) is calculated, another confounding variable such as z may influence the relationship between the two variables (x and y). In such cases, the effect of the third variable (z) can be adjusted by calculating the partial correlation coefficient. For this purpose, we need to calculate the correlation between the mutual variables separately (r_{zy}, r_{xz} and r_{xy}) and then calculate the partial correlation using the following formula:

$$r_{xy.z} = r_{xy|z} = \frac{r_{xy} - r_{xz} \times r_{zy}}{\sqrt{\left(1 - r_{xz}^2\right)} \times \sqrt{\left(1 - r_{zy}^2\right)}}$$

The test statistic for the partial correlation coefficient is the same as for the Pearson correlation coefficient, except that df $= n-3$.

Example 12.12 For the data in Table 12.1, obtain the correlation coefficient between the final grade of students and the number of student study hours after adjusting the number of absence sessions.

Solution: In this example, we consider the exam score as x, the number of study hours y, and the number of absence sessions as z. Like calculating Pearson correlation coefficient, their mutual correlation coefficient is equal to:

$$r_{xy} = 0.912, \quad r_{xz} = -0.964, \quad \text{and} \quad r_{zy} = -0.930$$

By substituting above values in the relevant formula, the adjusted correlation coefficient of the final grade and study hours for the number of student absence sessions is obtained.

$$r_{xy.z} = r_{xy|z} = \frac{r_{xy} - r_{xz} \times r_{zy}}{\sqrt{\left(1 - r_{xz}^2\right)} \times \sqrt{\left(1 - r_{zy}^2\right)}}$$

$$= \frac{0.912 - (-0.964) \times (-0.930)}{\sqrt{\left(1 - (-0.964)^2\right)} \times \sqrt{\left(1 - (-0.930)^2\right)}}$$

$$= \frac{0.01548}{0.09973}$$

$$= 0.158$$

It can be seen that after adjusting and controlling the number of student absence sessions, the correlation coefficient of the final grade and study hours, which was previously 0.912, decreased to 0.158.

12.2.12 Comparing Two Correlation Coefficients

How can two correlation coefficients be compared whether their differences are significant? These coefficients may be dependent, i.e., between one variable and two other variables (e.g., r_{xy} and r_{xz} in the same sample, or independent coefficients, i.e., between variable X and Y in two different samples ($r_{1_{xy}}$ and $r_{2_{xy}}$).

Dependent Correlations

If correlation coefficients are dependent, the hypothesis test of equality of two dependent correlation coefficients is used. The null and alternative hypotheses for this test are:

$$\begin{cases} H_0 : \rho_{xy} = \rho_{xz} \Rightarrow \rho_{xy} - \rho_{xz} = 0 \\ H_0 : \rho_{xy} \neq \rho_{xz} \Rightarrow \rho_{xy} - \rho_{xz} \neq 0 \end{cases}$$

The test statistic is defined as below:

$$t(r_{xy} - r_{xz}) = \frac{(r_{xy} - r_{xz}) \times \sqrt{(n-3)(1 + r_{yz})}}{2\left(1 - r_{xz}^2 - r_{yz}^2 - r_{xy}^2 + 2r_{xz}.r_{yz}.r_{yz}\right)}$$

where r_{xy} is the correlation coefficient between the two variables x and y, r_{xz} is the correlation coefficient between the two variables x and z, r_{yz} is the correlation coefficient between the two variables y and z and n is the sample size. This test statistic should be compared with the critical value of table t with degree of freedom $(n-3)$.

Example 12.13 The correlation coefficients between stiffness (x), pain (y), and activity level (z) in 24 elderly are $r_{xy} = 0.861$, $r_{yz} = 0.657$, and $r_{xz} = 0.824$, respectively. Is there a significant difference between the correlation coefficients of stiffness and pain (ρ_{xy}) with the correlation between pain and activity level (ρ_{xy})?

Solution: The critical value of t from t student table with degree of freedom df $=$ 24–3 $=$ 21 is 2.080 and test statistic is as below:

$$t\left(r_{xy} - r_{xz}\right) = \frac{\left(r_{xy} - r_{xz}\right) \times \sqrt{(n-3)\left(1+r_{yz}\right)}}{2\left(1 - r_{xz}^2 - r_{yz}^2 - r_{xy}^2 + 2r_{xz}.r_{yz}.r_{yz}\right)}$$

$$= \frac{(0.861 - 0.657) \times \sqrt{(24-3)(1+0.824)}}{2\left(1 - 0.824^2 - 0.657^2 - 0.861^2 + 2 \times 0.824 \times 0.657 \times 0.861\right)}$$

$$= \frac{1.2625}{0.1606}$$

$$= 7.862$$

Since the test statistics is greater than critical value (7.862 > 2.080), the null hypothesis is rejected and the two correlation coefficients are not equal.

Example 12.14 In a sample of 40 $(n = 40)$, the correlation coefficients between height of child (C) with father (F) is $r_{CF} = 0.604$, child with mother is $r_{CM} = 0.262$, and mother with father is $r_{MF} = 0.151$. Is there a significant difference between the correlation coefficients of child and father (ρ_{CF}) with the correlation between child and mother (ρ_{CM})?

Solution: The critical value of t from t student table with degree of freedom df $=$ 40–3 $=$ 37 is 2.026 and test statistic is as below:

$$t(r_{CM} - r_{CF}) = \frac{(r_{CM} - r_{CF}) \times \sqrt{(n-3)(1+r_{MF})}}{2\left(1 - r_{CM}^2 - r_{CF}^2 - r_{MF}^2 + 2r_{CM}.r_{CF}.r_{MF}\right)}$$

$$= \frac{(0.604 - 0.262) \times \sqrt{(40-3)(1+0.151)}}{2\left(1 - 0.604^2 - 0.262^2 - 0.151^2 + 2 \times 0.604 \times 0.262 \times 0.151\right)}$$

$$= \frac{2.2318}{1.1831}$$

$$= 1.886$$

Since the test statistics is greater than critical value (1.886 < 2.026), the null hypothesis is not rejected and the two correlation coefficients are equal.

Independent Correlations

If correlation coefficients are independent, the hypothesis test of equality of two independent correlation coefficients is used. The null and alternative hypotheses for this test are:

$$\begin{cases} H_0 : \rho_1 = \rho_2 \Rightarrow \rho_1 - \rho_2 = 0 \\ H_0 : \rho_1 \neq \rho_2 \Rightarrow \rho_1 - \rho_2 \neq 0 \end{cases}$$

Since it is a little difficult to calculate the combined distribution of two correlation coefficients, to test this hypothesis, we first transform both correlation coefficients using the z Fisher score as follows:

$$Z_1 = \frac{1}{2}\ln\left(\frac{1+r_1}{1-r_1}\right), \quad Z_2 = \frac{1}{2}\ln\left(\frac{1+r_2}{1-r_2}\right)$$

Note that the reverse transformation of Z_1 and Z_2 to r_1 of r_2 is obtained by:

$$r_1 = \frac{e^{Z_1} - e^{-Z_1}}{e^{Z_1} + e^{-Z_1}}, \quad r_2 = \frac{e^{Z_2} - e^{-Z_2}}{e^{Z_2} + e^{-Z_2}}$$

Because the combined distribution of Z_1 and Z_2 is approximately normal, the test statistic of z is used for it:

$$Z(r_1 - r_2) = \frac{Z_1 - Z_2}{\sqrt{\left(\frac{1}{n_1 - 3)} + \frac{1}{n_2 - 3}\right)}}$$

And it should be compared with the critical value of z score in the standard normal distribution.

Example 12.15 The correlation coefficients between stiffness (x) and activity (z) in 16 male elderly are $r_1 = 0.709$ and in 8 female elderly are $r_2 = 0.849$. Are the correlation coefficients different between the two groups?

Solution: The transformed values of r_1 and r_2 are equal to:

$$Z_1 = \frac{1}{2}\ln\left(\frac{1+r_1}{1-r_1}\right) = \frac{1}{2}\ln\left(\frac{1+0.709}{1-0.709}\right) = 0.8852$$

$$Z_2 = \frac{1}{2}\ln\left(\frac{1+r_2}{1-r_2}\right) = \frac{1}{2}\ln\left(\frac{1+0.849}{1-0.849}\right) = 1.2526$$

The test statistic for comparing the two correlation coefficients is equal to:

$$Z(r_1 - r_2) = \frac{Z_1 - Z_2}{\sqrt{\left(\frac{1}{n_1 - 3)} + \frac{1}{n_2 - 3}\right)}} = \frac{1.2526 - 0.8852}{\sqrt{\left(\frac{1}{16-3)} + \frac{1}{8-3}\right)}}$$

$$= \frac{0.3674}{0.526}$$

$$= 0.6984$$

Because the test statistic is less than the critical value of 1.96, the null hypothesis is not rejected, i.e., the difference in correlation coefficient between the two communities is not significant.

Example 12.16 The correlation coefficients between height of son (S) and father (F) in 20 adult males are $r_1 = r_{SF} = 0.241$, and correlation coefficients between height of daughter (D) and father (M) in 20 adult females are $r_2 = r_{DM} = 0.306$. Are the correlation coefficients different between the two groups?

Solution: The transformed values of r_1 and r_2 are equal to:

$$Z_1 = \frac{1}{2}\ln\left(\frac{1+r_1}{1-r_1}\right) = \frac{1}{2}\ln\left(\frac{1+0.241}{1-0.241}\right) = 0.246$$

$$Z_2 = \frac{1}{2}\ln\left(\frac{1+r_2}{1-r_2}\right) = \frac{1}{2}\ln\left(\frac{1+0.306}{1-0.306}\right) = 0.316$$

The test statistic for comparing the two correlation coefficients is equal to:

$$Z(r_1 - r_2) = \frac{Z_1 - Z_2}{\sqrt{\left(\frac{1}{n_1 - 3)} + \frac{1}{n_2 - 3}\right)}} = \frac{0.246 - 0.316}{\sqrt{\left(\frac{1}{20-3)} + \frac{1}{20-3}\right)}} = \frac{-0.070}{0.343} = -0.204$$

Because the test statistic is less than the critical value of 1.96 ($1.88 < 1.96$), the null hypothesis is not rejected, i.e., the difference in correlation coefficient between the two communities is not significant.

12.2.13 Precautions

Before calculating the Pearson correlation coefficient, first draw the scatter diagram of the two variables so that if there is a linear relationship between the two variables, the Pearson correlation coefficient is calculated.

When there is an outlier in the dataset, be cautious in applying the Pearson correlation coefficient. Because in these cases the value of the correlation coefficient changes by eliminating an outlier that may have been wrongly measured or taken from another population.

When the data source is several populations or sub-populations, it is necessary to be cautious in applying the coefficient of correlation. Because each sub-population may have a different correlation coefficient than the correlation coefficient of the whole population.

When one of the variables is pre-specified, it is not appropriate to use a correlation coefficient. For example, in measuring the response to different doses of a drug, the correlation coefficient cannot be clear, because the choice of a specific dose level of the drug is determined by the experimenter and is not random.

12.3   Regression

In medicine and health science, there are inherent relationships between two or more variables. Such relationships are occurring every day in the practice of health and medical fields. For example, blood pressure (BP) levels have a relationship with the body mass index (BMI), and the osteoporosis is associated with physical activity and dairy consumption, or tobacco consumption is related to many of cardiac and lung disorders.

The medical researchers have to depict and describe the relationship between these variables to be able to predict with some degree of precision, the value of one variable for an individual with the help of the other variables. That is, how much a unit increases or decreases in one variable, how much change it makes in another variable (Fig. 12.2). It is a question that correlation coefficient is clearly unable to answer and left the answer to the regression models. Because the correlation coefficient just indicates the strength and direction of the relationship between two variables with a single number between -1 and $+1$.

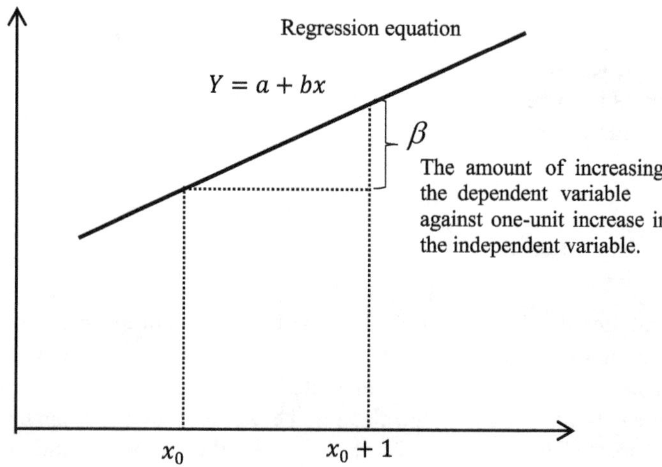

Fig. 12.2 Variations in variable Y per unit change in variable X

A way to describe the relationship between these two variables, and to predict the value of one variable for each individual from the known value of the other variable, is called regression, which was proposed by Francis Galton in the late-nineteenth century.

Therefore, the purposes of regression models can be summarized as follows:

To be able to predict the value of one variable from other variables that are associated to it.

Help to understand the mechanism of such relationships, particularly by studying the effect of a changing in the value of one variable if other conditions do not change.

For this purpose, a line should be fitted to the set of observed values that is closest to them, i.e., the vertical distance of each point with the fitted line should be minimize. The general equation of this line (regression) is as follows:

$$Y = a + bx + e$$

where Y is dependent variable, outcome or response variable, X is named independent or predictor variable, b is slope of regression line or regression coefficient, a is intercept and e is called error term or residual.

Typically, the above line minimizes the vertical distance between the observed data and the fitted line. There are several approaches to estimate a and b, but the "least squares method" is their most common one. Therefore, using this method to fit a regression line leads to minimal errors, which is the sum of the squares of the vertical distances of the observations from the line. Figure 12.3 shows difference between observed values from fitted regression line for systolic blood pressure and age in a sample of 11 girl school students (Table 12.12).

The values of b and a can be estimated from sampled data by the following equations:

$$b = \frac{\sum (x - \bar{x})(y - \bar{y})}{\sum (x - \bar{x})^2} = \frac{n . \sum xy - \left(\sum x\right)\left(\sum y\right)}{n . \sum x^2 - \left(\sum x\right)^2}$$

And

$$a = \frac{\left(\sum y\right)\left(\sum x^2\right) - \left(\sum x\right)\left(\sum xy\right)}{n . \sum x^2 - \left(\sum x\right)^2}$$

Or

$$a = \bar{y} - b . \bar{x}$$

The calculations of the above equation may be a bit complicated and difficult; to simplify the calculations, it is recommended to assign five columns for x, y, x^2, y^2, xy as in Table 12.2. Then, substitute the sum of each column directly in formula of the regression line.

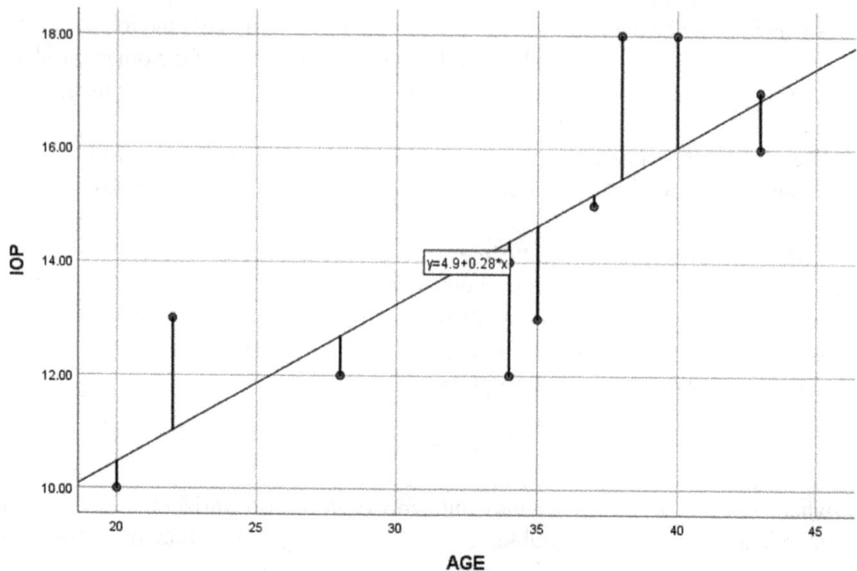

Fig. 12.3 Difference between observed values from the fitted regression line

Table 12.12 Age and systolic and diastolic blood pressures in millimeters of mercury, for 11 children girl students

Age	5	6	7	8	9	10	11	12	13	14	15
Systolic	99.8	102.6	103.0	105.4	106.8	107.2	110.6	111.5	114.0	114.3	115.0
Diastolic	66.9	67.4	67.8	69.6	69.9	69.8	71.5	71.6	72.7	74.1	74.8

12.3.1 Assumptions for Linear Regression

Like all parametric models, before using regression analysis, it is important to consider and evaluate four assumptions in order to validate the regression method.

The dependent variable (Y) should have a normal distribution for each value of the independent variable X (Fig. 12.4);

(The independent variable of X (predictor) doesn't need to be a random variable or have an approximately normal distribution (Unlike for correlation). The regression model is still valid even if the researcher chose specific values of X.)

The variance (or standard deviation) of Y should be the same for each value of X.

The relation between the two variables should be linear.

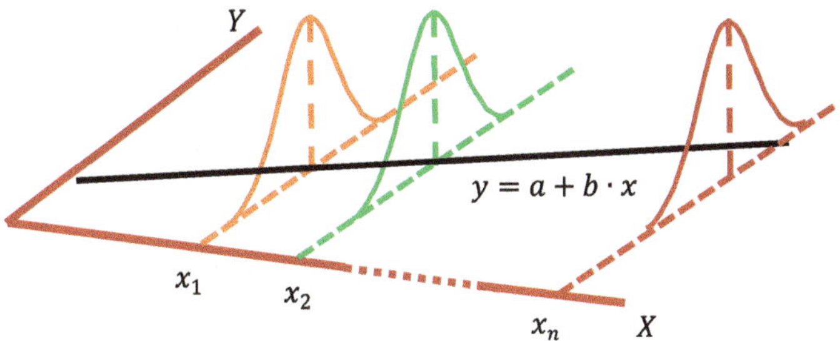

Fig. 12.4 Distribution of Y for each value of independent value of X

> **Tips**
>
> The normality assumption is evaluated based on the residuals and can be evaluated using a **QQ-plot** (plot 2). In the "ideal" normal distribution, observations lie well along the 45-degree line in the QQ-plot. The farther the observations are from this line, the farther the distribution is from normal.
>
> To investigate the homoscedasticity assumption, we need to ensure that there are no patterns in the residuals and that they are equally spread around the $y = 0$ line.
>
> To assess the linearity assumption, we make ensure that the residuals are not too far away from 0 (standardized values less than -2 or greater than 2 are deemed problematic).

Outliers have a strong effect on the regression equation, so the presence of outliers should be investigated.

Fortunately assess the first three assumptions is easy by plotting residuals against the x-values, because the residuals are informative and include the useful relevant information. If the residuals have a normal distribution (with a mean of zero), means that the three above assumptions are met. Then plot the residuals against the x-values. If the above plot shows that all the residuals are scattered around the horizontal line of mean (zero), it indicates that all the first three assumptions are met.

Example 12.17 Intraocular pressure (IOP), accommodation (ACC), and age of 11 patients are listed in Table 12.13. Find the regression line equation of IOP by age.

Table 12.13 Age, intraocular pressure (IOP), and accommodation (ACC) in a sample of 11 patients in optometry clinic

Age	20	22	28	34	34	35	37	38	40	43	43
IOP	10	13	12	12	14	13	15	18	18	17	16
ACC	9.5	11	10.5	6.5	7.5	8	5	6	5	2.5	3.5

Solution: Like correlation coefficient, we perform the calculations required for regression coefficients in Table 12.14 and substitute the sum of columns into the regression formula.

Now, we substitute values from the last row in the formula and estimate the regression coefficients of b and a.

$$b = \frac{n.\sum xy - \left(\sum x\right)\left(\sum y\right)}{n.\sum x^2 - \left(\sum x\right)^2} = \frac{11 \times 5539 - 374 \times 158}{11 \times 13316 - (374)^2} = \frac{1837}{6600} = 0.278$$

$$\bar{x} = \frac{374}{11} = 34, \quad \bar{y} = \frac{158}{11} = 14.3636$$

$$a = \bar{y} - b.\bar{x} = 14.3636 - 0.278 \times 34 = 4.900$$

Therefore, the equation of intraocular pressure (IOP) regression line in terms of age is as follows:

$$IOP = 4.900 + 0.278 \times Age$$

In order to predict the intraocular pressure (IOP) of a 30-year-old person, in the above equation we replace the value of 30 instead of age.

$$IOP = 4.900 + 0.278 \times Age$$
$$= 4.900 + 0.278 \times 30$$
$$= 13.24$$

Table 12.14 Necessary calculations for estimating regression coefficient for data on Table 12.11

Patient	Age (x)	IOP (y)	x^2	y^2	x.y
1	20	10	400	100	200
2	22	13	484	169	286
3	28	12	784	144	336
4	34	12	1156	144	408
5	34	14	1156	196	476
6	35	13	1225	169	455
7	37	15	1369	225	555
8	38	18	1444	324	684
9	40	18	1600	324	720
10	43	17	1849	289	731
11	43	16	1849	256	688
Sum	$\sum x = 374$	$\sum y = 158$	$\sum x^2 = 13316$	$\sum y^2 = 2340$	$\sum x.y = 5539$

12.3.2 Confidence Interval and Significant Test for Regression Coefficient

Regression coefficient b is an estimate of the population β, which is obtained from the sample data and is influenced by the sample size and sampling variations. Therefore, like other parameters, it can be tested with the value specified in null hypothesis (β_0). When we have only one independent variable in the regression model, the null and alternative hypotheses for regression coefficient test are as follows:

> **Tip**
>
> The values of the independent variable x, which is substituted in this equation to predict the dependent variable, must be in the range of the dataset (between minimum and maximum). For instance, in the example above, age between 20 and 43 should be chosen to predict intraocular pressure (IOP). Because the regression equation is obtained based on such data and if the data range changes, the regression line equation will also change.

$$\begin{cases} H_0 : \beta = \beta_0 \\ H_A : \beta \neq \beta_0 \end{cases}$$

The test statistic is:

$$t = \frac{\hat{\beta} - \beta_0}{\text{Se}(b)}$$

where the standard error of the regression coefficient is also equal to:

$$\text{Se}(b) = \frac{S_{\text{res}}}{S_x}$$

$$S_{\text{res}} = \sqrt{\frac{\left[\sum y^2 - \frac{(\sum y)^2}{n}\right] - b^2\left[\sum x^2 - \frac{(\sum x)^2}{n}\right]}{(n-2)}}$$

$$S_x = \sqrt{\left[\sum x^2 - \frac{(\sum x)^2}{n}\right]}$$

This test statistic should be compared with the critical value of the two tailed of the t distribution table and with the degree of freedom ($n-2$).

To find a 95% confidence interval for b by taking t standard errors on either side of the estimate.

$$b \pm t_{(1-\frac{\alpha}{2},n-2)} \times \mathrm{Se}(b)$$

Example 12.18 In Example 12.11 [intraocular pressure (IOP) and age of 11 patients] test whether the regression coefficient can be equal to 0? And find a 95% confidence interval for it.

Solution: In this example, degree of freedom is equal to df $= n-2 = 11-2 = 9$, $\beta_0 = 0$ and $\begin{cases} H_0 : \beta = \beta_0 \\ H_A : \beta \neq \beta_0 \end{cases}$. To find the standard error of residuals, we first calculate S_{res} and S_x.

$$S_{\mathrm{res}} = \sqrt{\frac{\left[\sum y^2 - \frac{(\sum y)^2}{n}\right] - b^2\left[\sum x^2 - \frac{(\sum x)^2}{n}\right]}{(n-2)}}$$

$$= \sqrt{\frac{\left[2340 - \frac{(158)^2}{11}\right] - 0.278^2\left[13316 - \frac{(374)^2}{11}\right]}{(11-2)}}$$

$$= 1.639$$

$$S_x = \sqrt{\left[\sum x^2 - \frac{(\sum x)^2}{n}\right]} = \sqrt{\left[13316 - \frac{(374)^2}{11}\right]} = 24.495$$

And

$$\mathrm{Se}(b) = \frac{S_{\mathrm{res}}}{S_x} = \frac{1.639}{24.495} = 0.067$$

Subsisting the above value gives us the test statistic.

$$t = \frac{\hat{\beta} - \beta_0}{\mathrm{Se}(b)} = \frac{0.278 - 0}{0.067} = 4.149$$

The above statistic is greater than the critical value of t with a degree of freedom of 9 and the null hypothesis is rejected.

$$4.149 > t_{(1-\frac{\alpha}{2},n-2)} = t_{(0.975,9)=2.262}$$

The 95% confidence interval for b is obtained as follows:

$$b \pm t_{(1-\frac{\alpha}{2},n-2)} \times Se(b) = 0.278 \pm 2.262 \times 0.067 = 0.278 \pm 0.151$$

$$= (0.127 \ \text{to} \ 0.429)$$

Because the above interval does not include the value specified in hypothesis zero (0), so there is a significant difference between the population β with zero.

12.3.3 Running Simple Linear Regression in SPSS

To calculate the simple linear regression in SPSS, we first select the Analyze\Regression\Linear path to open a window like the one below.

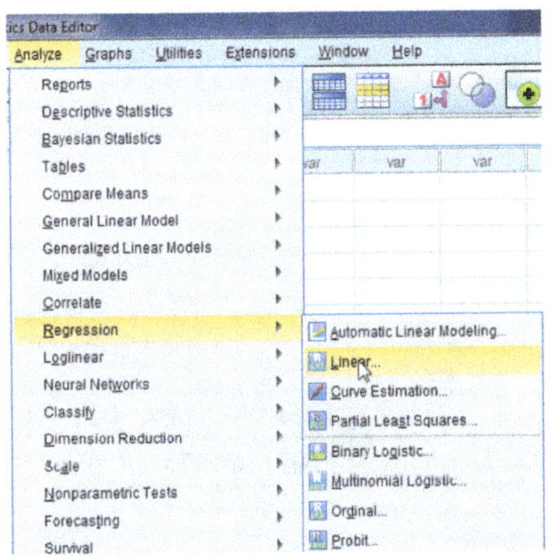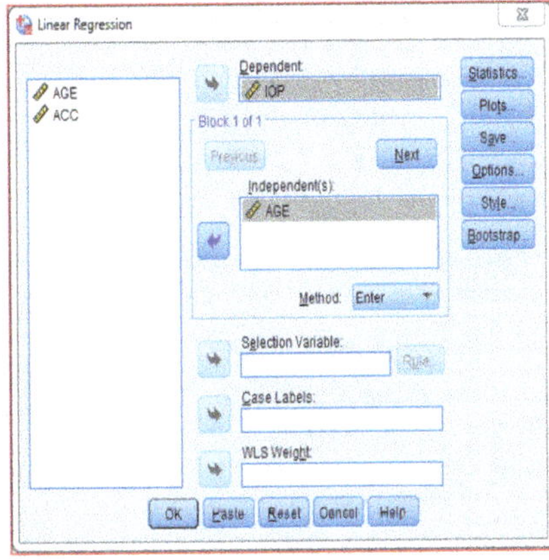

Select the independent variable of age from the left box and move to the Independent(s) box in the middle of the window. We also select the IOP-dependent variable from the same list and move to the Dependent box at the top of the middle of the window. To have the confidence interval of the regression coefficient, select the Statistics button at the top right to open another window. In this sub-window, we mark the Confidence interval option and close it by pressing the Continue button, and then we confirm the OK button to run the regression model and we will have the following output.

Unstandardized coefficients		Standardized coefficients		95.0% confidence interval for B			
B	Std. error	Beta		t	Sig.	Lower bound	Upper bound
(Constant)	4.900	2.323		2.110	0.064	−0.354	10.154
Age	0.278	0.067	0.812	4.169	0.002	0.127	0.429

As can be seen, the test statistic, regression coefficient, and confidence interval are equal to manual calculations.

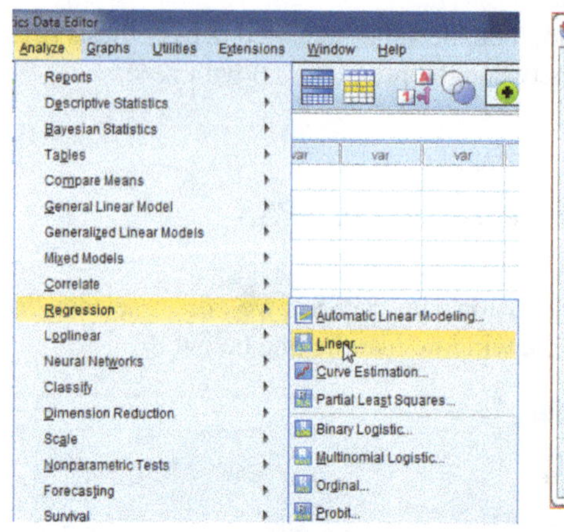
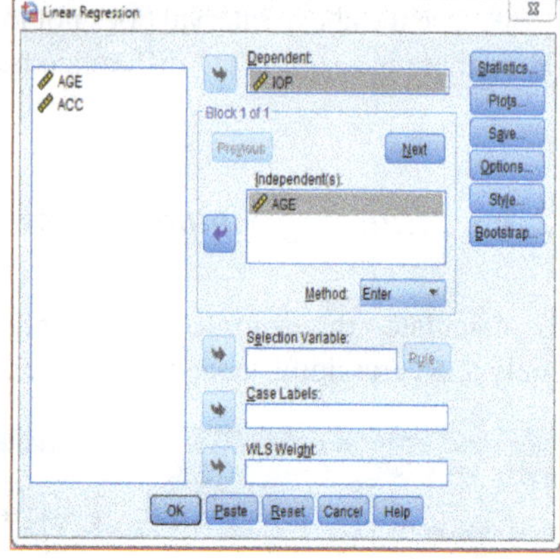

Unstandardized coefficients		Standardized coefficients	t	Sig	95.0% confidence interval for B	
B	Std. error	Beta			Lower bound	Upper bound
(Constant)	4.900	2.323		2.110	0.064	−0.354	10.154
AGE	0.278	0.067	0.812	4.169	0.002	0.127	0.429

Example 12.19 *Freezing of gait (FOG)* and Mini-Mental State Examination (MMSE) score in Parkinson's patients is in Table 12.15. Find the regression equation for FOG in terms of MMSE and calculate the test statistics and its confidence interval.

Solution: We summarize the required calculations in Table 12.16.

$$b = \frac{n. \sum xy - (\sum x)(\sum y)}{n. \sum x^2 - (\sum x)^2} = \frac{9 \times 1676 - 245 \times 63}{9 \times 6731 - (245)^2} = \frac{-351}{554} = -0.634$$

$$\bar{x} = \frac{245}{9} = 27.22, \bar{y} = \frac{63}{9} = 7.0$$

Table 12.15 *Freezing of gait (FOG)* and mini-mental state examination (MMSE) score in Parkinson's patients

MMSE	26	22	24	30	30	29	29	28	27
FOG	4	9	13	8	2	9	3	10	4

Table 12.16 Necessary calculations for estimating regression equation for data on Table 12.13

Patient	MMSE (x)	FOG (y)	x^2	y^2	$x.y$
1	26	5	676	25	130
2	22	9	484	81	198
3	24	13	576	169	312
4	30	8	900	64	240
5	30	2	900	4	60
6	29	9	841	81	261
7	29	3	841	9	87
8	28	10	784	100	280
9	27	4	729	16	108
Sum	$\sum x = 245$	$\sum y = 63$	$\sum x^2 = 6731$	$\sum y^2 = 549$	$\sum x.y = 1676$

$$a = \bar{y} - b.\bar{x} = 7.0 - (-0.634) \times 27.22 = 24.246$$

$$S_{res} = \sqrt{\frac{\left[\sum y^2 - \frac{(\sum y)^2}{n}\right] - b^2\left[\sum x^2 - \frac{(\sum x)^2}{n}\right]}{(n-2)}}$$

$$= \sqrt{\frac{\left[549 - \frac{(63)^2}{9}\right] - (-0.634)^2\left[6731 - \frac{(245)^2}{9}\right]}{(9-2)}}$$

$$= 3.449$$

$$S_x = \sqrt{\left[\sum x^2 - \frac{(\sum x)^2}{n}\right]} = \sqrt{\left[6731 - \frac{(245)^2}{9}\right]} = 7.846$$

and

$$Se(b) = \frac{S_{res}}{S_x} = \frac{3.449}{7.846} = 0.4396$$

Subsisting the above value gives us the test statistic.

$$t = \frac{\hat{\beta} - \beta_0}{Se(b)} = \frac{-0.634 - 0}{0.4396} = 1.442$$

The above statistic is less than the critical value of t with a degree of freedom of 7 and the null hypothesis is not rejected.

$$1.442 < t_{(1-\frac{\alpha}{2},n-2)} = t_{(0.975,9)=2.365}$$

The 95% confidence interval for b is obtained as follow:

$$b \pm t_{(1-\frac{\alpha}{2}, n-2)} \times Se(b) = -0.634 \pm 2.365 \times 0.4396$$
$$= -0.634 \pm 1.040$$
$$= (-1.674 \quad to \quad 0.406)$$

Because the above interval includes the value specified in hypothesis zero (0), so there is not a significant difference between the population β with zero.

12.4 Coefficient of Determination

As you're aware, the correlation coefficient assesses the linear relationship between two sets of data, indicating both similarity and dissimilarity. When squared (i.e., $R^2 = r^2$), it yields the coefficient of determination. This measure reveals the proportion of shared variance between two variables, often referred to as the common variance.

The coefficient of determination is valuable as it quantifies that how much of the variance in one variable (e.g., blood pressure) can be attributed to another variable (e.g., age). In essence, it represents the portion of the variance in the dependent variable (Y) that can be explained by a linear regression model involving the predictor variable (X). In simpler terms, it signifies the extent of shared variance between the two variables X and Y.

In general, a high R^2 value indicates that the model is a good fit for the data. The coefficient shows only the magnitude of linear association, not whether that association is statistically significant and it can never determine whether one variable is the cause of another variable.

The coefficient of determination takes values between 0 and 1. Since the coefficient of determination is a squared value, the direction of variations is meaningless. As shown in Fig. 12.4, a value of 0 indicates that the two variables have nothing in common to explain the variance of each other. A value of 1 says that the two variables are able to explain the entire variances for each other. An $R^2 = 0.36$ indicates that %36 of the variation in the outcome has been explained just by predictor included in the model.

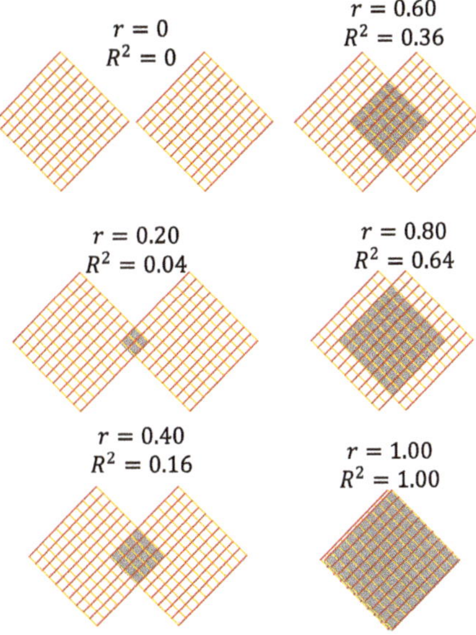

Subtracting the coefficient of determination from one, $(1 - R^2)$, gives the percentage of variance not shared between two variables, a quantity also called the *coefficient of alienation*.

12.5 Exercises

1. Body weight by kilograms and blood plasma volume (liters) of 8 healthy men were measured and are shown in Table 12.17.

   A. Is there a correlation between these two variables?
   B. If we consider the plasma volume as a dependent variable, find the regression line equation?
   C. From the regression line equation, predict plasma volume for people weighing 60, 65, and 70 kg.

Table 12.17 Body weight and blood plasma (liters) of 8 healthy men

Weight	58.0	70.0	74.0	63.5	62.0	70.5	71.0	66.3
Plasma volume	2.75	2.86	3.37	2.76	2.62	3.48	3.05	3.12

Hgb	PCV%	Age	Menopause
11.1	35	20	0
10.7	45	22	0
12.4	47	25	0
14.0	50	28	0
13.1	31	28	0
10.5	30	31	0
9.6	25	32	0
12.5	33	35	0
13.5	35	38	0
13.9	40	40	0
15.1	45	45	1
13.9	47	49	0
16.2	49	54	1
16.3	42	55	1
16,8	40	57	1
17.1	50	60	1
16.6	46	62	1
16.9	55	63	1
15.7	42	65	1
16.5	66	67	1

D. Can an individual plasma volume weighing 90 kg or 50 kg be estimated? Why?

2. Find the correlation between weight and arm circumference for the data in Table 9. 2. Then assume arm circumference as a dependent variable and obtain the arm circumference regression equation in terms of body weight. Then predict the value of the arm circumference for those who weigh 75 and 85 kg.

3. Is there a significant correlation between the two methods for the data in Table 11.15 (blood pressure of 16 patients)?

4. Draw a scatter diagram for data in Table 2.11. In which group is there a linear relationship between age and calcium value?

5. Calculate the correlation coefficient between age and calcium value for the data in Table 2.11 and for each group separately and test their significance.

6. The table below shows the level of hemoglobin (Hb), compressed cell volume (PCV), age and menopausal status (0 = not menopause and 1 = menopause).

A. Draw the scattergram of age against PCV and hemoglobin separately.
B. Estimate the correlation coefficient of age with PCV.
C. Is the estimated correlation coefficient significant?
D. Obtain hemoglobin level regression equation by age.
E. Is it permissible to consider the regression line coefficient as equal to zero?
F. Estimate point-biserial correlation coefficient between menopausal status and hemoglobin level.

    G. Estimate point-biserial correlation coefficient between PCV and hemoglobin level.

7. A psychiatrist assessed the level of depression and anxiety in 11 sample patients.

Anxiety	24	9	25	26	35	17	49	39	8	34	28
Depression	14	5	16	17	22	8	37	41	6	29	33

    A. Examine the scattergram of these two variables.
    B. Obtain the linear regression equation of anxiety in terms of depression.
    C. Can we consider the regression coefficient as equal to zero?
    D. Calculate the correlation coefficient of these two variables by Pearson, Spearman, and Kendall methods and test its equality with zero.
    E. If we consider anxiety as group 0 and depression as group 1, obtain the point-biserial correlation coefficient of two groups.

8. Table 12.18 shows the height of 20 adult girls and 20 adult boys and the Height of their mothers and fathers.

    A. First, obtain the correlation coefficient of height of children with father's height and mother's height separately (for 40 samples). Then compare these two correlation coefficients.

Table 12.18 Height of son, daughter, mother, and father

Gender	H. of daughter	H. of mother	H. of father	Gender	H. of son	H. of mother	H. of father
F	149	160	163	M	159	168	178
F	164	170	165	M	164	147	175
F	166	163	170	M	176	168	163
F	155	152	183	M	188	173	180
F	166	165	183	M	170	163	173
F	171	170	183	M	164	157	168
F	155	150	170	M	181	168	188
F	160	152	180	M	180	160	185
F	152	147	168	M	171	163	157
F	181	183	191	M	176	165	175
F	158	160	175	M	165	163	170
F	171	170	178	M	173	163	173
F	161	157	175	M	169	157	183
F	174	175	157	M	171	175	168
F	158	160	168	M	169	157	183
F	164	163	193	M	179	170	173
F	151	160	175	M	171	160	180
F	155	163	173	M	174	168	170
F	163	152	168	M	183	165	180
F	166	165	173	M	172	180	191

B. Find the correlation coefficient of height of adult girls with mother's height (20 samples) and also correlation coefficient of height of adult boys with father's height (20 samples) and scale them together.

Chapter 13
Analysis of Variance

13.1 Introduction

As we mentioned in Chap. 7, if the data is normally distributed, a two-sample t-test can be used to assess the significance of the difference between the means of two independent groups. To compare differences in the means of three or more independent groups simultaneously, an analysis of variance (ANOVA), which is a parametric test and an extension of the t-test, can be employed.

The purpose of the ANOVA is to compare the mean of the independent variable in three or more groups (as a dependent variable) simultaneously to see if the difference between the means of the groups is significant. Both a two-sample t-test and ANOVA can be used to compare the means of the two groups. The results of both tests will be the same, except that in the ANOVA instead of t statistic, it will give F statistic, which is the square of t statistic. But most researchers prefer the t-test because manual calculations are easier and more understandable, and they use ANOVA to compare the means of three groups and more.

To conduct an ANOVA, you need to have a categorical (ordinal or nominal) variable which includes at least two groups (e.g., marital status variable with the categories single, married, widow, and divorced) as the independent variable and a quantitative variable (e.g., no. of smoked cigarettes) as the dependent variable.

13.2 Multiple t-Tests or ANOVA

From the similarity of independent t-test results and ANOVA, this question may arise for novice researchers that why don't we just conduct several separate t-tests instead of an ANOVA test? To answer this question, suppose you want to compare cigarette smoking in terms of marital status; single (S), married (M), widow (W), and divorced (D). For this purpose, you have to repeat two independent samples t-tests six times:

© The Author(s), under exclusive license to Springer Nature Singapore Pte Ltd. 2024   321

S. H. Saneii and H. Doosti, *Practical Biostatistics for Medical and Health Sciences*,
https://doi.org/10.1007/978-981-97-3083-4_13

(1) singles with married ones, (2) singles with widows, (3) singles with divorced individuals, (4) marrieds with widows, (5) marrieds with divorced, and (6) widows with divorced.

Each time the t-test is performed, and the researcher decides whether there is a statistically significant difference between the means of the two groups. This decision is based on sampling variations and probability. Therefore, there is a slight chance you might be wrong every time you make a decision (probability of rejecting the true null hypothesis). That is, as the significance level increases, the confidence level comes down. The more tests are repeated, that is, the more decision-making about the significance of the t-tests, the greater the chances are that you will be wrong, and the lower the confidence level.

Now, if the probability of error in each test is equal to α and the probability of not making a mistake is equal to $1 - \alpha$ (confidence level), the probability of not rejecting the null hypothesis (the mean of the four groups are equal) is equal to the result of multiplying the confidence level of all six tests. More specifically, it can only be concluded that the mean of the 4 groups is equal if all six t-tests are not rejected. Because even if one of the comparisons is rejected, then the null hypothesis must be rejected. This means:

$$(1 - \alpha)(1 - \alpha)(1 - \alpha)(1 - \alpha)(1 - \alpha)(1 - \alpha)$$
$$= 0.95 \times 0.95 \times 0.95 \times 0.95 \times 0.95 \times 0.95$$
$$= 0.735.$$

This relationship shows that the more repeated t-tests are performed, the more likely the t-test to be significant (error) and the lower the confidence level. That is, decision-making comes with more error. This error is proportional to the number of t-tests summarized in Table 13.1.

In addition, as the number of groups to be compared increases, so does the number of t-tests. This increase is both time consuming and leads to computational error.

Considering the decrease in confidence level and increasing the significant level of α, a procedure should be employed to fix this problem (significant level) by adjusting the number of groups being compared. This procedure is called analysis of variance (ANOVA).

Table 13.1 Final error associated to the number of conducted t-tests on

No. of groups	2	3	4	5	6	7
No. of t-tests	1	3	6	10	15	21
Confidence level	0.95	0.902	0.735	0.599	0.463	0.341
Significant level	0.05	0.098	0.265	0.401	0.537	0.659

13.3 Mechanism of ANOVA

Analysis of variance (ANOVA) partitions the variability of individual data values into two separate components called "within group" and "between group" variations, which are called source of variations. Therefore, the total sum of squares (SS) is decomposed as ($SS_{Total} = SS_{Treatment} + SS_{Error} = SS_{Between} + SS_{Within}$), and similarly, the degree of freedom (df) can be partitioned as: $df_{Total} = df_{Between} + df_{within}$. Under the null hypothesis (all the population means are the same), it is expected that between group variance (variability between mean of groups) and within group variance (variability within each group) are influenced by chance and random variations. That is, their proportion must be equal to 1. Hence, the proportion of these two variances constitutes the test statistic (F). This statistic should be compared with the critical values in the corresponding F table with the degree of freedom for both numerator and denominator. The larger value of F, the greater and stronger evidence that the difference between means of groups is significant. This is why in ANOVA the study of the difference between the means is explained and performed by comparing the variances (instead of comparing the means).

In the particular case, when the number of groups studied is 2, the variance of the residuals in the ANOVA table is exactly equal to the pooled variance in the two-sample t-test, and the result of the comparison will be exactly equal to the two-sample t-test, because $F = t^2$.

In ANOVA method, the samples are divided into several groups and independent of each other, and equal sample size in each group is not mandatory. Therefore, the analysis of variance observations is considered in Table 13.2.

Table 13.2 Position of individuals studied for analysis of variance

		Different treatments					
		Group 1	**Group 2**		**Group j**		**Group k**
Subjects	Sample 1	y_{11}	y_{12}	$\cdots$	y_{1j}	$\cdots$	y_{1k}
	Sample 1	y_{21}	y_{22}	$\cdots$	y_{2j}	$\cdots$	y_{2k}
		$\vdots$	$\vdots$	$\vdots$	$\vdots$	$\vdots$	$\vdots$
	Sample i	y_{i1}	y_{i2}	$\cdots$	y_{ij}	$\cdots$	y_{ik}
		$\vdots$	$\vdots$	$\vdots$	$\vdots$	$\vdots$	$\vdots$
	Sample n	y_{in_1}	y_{in_2}	$\cdots$	y_{in_j}	$\cdots$	y_{in_k}

$N = n_1 + n_2 + \cdots + n_k$

> **Tip**
>
> One of the first applications of ANOVA was agricultural experiments that were performed in different plots of farmland under the treatment of different fertilizers, seed types, insecticides, etc. For this reason, the term "treatment" is used in ANOVA instead of between groups.

To better understand the concept of variance partitioning, between group and within group variations pay attention to the following example.

Example 13.1 In a study, 16 people with high blood pressure participated in a systolic blood pressure reduction treatment program. After one month, the reduction on their blood pressure is recorded in Table 13.3. The researcher has asked the statistician to analyze and interpret this data.

Solution: Obviously, the statistician must first find the mean and variance of these 16 subjects:

$$\bar{y} = \frac{\sum y}{n} = \frac{146}{16} = 9.125,$$

$$SS = \sum (y_i - \bar{y})^2$$
$$= (9 - 9.125)^2 + (11 - 9.125)^2 + \cdots + (13 - 9.125)^2 + (9 - 9.125)^2$$
$$= 113.75.$$

If the sum of the squares is divided by the total degree of freedom df $= 16-1 = 15$, the total variance is obtained.

In response to a statistician's question, the researcher said that these 16 subjects have participated in three different treatment programs, and each row of the above table is related to each treatment program.

The statistician then found it necessary to calculate the mean and sum of squares of each treatment program (row) separately.

Thus, the mean and sum of the squares of the first row are:

$$y_1 = \frac{\sum y_{i1}}{n} = \frac{66}{6} = 11,$$

Table 13.3 Reduction in systolic blood pressure in 16 subjects

9	11	10	13	12	11
6	8	7	4	5	–
7	10	11	13	9	–

$$SS_1 = \sum (y_i - \bar{y})^2$$
$$= (9-11)^2 + (11-11)^2 + (10-11)^2 + (13-11)^2 + (12-11)^2 + (11-11)^2$$
$$= 10.$$

The mean and sum of the squares of the second row are:

$$y_2 = \frac{\sum y_{i2}}{n} = \frac{30}{5} = 6,$$

$$SS_2 = \sum (y_i - \bar{y})^2$$
$$= (6-6)^2 + (8-6)^2 + (7-6)^2 + (4-6)^2 + (5-6)^2$$
$$= 10.$$

The mean and sum of the squares of the third row are:

$$y_3 = \frac{\sum y_{i3}}{n} = \frac{50}{5} = 10,$$

$$SS_3 = \sum (y_i - \bar{y})^2$$
$$= (7-10)^2 + (10-10)^2 + (11-10)^2 + (13-10)^2 + (9-10)^2$$
$$= 20.$$

S/he then added the sum of squares in each row to get the sum of squares of the within groups.

$$SS_W = SS_1 + SS_2 + SS_3 = 10 + 10 + 20 = 40.$$

If this sum of squares is divided by within the group's degree of freedom df $=$ $(6-1) + (5-1) + (5-1) = 13$, the within group variance is obtained.

In order to obtain the between group's sum of squares and the mean, he assumed that in the first group, there are 6 subjects with a mean of 11, in the second group, there are 5 subjects with a mean of 6, and in the third group, there are 5 subjects with a mean of 10. Therefore, the weighted mean of the groups is identical to the grand mean:

$$\bar{y} = \frac{(6 \times 11) + (5 \times 6) + (5 \times 10)}{6 + 5 + 5} = \frac{146}{16} = 9.125.$$

And the sum of squares for the means of these three groups (11, 6 and 10) is obtained as follows:

$$SS_B = \sum n_j (\bar{y}_j - \bar{y})^2$$
$$= 6 \times (11 - 9.125)^2 + 5 \times (6 - 9.125)^2 + 5 \times (10 - 9.125)^2$$
$$= 73.75.$$

If this sum of squares is also divided by the degree of freedom between the groups $df = 3 - 1 = 2$, the between group variance is obtained.

Thus, it can be seen that the total sum of squares was divided into two parts between groups and within groups.

This example will be completely solved later by the ANOVA technique.

13.4   Commonly Used Definitions in ANOVA

There are several technical terms that we encounter frequently when calculating and performing the ANOVA test. A brief explanation of few terms will be provided:

Sum of Squares (SS): Sum of squares (SS) is a statistical term used in regression analysis and ANOVA to identify the dispersion of data points as well as how well the data can fit the model in regression analysis. The sum of squares got its name because it first squares the difference of each observation from the grand mean, and then add up all squares ($SS = \sum (x_i - \bar{x})^2$). Therefore, the sum of squares is a measure of deviation from the mean.

Mean of Squares (MS): Mean squares represent an estimate of population variance and are calculated by dividing the sum of squares by the degrees of freedom.

Degrees of Freedom (df): Degrees of freedom refers to the numbers of independent values in the final calculation of a statistic that are logically free **and can have different values in a dataset**.

F **Statistic**: F statistic is the proportion of **the two variances to each other**. In ANOVA, F statistic is obtained by **dividing the mean squares between groups (MS_B) by the mean squares within groups (MS_W)**.

Source of Variations: In ANOVA, the total variations (variance) are divided into two components "between groups" and "within groups" in order to obtain F test statistic by dividing them. These two components are called the "source of variations".

Between Group Variability: It refers to variations between the distributions of individual groups (or levels) and measures variations in groups relative to the grand mean. If the distributions overlap or are closed, the grand mean will be similar to the individual means. If the distributions are close or overlapping, the total mean will be similar to the group means, while if the distributions are far apart, the difference between the mean and the total mean will be large.

Within Group Variability: It refers to the total deviations of the observations of each particular group (or levels) from the mean of that group as not all the values within each group are the same. The sum of the deviations of each observation from the mean of its own group shows the within group variations.

Error: When a statistical model is fitted to the observed data, it does not fit the actual data completely and generates a residual called error. But because ANOVA determines the amount of variations that can be attributed to the factor or treatment, that part of the variations that cannot be attributed to factors or groups are called errors or within the group variations.

13.5 Assumptions for One-Way ANOVA

A part of the process of implementing parameter models, including ANOVA, is checking and controlling the data you want to analyze to make sure that particular model (say, ANOVA) can really be used for that data. Regarding one-way ANOVA, it should also be checked whether the data is suitable for use, and this means that it meets the requirements and assumptions of ANOVA. To perform one-way analysis of variance, the following 5 assumptions must be considered:

Tips

One-way analysis of variance can be used with both balanced (equal sample sizes for all groups) and unbalanced (unequal sample sizes for all groups) designs. An advantage of balanced design is much less sensitive to violations of the assumption of homogeneity of variances than an unbalanced one.

It is recommended to use a balanced design whenever possible. When a balanced design is impossible to achieve, a slightly unbalanced design is preferable to a severely unbalanced one.

The **dependent variable** should be measured at quantitative scale (**interval** or **ratio level**) and **independent variable** should include **three or more categorical independent groups**.

You should have **independence of observations**, which means that there is no relationship between the observations in each group or between the groups themselves.

The dependent variable (Y) should be approximately normally distributed for each category of the independent variable (X). You can test easily for normality using the Shapiro–Wilks or Kolmogorov–Smirnov tests of normality using SPSS Statistics.

The variance of the dependent variable (Y) must be approximately equal for all categories of independent variables (X), which is called homogeneity of variances. You can test this assumption in SPSS statistics using Levene's test for homogeneity of variances.

There should be **no significant outliers**.

13.6 Hypotheses in ANOVA

In ANOVA, where the means of at least three groups are compared, the null hypothesis is usually explained as "the mean of the populations of these several groups are equal". However, the alternative hypothesis cannot be expressed as "all the means of several groups are different", but it must be said that "at least one of the means of these several groups is different from the others". Therefore, the null hypothesis (H_0) and the opposite hypothesis (H_A) are stated as follows:

$$\begin{cases} H_0 : \mu_1 = \mu_2 = \ldots = \mu_k \\ H_A : \mu_i \neq \mu_j : for \, \forall \, i \, \text{and} \, j \end{cases}$$

For a particular case where three groups are compared, the alternative hypothesis is more meaningful as follows:

$$H_A : \mu_1 \neq \mu_2 \quad \text{or} \quad \mu_1 \neq \mu_3 \quad \text{or} \quad \mu_2 \neq \mu_3.$$

In this test, only if the mean of one group is different from the other, then the whole null hypothesis will be rejected.

13.7 The ANOVA Table

The ANOVA is an abbreviation for **An**alysis **of Va**riance and, as mentioned, partitions and breaks down the total variance. Explaining the details of the one-way analysis of variance table (Table 13.4) can be useful to the reader's understanding of why the total variance partitioning is used to "compare mean of several groups".

$$CF = \frac{\left(\sum \sum y_{ij} \right)^2}{N},$$

$$SS_T = \sum_{i=1}^{k} \sum_{j=1}^{n_j} \left(y_{ij} - \bar{y} \right)^2 = \sum \sum y_{ij}^2 - CF,$$

Table 13.4 Main table of one-way ANOVA

Source of variation	Sum of square (SS)	Degree of freedom (df)	Mean of square (MS)	F
Intergroup, between	SS_B	$K - 1$	$MS_B = \frac{SS_B}{K-1}$	$\frac{MS_B}{MS_W}$
Intragroup, within	SS_W	$N - K$	$MS_W = \frac{SS_W}{N-K}$	
Total	SS_T	$N - 1$		$-$

$$SS_B = \sum_{j=1}^{k} n_j \left(\bar{y}_j - \bar{y}\right)^2 = \sum_{j=1}^{k} \frac{\left(\sum_{j=1}^{n_j} y_{ij}\right)^2}{j} - CF,$$

$$SS_W = \sum_{i=1}^{k} \sum_{j=1}^{n_j} \left(y_{ij} - \bar{y}_i\right)^2 = \sum \sum y_{ij}^2 - \sum_{j=1}^{k} \frac{\left(\sum_{j=1}^{n_j} y_{ij}\right)^2}{n_j}.$$

In this table, $\bar{y}_j$ and n_j are the mean and number of observations of jth, $\bar{y}$, N are the grand mean and total number of samples for all groups, K is the number of groups being compared, y_{ij} is the observed value of the ith sample from jth group, and the F is test statistic obtained from the proportion of two variances ($F = \frac{MS_B}{MS_W}$).

In this table, the significance level is determined using the p-value. To obtain it, the position of the statistic in the distribution to which it belongs must be known. In other words, there must be a distribution that is a criterion. This distribution is called the F distribution, the first letter of the famous statistician Ronald Fisher. That is why the F test is another name for the ANOVA test.

As mentioned earlier, the deviation of each observation from the grand mean is divided into two parts: "deviation of each observation from the mean of its own group" and "deviation of the mean of each group from the grand mean" as the following formula and Figs. 13.1 and 13.2.

$$y_{ij} - \bar{y} = y_{ij} + \left[-\bar{y}_i + \bar{y}_i\right] - \bar{y} = \left(y_{ij} - \bar{y}_i\right) + \left(\bar{y}_i - \bar{y}\right).$$

In the right Fig. 13.2, the grand mean of population is known, but it is not enough to explain all points (values) of the population alone. More accurately, based on the common characteristics of points with each other, three separate groups can be

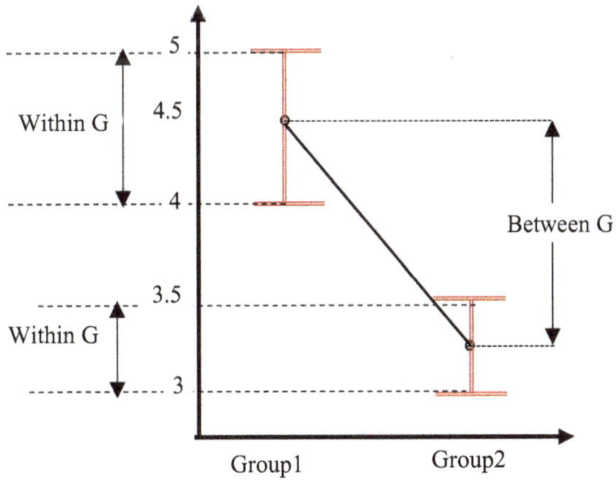

Fig. 13.1 Geometric representation of between group and within group deviations

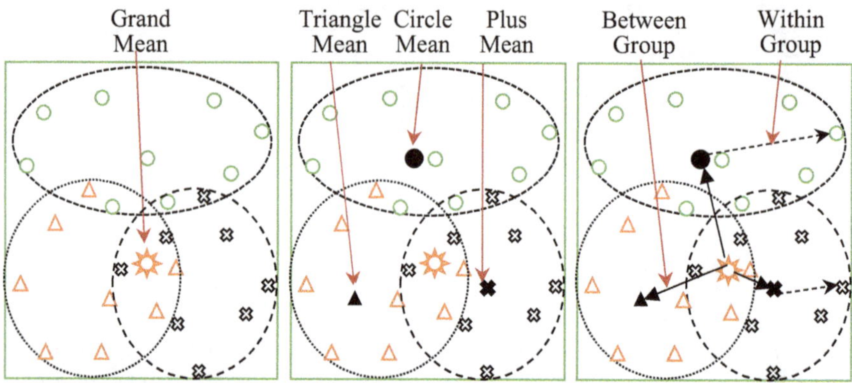

Fig. 13.2 Schematic representation of between and within group deviations

identified to better describe the population. Therefore, as shown in Fig. 13.2 (middle), the population is partitioned around three means. The whole population can now be explained in terms of the mean of each group.

In the analysis of variance, the mean deviation of each group from the grand mean (line arrow) is squared $((\bar{y}_i - \bar{y})^2)$ and multiplied by the sample size in each group (n_j) and is divided by the degree of freedom between groups $(k - 1)$ to obtain the between group variance (MS_B) and put it in the nominator of the fraction F. Then, we squared the data deviation of each group from the mean of its own group (dotted arrow) $((y_{ij} - \bar{y}_i)^2)$ and divide their sum by the degree of freedom within the group to obtain the intragroup variance (MS_W) and we put it in denominator of the fraction F.

The test statistic obtained from the ratio of between groups variance to within group variance has an F distribution (Fig. 13.3). If the calculated F is greater than the critical value of the 0.05 with degree of freedom $(df = (n - k, k - 1))$, the null hypothesis is rejected. That is, at least the mean of one group is different from the means of the other groups.

Fig. 13.3 Changes in the F distribution in terms of the degree of freedom of the nominator and the denominator of the fraction

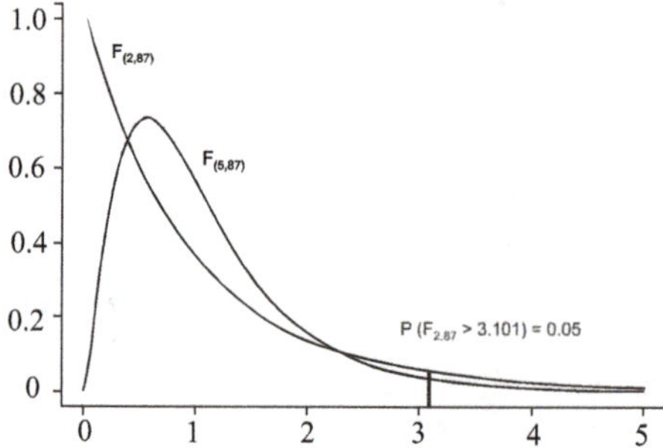

Example 13.2 A study was designed to compare the effect of three treatments on the mobility score of physiotherapy patients. In this study, 24 patients were randomly assigned to three groups: control (A), just physiotherapy (B), and physiotherapy with counseling (C). The results of this study are presented in Table 13.5. Is the mobility score of these 3 groups equal?

Solution: In this example $n_1 = n_2 = n_3 = 8$, $n = n_1 + n_2 + n_3 = 8 + 8 + 8 = 24$ and required calculations for the implementation of ANOVA are given on the right side of Table 9.5.

First, we obtain the correction factor (CF), which facilitates the rest of the computational process Table 13.6.

$$CF = \frac{\left(\sum\sum y_{ij}\right)^2}{n} = \frac{(1008)^2}{24} = 42336,$$

$$SS_T = \sum\sum y_{ij}^2 - CF = 43464 - 42336 = 1128,$$

$$SS_B - \sum \frac{\left(\sum_{j=1}^{n_j} y_{ij}\right)^2}{n_i} - CF = \frac{(280)^2}{8} + \frac{(360)^2}{8} + \frac{(368)^2}{8} - 42336 = 592,$$

$$SS_W = SS_T - SS_B = 1128 - 592 = 536.$$

We now substitute the above-calculated values in the ANOVA table as below:

The calculated F statistic (11.597) is greater than the critical value of $F(0.95, 2, 23) = 4.28$ in the F distribution tables, so the null hypothesis that the mean of the groups is equal is rejected.

Table 13.5 Mobility score in 3 groups

Sample	A	B	C	A^2	B^2	C^2
1	35	38	47	1225	1444	2209
2	38	43	53	1444	1849	2809
3	42	45	42	1764	2025	1764
4	34	52	45	1156	2704	2025
5	28	40	46	784	1600	2116
6	39	46	37	1521	2116	1369
7	34	54	48	1156	2916	2304
8	30	42	50	900	1764	2500
Sum	$\sum y_{i1} = 280$	$\sum y_{i2} = 360$	$\sum y_{i3} = 368$	9950	16418	17096
	$\sum\sum y_{ij} = 1008$			$\sum\sum y_{ij}^2 = 43464$		

Table 13.6 One-way ANOVA table for mobility score

Source of variation	Sum of square (SS)	Degree of freedom (DF)	Mean of square (MS)	F
Between	592	2	296	11.597
Within	536	21	25.52	–
Total	1128	23		–

Table 13.7 Blood pressure reduction in 16 subjects after 3 treatments

Medication	Exercise	Herbal diet
9	6	7
11	8	10
10	7	11
13	4	13
12	5	9
11	–	–

Example 13.3 A research project is designed to investigate the effect of 3 different treatments on reduction of systolic the blood pressure of individuals diagnosed with high blood pressure. They are randomly allocated to three groups; (1) medication, (2) exercises, and (3) an herbal diet. After a month, the reduction in each person's blood pressure is recorded and is shown in Table 13.7. Is there difference among the means of 3 treatments?

Solution: In this example $n_1 = 6, n_2 = 5, n_3 = 5, n = n_1 + n_2 + n_3 = 6 + 5 + 5 = 16$ and required calculations for the implementation of ANOVA are given on the right side of Table 13.8.

First, we obtain the correction factor (CF), which facilitates the rest of the computational process Table 13.9.

Table 13.8 Necessary calculation for ANOVA

Sample	Med	Exe	Herb	Med^2	Exe^2	Herb^2
1	9	6	7	81	36	49
2	11	8	10	121	64	100
3	10	7	11	100	49	121
4	13	4	13	69	16	169
5	12	5	9	144	25	81
6	11	–	–	121	–	–
Sum	$\sum y_{i1} = 66$	$\sum y_{i2} = 30$	$\sum y_{i3} = 50$	736	190	520
	$\sum\sum y_{ij} = 146$			$\sum\sum y_{ij}^2 = 1446$		

Table 13.9 One-way ANOVA table for mobility score

Source of variation	Sum of square (SS)	Degree of freedom (df)	Mean of square (MS)	F
Between	73.75	2	36.875	11.984
Within	40	13	3.077	–
Total	113.75	15		–

$$CF = \frac{\left(\sum\sum y_{ij}\right)^2}{n} = \frac{(146)^2}{16} = 1332.25,$$

$$SS_T = \sum\sum y_{ij}^2 - CF = 1446 - 1332.25 = 113.75,$$

$$SS_B = \sum \frac{\left(\sum_{i=1}^{n_j} y_{ij}\right)^2}{n_j} - CF = \frac{(66)^2}{6} + \frac{(30)^2}{5} + \frac{(50)^2}{5} - 1332.25 = 73.75,$$

$$SS_W = SS_T - SS_B = 113.75 - 73.75 = 40.$$

We now substitute the above-calculated values in the ANOVA table as below:

The calculated F statistic (11.984) is greater than the critical value of $F(0.95, 2, 13) = 6.70$ in the F distribution tables, so the null hypothesis that the mean of the groups is equal is rejected.

13.8 Running ANOVA on SPSS

To run a one-way analysis of variance in SPSS, all quantitative data (dependent variable) must be entered in one column and the names of groups (independent variable or factor) in another column. Then select the path of Analyze\Compare Means\One-way ANOVA to open a windows like Fig. 13.4.

Select the name of the dependent variable (in this example Mobility) and move it to the right box titled Dependent List. We move the name of the multilevel group or factor (say, treatment) that is qualitative (nominal or ordinal) to the Factor box. Then by selecting options, another window will open. In this window, by selecting the option of "Homogeneity of variance test", Levene statistic is calculated in order to test the equality of the variance of the groups. Also, by selecting the "Fixed and random effects" option, it calculates the standard error, standard deviation, 95% confidence interval for the fixed effects model.

The most important output of SPSS software for ANOVA is the following two tables.

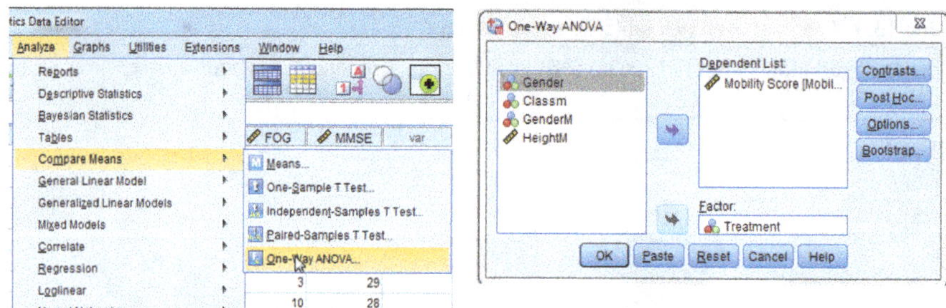

Fig. 13.4 Path and main windows of one-way ANOVA in SPSS

Test of homogeneity of variances

		Levene statistic	df1	df2	Sig.
Mobility score	Based on mean	0.161	2	21	0.852
	Based on median	0.152	2	21	0.860
	Based on median and with adjusted df	0.152	2	20.444	0.860
	Based on trimmed mean	0.161	2	21	0.852

The p-value in this test was $p = 0.852$ and is more than the significance level of 0.05. Therefore, the hypothesis of the p-value in this test was 0.852 $p = 0.852$ and more than the significance level of 0.05. Therefore, the hypothesis of equal group variance is not rejected and the condition of homogeneity of variances for ANOVA is established.

The main table of ANOVA will be as follows

ANOVA

Mobility score

	Sum of squares	df	Mean square	F	Sig.
Between groups	**592.000**	2	296.000	11.597	.000
Within groups	**536.000**	21	25.524		
Total	**1128.000**	23			

The calculations and conclusions are exactly the same as the manual calculations performed above.

The main table of ANOVA for Example 13.2 is as follows

ANOVA

Reduction

	Sum of squares	df	Mean square	F	Sig.
Between groups	73.750	2	36.875	11.984	.001
Within groups	40.000	13	3.077		
Total	113.750	15			

13.9 Identifying Which Means is Different

Analysis of variance (ANOVA) is the most common method of comparing the means of several groups. After conducting ANOVA, if there is no reason to reject the null hypothesis, there will be no more questions and the research question will be fully answered. However, if the rejection of the null hypothesis and the significance of the mean of the groups are concluded, according to ANOVA, it is not possible to say which particular groups are different from each other. In order to reach the full answer to the question, the researcher needs to know which pair of groups led to such a result. Is only one group different from the other groups or are the groups significantly different from each other?

For this purpose, empirical (informal) methods such as constructing confidence interval and boxplots or argumentative (formal) methods such as multiple comparison tests and range tests can be used.

Calculate confidence interval estimates of the means for each of the different groups, and then check if one or more of them does overlap with the others.

Construct boxplots for different groups, and then examine if one or more of them does overlap with the others.

Range tests allow user to identify subsets of means that are not significantly different from each other.

Multiple comparison tests (such as Bonferroni, Duncan, Tukey, Scheffe, Dunnet, Fisher LSD, and Student–Newman–Keuls) using pairs of means make adjustments to overcome the problem of increasing a significance level when the number of tests increases. These tests are usually selected based on whether or not the variance of the groups compared is equal, whether the sample size of the groups is equal or not, and whether the number of comparisons is low or high.

Example 13.4 Subjects have to press a button when they hear a signal. Three treatment signals with different intensities (low, moderate, and severe) are given to hearing problem patients. The time elapsed between presentation of the signal and button is pushed by patients in hundredths of a second are recorded in the table below. Are these three treatments different? If so, which groups are significantly different from the others?

Signal	Observations				
Low	27	24	28	21	32
Moderate	19	14	15	13	–
Severe	21	18	22	14	16

Solution: In this example $n_1 = 5, n_2 = 4, n_3 = 5, n = n_1 + n_2 + n_3 = 5 + 4 + 5 = 14$ and required calculations for the implementation of ANOVA are given on the right side of Table 13.10.

First, we obtain the correction factor (CF), which facilitates the rest of the computational process.

$$CF = \frac{\left(\sum\sum y_{ij}\right)^2}{n} = \frac{(284)^2}{14} = 5761.143,$$

$$SS_T = \sum\sum y_{ij}^2 - CF = 6206 - 5761.143 = 444.857,$$

$$SS_B = \sum \frac{\left(\sum_{i=1}^{n_j} y_{ij}\right)^2}{n_i} - CF$$

$$= \frac{(132)^2}{5} + \frac{(61)^2}{4} + \frac{(91)^2}{5} - 5801.786$$

$$= 310.107,$$

$$SS_W = SS_T - SS_B = 444.857 - 310.107 = 134.75.$$

We now substitute the above-calculated values in the ANOVA table as below (Table 13.11):

The calculated F statistic (12.657) is greater than the critical value of $F(0.95, 2, 11) = 7.21$ in the F distribution tables, so the null hypothesis that the mean of the groups is equal is rejected Table 13.11.

Table 13.10 Necessary calculations for ANOVA

Signal	Observations					Sum	
Low	27	24	28	21	32	$\sum y_{i1} = 132$	$\sum\sum y_{ij} = 284$
Moderate	19	14	15	13	–	$\sum y_{i2} = 61$	
Severe	21	18	22	14	16	$\sum y_{i3} = 91$	
$(Low)^2$	729	576	784	441	1024	3554	$\sum\sum y_{ij}^2 = 6206$
$(Moderate)^2$	361	196	225	169	–	951	
$(Severe)^2$	441	324	484	196	256	1701	

Table 13.11 One-way ANOVA table for mobility score

Source of variation	Sum of square (SS)	Degree of freedom (df)	Mean of square (MS)	F
Between	310.107	2	155.054	12.657
Within	134.750	11	12.250	–
Total	444.857	13		–

The SPSS output for this comparison is shown below:

ANOVA

MlSec

	Sum of squares	df	Mean square	*F*	Sig.
Between groups	310.107	2	155.054	12.657	0.001
Within groups	134.750	11	12.250		
Total	444.857	13			

In order to identify which groups' mean was different from the others, we first draw the boxplot for all three groups.

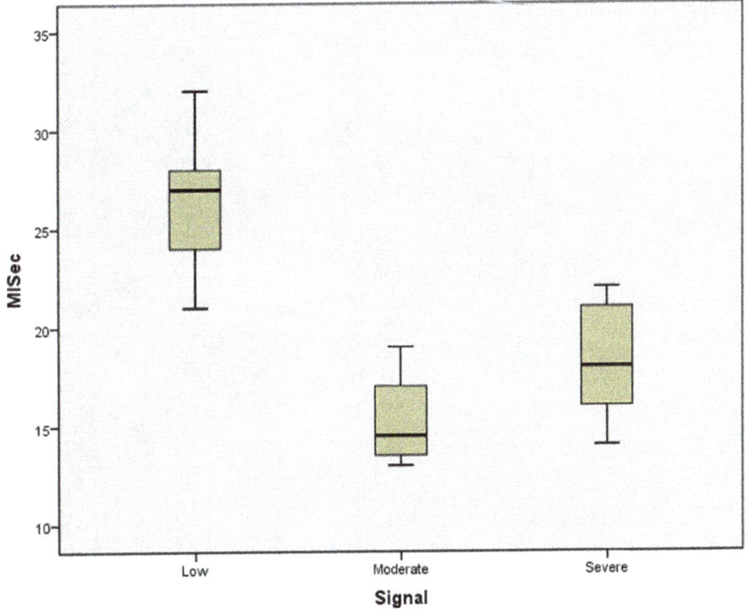

The 95% confidence interval for these three groups is:

Descriptives

	Signal		Lower	Upper
MlSec	Low	95% confidence interval for mean	21.24	31.56
	Moderate	95% confidence interval for mean	11.07	19.43
	Severe	95% confidence interval for mean	14.04	22.36

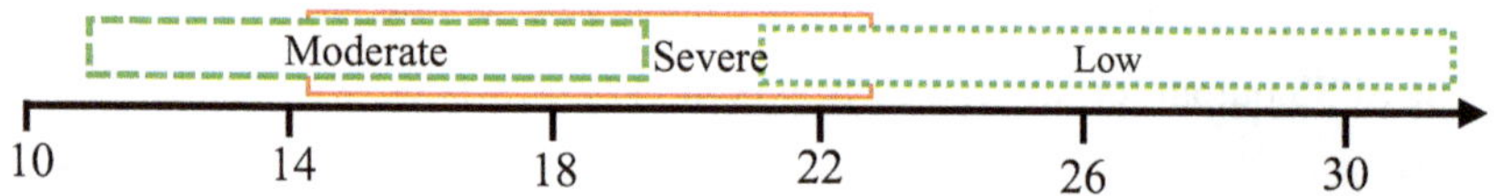

As can be seen from the diagrams above, the difference between low signal and medium signal and with high signal is significant.

Tukey, Fisher LSD, and Bonferroni post hoc tests also show differences.

Multiple comparisons

Dependent variable: MlSec

	(*I*) Signal	(*J*) Signal	Mean difference (*I−J*)	Std. error	Sig.	95% confidence interval	
						Lower	Upper
Tukey HSD	Low	Moderate	11.150[*]	2.348	.002	4.81	17.49
		Severe	8.200[*]	2.214	.009	2.22	14.18
	Moderate	Low	−11.150[*]	2.348	.002	−17.49	−4.81
		Severe	−2.950	2.348	.447	−9.29	3.39
	Severe	Low	−8.200[*]	2.214	.009	−14.18	−2.22
		Moderate	2.950	2.348	.447	−3.39	9.29
LSD	Low	Moderate	11.150[*]	2.348	.001	5.98	16.32
		Severe	8.200[*]	2.214	.003	3.33	13.07
	Moderate	Low	−11.150[*]	2.348	.001	−16.32	−5.98
		Severe	−2.950	2.348	.235	−8.12	2.22
	Severe	Low	−8.200[*]	2.214	.003	−13.07	−3.33
		Moderate	2.950	2.348	.235	−2.22	8.12
Bonferroni	Low	Moderate	11.150[*]	2.348	.002	4.53	17.77
		Severe	8.200[*]	2.214	.010	1.96	14.44
	Moderate	Low	−11.150[*]	2.348	.002	−17.77	−4.53
		Severe	−2.950	2.348	.705	−9.57	3.67
	Severe	Low	−8.200[*]	2.214	.010	−14.44	−1.96
		Moderate	2.950	2.348	.705	−3.67	9.57

[*]. The mean difference is significant at the 0.05 level.

13.10 Exercises

1. The pulse rates (beats per minute) for 3 age group of men are shown in Table 13.12. Are the difference between the mean of these groups for men significant?
2. The pulse rates (beats per minute) for 4 age groups of women are shown in Table 13.13. Are the difference between the mean of these groups for women significant?
3. A college instructor conducted an experiment to determine whether the teaching method affects the amount that students learn. Three instructional methods were used in the study: a traditional lecture (in person) method, an online method and a third method that combined in person, and online method. Two classes were taught with each method. Four students were sampled from each class. Their final exam scores are shown in Table 13.14. Compare the difference between 3 teaching methods.

   A researcher investigated five different techniques to lower the blood pressure of individuals with high blood pressure. The subjects are randomly assigned to 5 groups; the first one takes medication, the second one exercises, the third group follows a special diet, the fourth group takes medication and exercises,

Table 13.12 Pulse rates (beats per minute) for 3 age groups of men

Age groups	Pulse rates (beats per minute)
20–30	80–66–84–82–104–82–78–80–70
31–50	80–74–86–96–98–88–82–72–66–80
51–80	100–72–90–86–64–72–94–82–72–72

Table 13.13 Final exam scores for 3 teaching methods

Age groups	Pulse rates (beats per minute)
20–30	64–78–72–71–63–65–70–73–64
31–50	76–80–90–58–74–96–72–58–66–80–92
51–70	52–52–82–82–60–52–74
71+	54–72–62–55–61–59

Table 13.14 Pulse rates (beats per minute) for 4 age groups of women

Teaching methods		
In person	Online	Combination
62–74–68–69–75–77–82–74	66–66–81–75–76–84–90–93	87–75–69–84–75–87–89–94

and the fifth group follows diet and exercises. After a month, the reduction in each person's blood pressure is recorded. At sig. level of 0.05, test the claim that there is no difference among the means of five methods. The data is shown in Table 13.15.

4. Number of outpatient visit to a private clinic for last 15 days is shown in Table 13.16. The research question is "Do people with private health insurance visit their physicians more frequently than people with no insurance or other types of insurance"? This question will be answered by using data from a study of 57 adult attendees of a community-based health fair.

Table 13.15 Pulse rates (beats per minute) for 4 age groups of women

Teaching methods

Medication	Exercises	Diet	Diet+ exercise	Medication +exercise
10	6	5	3	9
12	8	9	6	9
9	3	12	7	8
13	0	8	6	11
15	2	4	2	8
–	–		1	6

Table 13.16 Number of outpatient visit to private clinic

No insurance	Private insurance	Social insurance	Military insurance
0	1	2	1
1	2	2	2
1	3	3	2
1	3	3	3
2	4	4	3
2	4	4	4
2	4	4	4
2	5	4	4
3	5	5	5
3	5	5	5
3	6	6	5
4	6	7	6
4	7	8	7
5	8	–	8
6	9	–	–

Appendix A
Selected Statistical Function in Excel

Function in excel	Descriptions
AVERAGE	Mean/average (arithmetic mean) of specified data
BINOMDIST	Returns the individual term binomial distribution probability
CHIDIST	Returns the right-tailed probability of the Chi-squared distribution
CHIINV	Returns the inverse of the right-tailed probability of the Chi-squared distribution. If probability = CHIDIST(x,…), then CHIINV(probability,…) = x. Use this function to compare observed results with expected ones in order to decide whether your original hypothesis is valid
CHITEST	Returns the test for independence. CHITEST returns the value from the Chi-squared (χ^2) distribution for the statistic and the appropriate degrees of freedom
CONFIDENCE.NORM	Returns the confidence interval for a population mean, using a normal distribution. The confidence interval is a range of values
CONFIDENCE.T	Returns the confidence interval for a population mean, using a Student's t-distribution
CORREL	Returns the correlation coefficient of the array1 and array2 cell ranges. Use the correlation coefficient to determine the relationship between two properties
CRITBINOM	Returns the smallest value for which the cumulative binomial distribution is greater than or equal to a criterion value. Use this function for quality assurance applications
FORECAST	Calculates, or predicts, a future value by using existing values. The predicted value is a y-value for a given x-value. The known values are existing x-values and y-values, and the new value is predicted by using linear regression
FTEST	An F test returns the two-tailed probability that the variances in array1 and array2 are not significantly different. Use this function to determine whether two samples have different variances

(continued)

(continued)

Function in excel	Descriptions
INTERCEPT	Calculates the point at which a line will intersect the Y-axis by using existing x-values and y-values. The intercept point is based on a best-fit regression line plotted through the known x-values and known y-values. Use the INTERCEPT function when you want to determine the value of the dependent variable when the independent variable is 0 (zero)
KURT	Returns the kurtosis of a dataset. Kurtosis characterizes the relative peakedness or flatness of a distribution compared with the normal distribution. Positive kurtosis indicates a relatively peaked distribution. Negative kurtosis indicates a relatively flat distribution
LOG	Returns the logarithm of a number to the base you specify
MAX	Returns the largest value in a set of values
MEDIAN	Returns the median of the given numbers. The median is the number in the middle of a set of numbers (50% lower than and 50% upper than)
MIN	Returns the smallest number in a set of values
MODE	Returns the most frequently occurring, or repetitive, value in an array or range of data
NORMDIST	Returns the normal distribution for the specified mean and standard deviation. This function has a very wide range of applications in statistics, including hypothesis testing
NORMINV	Returns the inverse of the normal cumulative distribution for the specified mean and standard deviation
NORMSDIST	Returns the standard normal cumulative distribution function. The distribution has a mean of 0 (zero) and a standard deviation of one. Use this function in place of a table of standard normal curve areas
NORMSINV	Returns the inverse of the standard normal cumulative distribution. The distribution has a mean of zero and a standard deviation of one
PERCENTILE	Returns the k-th percentile of values in a range. You can use this function to establish a threshold of acceptance. For example, you can decide to examine candidates who score above the 90th percentile
Poisson.DIST	Returns the Poisson distribution. A common application of the Poisson distribution is predicting the number of events over a specific time, such as the number of cars arriving at a toll plaza in 1 min
QUARTILE	Returns the quartile of a dataset. Quartiles often are used in sales and survey data to divide populations into groups. For example, you can use QUARTILE to find the top 25 percent of incomes in a population
RSQ	Returns the square of the Pearson product moment correlation coefficient through data points in known_y's and known_x's. For more information, see the PEARSON function. The r-squared value can be interpreted as the proportion of the variance in y attributable to the variance in x

(continued)

(continued)

Function in excel	Descriptions
SKEW	Returns the skewness of a distribution. Skewness characterizes the degree of asymmetry of a distribution around its mean. Positive skewness indicates a distribution with an asymmetric tail extending toward more positive values. Negative skewness indicates a distribution with an asymmetric tail extending toward more negative values
SLOPE	Returns the slope of the linear regression line through data points in known_y's and known_x's. The slope is the vertical distance divided by the horizontal distance between any two points on the line, which is the rate of change along the regression line
SQRT	Returns a positive square root
STANDARDIZE	Returns a normalized value from a distribution characterized by mean and standard_dev
STDEV.P	Calculates standard deviation based on the entire population given as arguments. The standard deviation is a measure of how widely values are dispersed from the average value (the mean)
STDEV.S	Estimates standard deviation based on a sample data. The standard deviation is a measure of how widely values are dispersed from the average value (the mean)
SUM	The SUM function adds all the numbers that you specify as arguments. Each argument can be a range (range: two or more cells on a sheet. The cells in a range can be adjacent or non-adjacent), a cell reference (for example, B3), an array, a formula (for example, SUM(A1:A5) adds all the numbers that are contained in cells A1 through A5. For another example, SUM(A1, A3, A5) adds the numbers that are contained in cells A1, A3, and A5)
TDIST	Returns the Percentage Points (probability) for the Student t-distribution where a numeric value (x) is a calculated value of t for which the Percentage Points are to be computed. The t-distribution is used in the hypothesis testing of small sample datasets. Use this function in place of a table of critical values for the t-distribution
TINV	Returns the two-tailed inverse of the Student's t distribution
TTEST	Returns the probability associated with a Student's t-test. Use TTEST to determine whether two samples are likely to have come from the same two underlying populations that have the same mean
TREND	Returns values along a linear trend. Fits a straight line (using the method of least squares) to the arrays known_y's and known_x's. Returns the y-values along that line for the array of new_x's that you specify
VAR.P	Calculates variance based on the entire population data

(continued)

(continued)

Function in excel	Descriptions
VAR.S	Estimates variance based on a sample data
ZTEST	Returns the one-tailed p-value of a z test. For a given hypothesized population mean, $\mu 0$, ZTEST returns the probability that the sample mean would be greater than the average of observations in the dataset (array)—that is, the observed sample mean

Appendix B
Common Statistical Tables

Table B.1 Area under the standard normal curve between **0 (mean)** and z

z	0.00	0.01	0.02	0.03	0.04	0.05	0.06	0.07	0.08	0.09
0.0	0.0000	0.0040	0.0080	0.0120	0.0160	0.0199	0.0239	0.0279	0.0319	0.0359
0.1	0.0398	0.0438	0.0478	0.0517	0.0557	0.0596	0.0636	0.0675	0.0714	0.0753
0.2	0.0793	0.0832	0.0871	0.0910	0.0948	0.0987	0.1026	0.1064	0.1103	0.1141
0.3	0.1179	0.1217	0.1255	0.1293	0.1331	0.1368	0.1406	0.1443	0.1480	0.1517
0.4	0.1554	0.1591	0.1628	0.1664	0.1700	0.1736	0.1772	0.1808	0.1844	0.1879
0.5	0.1915	0.1950	0.1985	0.2019	0.2054	0.2088	0.2123	0.2157	0.2190	0.2224
0.6	0.2257	0.2291	0.2324	0.2357	0.2389	0.2422	0.2454	0.2486	0.2517	0.2549
0.7	0.2580	0.2611	0.2642	0.2673	0.2704	0.2734	0.2764	0.2794	0.2823	0.2852
0.8	0.2881	0.2910	0.2939	0.2967	0.2995	0.3023	0.3051	0.3078	0.3106	0.3133
0.9	0.3159	0.3186	0.3212	0.3238	0.3264	0.3289	0.3315	0.3340	0.3365	0.3389
1.0	0.3413	0.3438	0.3461	0.3485	0.3508	0.3531	0.3554	0.3577	0.3599	0.3621
1.1	0.3643	0.3665	0.3686	0.3708	0.3729	0.3749	0.3770	0.3790	0.3810	0.3830
1.2	0.3849	0.3869	0.3888	0.3907	0.3925	0.3944	0.3962	0.3980	0.3997	0.4015
1.3	0.4032	0.4049	0.4066	0.4082	0.4099	0.4115	0.4131	0.4147	0.4162	0.4177
1.4	0.4192	0.4207	0.4222	0.4236	0.4251	0.4265	0.4279	0.4292	0.4306	0.4319
1.5	0.4332	0.4345	0.4357	0.4370	0.4382	0.4394	0.4406	0.4418	0.4429	0.4441
1.6	0.4452	0.4463	0.4474	0.4484	0.4495	0.4505	0.4515	0.4525	0.4535	0.4545
1.7	0.4554	0.4564	0.4573	0.4582	0.4591	0.4599	0.4608	0.4616	0.4625	0.4633

(continued)

© The Editor(s) (if applicable) and The Author(s), under exclusive license
to Springer Nature Singapore Pte Ltd. 2024
S. H. Saneii and H. Doosti, *Practical Biostatistics for Medical and Health Sciences*,
https://doi.org/10.1007/978-981-97-3083-4

Table B.1 (continued)

z	0.00	0.01	0.02	0.03	0.04	0.05	0.06	0.07	0.08	0.09
1.8	0.4641	0.4649	0.4656	0.4664	0.4671	0.4678	0.4686	0.4693	0.4699	0.4706
1.9	0.4713	0.4719	0.4726	0.4732	0.4738	0.4744	0.4750	0.4756	0.4761	0.4767
2.0	0.4772	0.4778	0.4783	0.4788	0.4793	0.4798	0.4803	0.4808	0.4812	0.4817
2.1	0.4821	0.4826	0.4830	0.4834	0.4838	0.4842	0.4846	0.4850	0.4854	0.4857
2.2	0.4861	0.4864	0.4868	0.4871	0.4875	0.4878	0.4881	0.4884	0.4887	0.4890
2.3	0.4893	0.4896	0.4898	0.4901	0.4904	0.4906	0.4909	0.4911	0.4913	0.4916
2.4	0.4918	0.4920	0.4922	0.4925	0.4927	0.4929	0.4931	0.4932	0.4934	0.4936
2.5	0.4938	0.4940	0.4941	0.4943	0.4945	0.4946	0.4948	0.4949	0.4951	0.4952
2.6	0.4953	0.4955	0.4956	0.4957	0.4959	0.4960	0.4961	0.4962	0.4963	0.4964
2.7	0.4965	0.4966	0.4967	0.4968	0.4969	0.4970	0.4971	0.4972	0.4973	0.4974
2.8	0.4974	0.4975	0.4976	0.4977	0.4977	0.4978	0.4979	0.4979	0.4980	0.4981
2.9	0.4981	0.4982	0.4982	0.4983	0.4984	0.4984	0.4985	0.4985	0.4986	0.4986
3.0	0.4987	0.4987	0.4987	0.4988	0.4988	0.4989	0.4989	0.4989	0.4990	0.4990
3.1	0.4990	0.4991	0.4991	0.4991	0.4992	0.4992	0.4992	0.4992	0.4993	0.4993
3.2	0.4993	0.4993	0.4994	0.4994	0.4994	0.4994	0.4994	0.4995	0.4995	0.4995
3.3	0.4995	0.4995	0.4995	0.4996	0.4996	0.4996	0.4996	0.4996	0.4996	0.4997
3.4	0.4997	0.4997	0.4997	0.4997	0.4997	0.4997	0.4997	0.4997	0.4997	0.4998

Negative values can be calculated based on symmetrical

Table B.2 Percentage points of the standard normal distribution

P values	Percentage points	
	Two-tailed	One-tailed
0.5	0.67	0
0.4	0.84	0.25
0.3	1.04	0.52
0.2	1.28	0.84
0.1	1.64	1.28
0.05	1.96	1.64
0.02	2.33	2.05
0.01	2.58	2.33
0.005	2.81	2.58
0.002	3.09	2.88
0.001	3.29	3.09
0.0001	3.89	3.72

Table B.3 Percentage points of student's *t*-distribution

df	One tail, *A*	0.10	0.05	0.025	0.01	0.005
	Two tails, *A*	0.20	0.10	0.05	0.02	0.01
1		3.078	6.314	12.706	31.821	63.657
2		1.886	2.920	4.303	6.965	9.925
3		1.638	2.353	3.182	4.541	5.841
4		1.533	2.132	2.776	3.747	4.604
5		1.476	2.015	2.571	3.365	4.032
6		1.440	1.943	2.447	3.143	3.707
7		1.415	1.895	2.365	2.998	3.499
8		1.397	1.860	2.306	2.896	3.355
9		1.383	1.833	2.262	2.821	3.250
10		1.372	1.812	2.228	2.764	3.169
11		1.363	1.796	2.201	2.718	3.106
12		1.356	1.782	2.179	2.681	3.055
13		1.350	1.771	2.160	2.650	3.012
14		1.345	1.761	2.145	2.624	2.977
15		1.341	1.753	2.131	2.602	2.947
16		1.337	1.746	2.120	2.583	2.921
17		1.333	1.740	2.110	2.567	2.898
18		1.330	1.734	2.101	2.552	2.878
19		1.328	1.729	2.093	2.539	2.861
20		1.325	1.725	2.086	2.528	2.845
21		1.323	1.721	2.080	2.518	2.831
22		1.321	1.717	2.074	2.508	2.819
23		1.319	1.714	2.069	2.500	2.807
24		1.318	1.711	2.064	2.492	2.797
25		1.316	1.708	2.060	2.485	2.787
26		1.315	1.706	2.056	2.479	2.779
27		1.314	1.703	2.052	2.473	2.771
28		1.313	1.701	2.048	2.467	2.763
29		1.311	1.699	2.045	2.462	2.756
30		1.310	1.697	2.042	2.457	2.750
32		1.309	1.694	2.037	2.449	2.738
34		1.307	1.691	2.032	2.441	2.728

(continued)

Table B.3 (continued)

36		1.306	1.688	2.028	2.434	2.719
38		1.304	1.686	2.024	2.429	2.712
40		1.303	1.684	2.021	2.423	2.704
45		1.301	1.679	2.014	2.412	2.690
50		1.299	1.676	2.009	2.403	2.678
55		1.297	1.673	2.004	2.396	2.668
60		1.296	1.671	2.000	2.390	2.660
65		1.295	1.669	1.997	2.385	2.654
70		1.294	1.667	1.994	2.381	2.648
75		1.293	1.665	1.992	2.377	2.643
80		1.292	1.664	1.990	2.374	2.639
90		1.291	1.662	1.987	2.368	2.632
100		1.290	1.660	1.984	2.364	2.626
500		1.283	1.648	1.965	2.334	2.586
1000		1.282	1.646	1.962	2.330	2.581

Table B.4 Percentage points of the F distribution: df1 for numerator and df2 for denominator

df2	Significant level = 0.01 df1 for numerator												
	1	2	3	4	5	6	7	8	9	10	12	15	20
1	4052	4999	5403	5625	5764	5859	5928	5982	6022	6056	6106	6157	6209
2	98.50	99.00	99.17	99.25	99.30	99.33	99.36	99.37	99.39	99.40	99.42	99.43	99.45
3	34.12	30.82	29.46	28.71	28.24	27.91	27.67	27.49	27.35	27.23	27.05	26.87	26.69
4	21.20	18.00	16.69	15.98	15.52	15.21	14.98	14.80	14.66	14.55	14.37	14.20	14.02
5	16.26	13.27	12.06	11.39	10.97	10.67	10.46	10.29	10.16	10.05	9.89	9.72	9.55
6	13.75	10.92	9.78	9.15	8.75	8.47	8.26	8.10	7.98	7.87	7.72	7.56	7.40
7	12.25	9.55	8.45	7.85	7.46	7.19	6.99	6.84	6.72	6.62	6.47	6.31	6.16
8	11.26	8.65	7.59	7.01	6.63	6.37	6.18	6.03	5.91	5.81	5.67	5.52	5.36
9	10.56	8.02	6.99	6.42	6.06	5.80	5.61	5.47	5.35	5.26	5.11	4.96	4.81
10	10.04	7.56	6.55	5.99	5.64	5.39	5.20	5.06	4.94	4.85	4.71	4.56	4.41
11	9.65	7.21	6.22	5.67	5.32	5.07	4.89	4.74	4.63	4.54	4.40	4.25	4.10
12	9.33	6.93	5.95	5.41	5.06	4.82	4.64	4.50	4.39	4.30	4.16	4.01	3.86
13	9.07	6.70	5.74	5.21	4.86	4.62	4.44	4.30	4.19	4.10	3.96	3.82	3.66
14	8.86	6.51	5.56	5.04	4.69	4.46	4.28	4.14	4.03	3.94	3.80	3.66	3.51
15	8.68	6.36	5.42	4.89	4.56	4.32	4.14	4.00	3.89	3.80	3.67	3.52	3.37
16	8.53	6.23	5.29	4.77	4.44	4.20	4.03	3.89	3.78	3.69	3.55	3.41	3.26
17	8.40	6.11	5.18	4.67	4.34	4.10	3.93	3.79	3.68	3.59	3.46	3.31	3.16
18	8.29	6.01	5.09	4.58	4.25	4.01	3.84	3.71	3.60	3.51	3.37	3.23	3.08
19	8.18	5.93	5.01	4.50	4.17	3.94	3.77	3.63	3.52	3.43	3.30	3.15	3.00
20	8.10	5.85	4.94	4.43	4.10	3.87	3.70	3.56	3.46	3.37	3.23	3.09	2.94
21	8.02	5.78	4.87	4.37	4.04	3.81	3.64	3.51	3.40	3.31	3.17	3.03	2.88

(continued)

Table B.4 (continued)

df$_2$	Significant level = 0.01 df1 for numerator												
	1	2	3	4	5	6	7	8	9	10	12	15	20
22	7.95	5.72	4.82	4.31	3.99	3.76	3.59	3.45	3.35	3.26	3.12	2.98	2.83
23	7.88	5.66	4.76	4.26	3.94	3.71	3.54	3.41	3.30	3.21	3.07	2.93	2.78
24	7.82	5.61	4.72	4.22	3.90	3.67	3.50	3.36	3.26	3.17	3.03	2.89	2.74
25	7.77	5.57	4.68	4.18	3.85	3.63	3.46	3.32	3.22	3.13	2.99	2.85	2.70
26	7.72	5.53	4.64	4.14	3.82	3.59	3.42	3.29	3.18	3.09	2.96	2.81	2.66
27	7.68	5.49	4.60	4.11	3.78	3.56	3.39	3.26	3.15	3.06	2.93	2.78	2.63
28	7.64	5.45	4.57	4.07	3.75	3.53	3.36	3.23	3.12	3.03	2.90	2.75	2.60
29	7.60	5.42	4.54	4.04	3.73	3.50	3.33	3.20	3.09	3.00	2.87	2.73	2.57
30	7.56	5.39	4.51	4.02	3.70	3.47	3.30	3.17	3.07	2.98	2.84	2.70	2.55
40	7.31	5.18	4.31	3.83	3.51	3.29	3.12	2.99	2.89	2.80	2.66	2.52	2.37
60	7.08	4.98	4.13	3.65	3.34	3.12	2.95	2.82	2.72	2.63	2.50	2.35	2.20
120	6.85	4.79	3.95	3.48	3.17	2.96	2.79	2.66	2.56	2.47	2.34	2.19	2.03
∞	6.63	4.61	3.78	3.32	3.02	2.80	2.64	2.51	2.41	2.32	2.18	2.04	1.88

df$_2$	Significant level = 0.05 df1 for numerator												
	1	2	3	4	5	6	7	8	9	10	12	15	20
1	161.4	199.5	215.7	224.6	230.2	234.0	236.8	238.9	240.5	241.9	243.9	245.9	248.0
2	18.51	19.00	19.16	19.25	19.30	19.33	19.35	19.37	19.38	19.40	19.41	19.43	19.45
3	10.13	9.55	9.28	9.12	9.01	8.94	8.89	8.85	8.81	8.79	8.74	8.70	8.66
4	7.71	6.94	6.59	6.39	6.26	6.16	6.09	6.04	6.00	5.96	5.91	5.86	5.80
5	6.61	5.79	5.41	5.19	5.05	4.95	4.88	4.82	4.77	4.74	4.68	4.62	4.56
6	5.99	5.14	4.76	4.53	4.39	4.28	4.21	4.15	4.10	4.06	4.00	3.94	3.87
7	5.59	4.74	4.35	4.12	3.97	3.87	3.79	3.73	3.68	3.64	3.57	3.51	3.44

(continued)

Table B.4 (continued)

df_2	Significant level = 0.05 df1 for numerator												
	1	2	3	4	5	6	7	8	9	10	12	15	20
8	5.32	4.46	4.07	3.84	3.69	3.58	3.50	3.44	3.39	3.35	3.28	3.22	3.15
9	5.12	4.26	3.86	3.63	3.48	3.37	3.29	3.23	3.18	3.14	3.07	3.01	2.94
10	4.96	4.10	3.71	3.48	3.33	3.22	3.14	3.07	3.02	2.98	2.91	2.85	2.77
11	4.84	3.98	3.59	3.36	3.20	3.09	3.01	2.95	2.90	2.85	2.79	2.72	2.65
12	4.75	3.89	3.49	3.26	3.11	3.00	2.91	2.85	2.80	2.75	2.69	2.62	2.54
13	4.67	3.81	3.41	3.18	3.03	2.92	2.83	2.77	2.71	2.67	2.60	2.53	2.46
14	4.60	3.74	3.34	3.11	2.96	2.85	2.76	2.70	2.65	2.60	2.53	2.46	2.39
15	4.54	3.68	3.29	3.06	2.90	2.79	2.71	2.64	2.59	2.54	2.48	2.40	2.33
16	4.49	3.63	3.24	3.01	2.85	2.74	2.66	2.59	2.54	2.49	2.42	2.35	2.28
17	4.45	3.59	3.20	2.96	2.81	2.70	2.61	2.55	2.49	2.45	2.38	2.31	2.23
18	4.41	3.55	3.16	2.93	2.77	2.66	2.58	2.51	2.46	2.41	2.34	2.27	2.19
19	4.38	3.52	3.13	2.90	2.74	2.63	2.54	2.48	2.42	2.38	2.31	2.23	2.16
20	4.35	3.49	3.10	2.87	2.71	2.60	2.51	2.45	2.39	2.35	2.28	2.20	2.12
21	4.32	3.47	3.07	2.84	2.68	2.57	2.49	2.42	2.37	2.32	2.25	2.18	2.10
22	4.30	3.44	3.05	2.82	2.66	2.55	2.46	2.40	2.34	2.30	2.23	2.15	2.07
23	4.28	3.42	3.03	2.80	2.64	2.53	2.44	2.37	2.32	2.27	2.20	2.13	2.05
24	4.26	3.40	3.01	2.78	2.62	2.51	2.42	2.36	2.30	2.25	2.18	2.11	2.03
25	4.24	3.39	2.99	2.76	2.60	2.49	2.40	2.34	2.28	2.24	2.16	2.09	2.01
26	4.23	3.37	2.98	2.74	2.59	2.47	2.39	2.32	2.27	2.22	2.15	2.07	1.99
27	4.21	3.35	2.96	2.73	2.57	2.46	2.37	2.31	2.25	2.20	2.13	2.06	1.97
28	4.20	3.34	2.95	2.71	2.56	2.45	2.36	2.29	2.24	2.19	2.12	2.04	1.96
29	4.18	3.33	2.93	2.70	2.55	2.43	2.35	2.28	2.22	2.18	2.10	2.03	1.94

(continued)

Table B.4 (continued)

df_2	Significant level = 0.05 df1 for numerator												
	1	2	3	4	5	6	7	8	9	10	12	15	20
30	4.17	3.32	2.92	2.69	2.53	2.42	2.33	2.27	2.21	2.16	2.09	2.01	1.93
40	4.08	3.23	2.84	2.61	2.45	2.34	2.25	2.18	2.12	2.08	2.00	1.92	1.84
60	4.00	3.15	2.76	2.53	2.37	2.25	2.17	2.10	2.04	1.99	1.92	1.84	1.75
120	3.92	3.07	2.68	2.45	2.29	2.17	2.09	2.02	1.96	1.91	1.83	1.75	1.66
∞	3.84	3.00	2.60	2.37	2.21	2.10	2.01	1.94	1.88	1.83	1.75	1.67	1.57

Table B.5 Random numbers

10,480	15,011	01,536	02,011	81,647	91,646	67,179	14,194	62,590	36,207	20,969	99,570	91,291	90,700
22,368	46,573	25,595	85,393	30,995	89,198	27,982	53,402	93,965	34,095	52,666	19,174	39,615	99,505
24,130	48,360	22,527	97,265	76,393	64,809	15,179	24,830	49,340	32,081	30,680	19,655	63,348	58,629
42,167	93,093	06,243	61,680	07,856	16,376	39,440	53,537	71,341	57,004	00,849	74,917	97,758	16,379
37,570	39,975	81,837	16,656	06,121	91,782	60,468	81,305	49,684	60,672	14,110	06,927	01,263	54,613
77,921	06,907	11,008	42,751	27,756	53,498	18,602	70,659	90,655	15,053	21,916	81,825	44,394	42,880
99,562	72,905	56,420	69,994	98,872	31,016	71,194	18,738	44,013	48,840	63,213	21,069	10,634	12,952
96,301	91,977	05,463	07,972	18,876	20,922	94,595	56,869	69,014	60,045	18,425	84,903	42,508	32,307
89,579	14,342	63,661	10,281	17,453	18,103	57,740	84,378	25,331	12,566	58,678	44,947	05,584	56,941
85,475	36,857	43,342	53,988	53,060	59,533	38,867	62,300	08,158	17,983	16,439	11,458	18,593	64,952
28,918	69,578	88,231	33,276	70,997	79,936	56,865	05,859	90,106	31,595	01,547	85,590	91,610	78,188
63,553	40,961	48,235	03,427	49,626	69,445	18,663	72,695	52,180	20,847	12,234	90,511	33,703	90,322
09,429	93,969	52,636	92,737	88,974	33,488	36,320	17,617	30,015	08,272	84,115	27,156	30,613	74,952
10,365	61,129	87,529	85,689	48,237	52,267	67,689	93,394	01,511	26,358	85,104	20,285	29,975	89,868
07,119	97,336	71,048	08,178	77,233	13,916	47,564	81,056	97,735	85,977	29,372	74,461	28,551	90,707
51,085	12,765	51,821	51,259	77,452	16,308	60,756	92,144	49,442	53,900	70,960	63,990	75,601	40,719
02,368	21,382	52,404	60,268	89,368	19,885	55,322	44,819	01,188	65,255	64,835	44,919	05,944	55,157
01,011	54,092	33,362	94,904	31,273	04,146	18,594	29,852	71,585	85,030	51,132	01,915	92,747	64,951
52,162	53,916	46,369	58,586	23,216	14,513	83,149	98,736	23,495	64,350	94,738	17,752	35,156	35,749
07,056	97,628	33,787	09,998	42,698	06,691	76,988	13,602	51,851	46,104	88,916	19,509	25,625	58,104
48,663	91,245	85,828	14,346	09,172	30,168	90,229	04,734	59,193	22,178	30,421	61,666	99,904	32,812
54,164	58,492	22,421	74,103	47,070	25,306	76,468	26,384	58,151	06,646	21,524	15,227	96,909	44,592
32,639	32,363	05,597	24,200	13,363	38,005	94,342	28,728	35,806	06,912	17,012	64,161	18,296	22,851

(continued)

Table B.5 (continued)

29,334	27,001	87,637	87,308	58,731	00,256	45,834	15,398	46,557	41,135	10,367	07,684	36,188	18,510
02,488	33,062	28,834	07,351	19,731	92,420	60,952	61,280	50,001	67,658	32,586	86,679	50,720	94,953
81,525	72,295	04,839	96,423	24,878	82,651	66,566	14,778	76,797	14,780	13,300	87,074	79,666	95,725
29,676	20,591	68,086	26,432	46,901	20,849	89,768	81,536	86,645	12,659	92,259	57,102	80,428	25,280
00,742	57,392	39,064	66,432	84,673	40,027	32,832	61,362	98,947	96,067	64,760	64,584	96,096	98,253
05,366	04,213	25,669	26,422	44,407	44,048	37,937	63,904	45,766	66,134	75,470	66,520	34,693	90,449
91,921	26,418	64,117	94,305	26,766	25,940	39,972	22,209	71,500	64,568	91,402	42,416	07,844	69,618
00,582	04,711	87,917	77,341	42,206	35,126	74,087	99,547	81,817	42,607	43,808	76,655	62,028	76,630
00,725	69,884	62,797	56,170	86,324	88,072	76,222	36,086	84,637	93,161	76,038	65,855	77,919	88,006
69,011	65,797	95,876	55,293	18,988	27,354	26,575	08,625	40,801	59,920	29,841	80,150	12,777	48,501
25,976	57,948	29,888	88,604	67,917	48,708	18,912	82,271	65,424	69,774	33,611	54,262	85,963	03,547
09,763	83,473	73,577	12,908	30,883	18,317	28,290	35,797	05,998	41,688	34,952	37,888	38,917	88,050
91,567	42,595	27,958	30,134	04,024	86,385	29,880	99,730	55,536	84,855	29,080	09,250	79,656	73,211
17,955	56,349	90,999	49,127	20,044	59,931	06,115	20,542	18,059	02,008	73,708	83,517	36,103	42,791
46,503	18,584	18,845	49,618	02,304	51,038	20,655	58,727	28,168	15,475	56,942	53,389	20,562	87,338
92,157	89,634	94,824	78,171	84,610	82,834	09,922	25,417	44,137	48,413	25,555	21,246	35,509	20,468
14,577	62,765	35,605	81,263	39,667	47,358	56,873	56,307	61,607	49,518	89,656	20,103	77,490	18,062
98,427	07,523	33,362	64,270	01,638	92,477	66,969	98,420	04,880	45,585	46,565	04,102	46,880	45,709
53,976	54,914	06,990	67,245	68,350	82,948	11,398	42,878	80,287	88,267	47,363	46,634	06,541	97,809
15,664	10,493	20,492	38,391	91,132	21,999	59,516	81,652	27,195	48,223	46,751	22,923	32,261	85,653

Table B.6 Percentage points of Chi-squared distribution

df	0.995	0.99	0.975	0.95	0.90	0.10	0.05	0.025	0.01	0.005
1	–	–	0.001	0.004	0.016	2.706	3.841	5.024	6.635	7.879
2	0.010	0.020	0.051	0.103	0.211	4.605	5.991	7.378	9.210	10.597
3	0.072	0.115	0.216	0.352	0.584	6.251	7.815	9.348	11.345	12.838
4	0.207	0.297	0.484	0.711	1.064	7.779	9.488	11.143	13.277	14.860
5	0.412	0.554	0.831	1.145	1.610	9.236	11.071	12.833	15.086	16.750
6	0.676	0.872	1.237	1.635	2.204	10.645	12.592	14.449	16.812	18.548
7	0.989	1.239	1.690	2.167	2.833	12.017	14.067	16.013	18.475	20.278
8	1.344	1.646	2.180	2.733	3.490	13.362	15.507	17.535	20.090	21.955
9	1.735	2.088	2.700	3.325	4.168	14.684	16.919	19.023	21.666	23.589
10	2.156	2.558	3.247	3.940	4.865	15.987	18.307	20.483	23.209	25.188
11	2.603	3.053	3.816	4.575	5.578	17.275	19.675	21.920	24.725	26.757
12	3.074	3.571	4.404	5.226	6.304	18.549	21.026	23.337	26.217	28.299
13	3.565	4.107	5.009	5.892	7.042	19.812	22.362	24.736	27.688	29.819
14	4.075	4.660	5.629	6.571	7.790	21.064	23.685	26.119	29.141	31.319
15	4.601	5.229	6.262	7.261	8.547	22.307	24.996	27.488	30.578	32.801
16	5.142	5.812	6.908	7.962	9.312	23.542	26.296	28.845	32.000	34.267
17	5.697	6.408	7.564	8.672	10.085	24.769	27.587	30.191	33.409	35.718
18	6.265	7.015	8.231	9.390	10.865	25.989	28.869	31.526	34.805	37.156
19	6.844	7.633	8.907	10.117	11.651	27.204	30.144	32.852	36.191	38.582
20	7.434	8.260	9.591	10.851	12.443	28.412	31.410	34.170	37.566	39.997
21	8.034	8.897	10.283	11.591	13.240	29.615	32.671	35.479	38.932	41.401
22	8.643	9.542	10.982	12.338	14.042	30.813	33.924	36.781	40.289	42.796
23	9.262	10.196	11.689	13.091	14.848	32.007	35.172	38.076	41.638	44.181
24	9.886	10.856	12.401	13.848	15.659	33.196	36.415	39.364	42.980	45.559
25	10.520	11.524	13.120	14.611	16.473	34.382	37.652	40.646	44.314	46.928
26	11.160	12.198	13.844	15.379	17.292	35.563	38.885	41.923	45.642	48.290
27	11.808	12.879	14.573	16.151	18.114	36.741	40.113	43.194	46.963	49.645
28	12.461	13.565	15.308	16.928	18.939	37.916	41.337	44.461	48.278	50.993
29	13.121	14.257	16.047	17.708	19.768	39.087	42.557	45.722	49.588	52.336
30	13.787	14.954	16.791	18.493	20.599	40.256	43.773	46.979	50.892	53.672
40	20.707	22.164	24.433	26.509	29.051	51.805	55.758	59.342	63.691	66.766
50	27.991	29.707	32.357	34.764	37.689	63.167	67.505	71.420	76.154	79.490
60	35.534	37.485	40.482	43.188	46.459	74.397	79.082	83.298	88.379	91.952
70	43.275	45.442	48.758	51.739	55.329	85.527	90.531	95.023	100.425	104.215
80	51.172	53.540	57.153	60.391	64.278	96.578	101.879	106.629	112.329	116.321
90	59.196	61.754	65.647	69.126	73.291	107.565	113.145	118.136	124.116	128.299

Table B.7 Critical values of the Spearman's ranked correlation coefficient (r)

n	0.10	0.05	0.02	0.01
5	0.900	–	–	–
6	0.829	0.886	0.943	–
7	0.714	0.786	0.893	0.929
8	0.643	0.738	0.833	0.881
9	0.600	0.700	0.783	0.833
10	0.564	0.648	0.745	0.794
11	0.536	0.618	0.709	0.818
12	0.497	0.591	0.703	0.780
13	0.475	0.566	0.673	0.745
14	0.457	0.545	0.646	0.716
15	0.441	0.525	0.623	0.689
16	0.425	0.507	0.601	0.666
17	0.412	0.490	0.582	0.645
18	0.399	0.476	0.564	0.625
19	0.388	0.462	0.549	0.608
20	0.377	0.450	0.534	0.591
21	0.368	0.438	0.521	0.576
22	0.359	0.428	0.508	0.562
23	0.351	0.418	0.496	0.549
24	0.343	0.409	0.485	0.537
25	0.336	0.400	0.475	0.526
26	0.329	0.392	0.465	0.515
27	0.323	0.385	0.456	0.505
28	0.317	0.377	0.488	0.496
29	0.311	0.370	0.440	0.487
30	0.305	0.364	0.432	0.478

The null hypothesis is rejected if the calculated Spearman rank correlation is greater than the given value in the table. Ignore the sign of negative coefficient correlations

Table B.8 Critical values of the Pearson's correlation coefficient (r)

N	One-tailed α						
	0.10	0.05	0.025	0.01	0.005	0.001	0.0005
	Two-tailed α						
	0.20	0.10	0.05	0.02	0.01	0.002	0.001
3	0.951	0.988	0.997	1.00	1.00	1.00	1.00
4	0.800	0.900	0.950	0.980	0.990	0.998	0.999
5	0.687	0.805	0.878	0.934	0.959	0.986	0.991
6	0.608	0.729	0.811	0.882	0.917	0.963	0.974
7	0.551	0.669	0.755	0.833	0.875	0.935	0.951
8	0.507	0.622	0.707	0.789	0.834	0.905	0.925
9	0.472	0.582	0.666	0.750	0.798	0.875	0.898
10	0.443	0.549	0.632	0.716	0.765	0.847	0.872
11	0.419	0.522	0.602	0.685	0.735	0.820	0.847
12	0.398	0.497	0.576	0.658	0.708	0.795	0.823
13	0.380	0.476	0.553	0.634	0.684	0.772	0.801
14	0.365	0.458	0.533	0.612	0.661	0.750	0.780
15	0.351	0.441	0.514	0.592	0.641	0.730	0.760
16	0.338	0.426	0.497	0.574	0.623	0.711	0.742
17	0.327	0.412	0.482	0.558	0.606	0.694	0.725
18	0.317	0.400	0.468	0.543	0.590	0.678	0.708
19	0.308	0.389	0.456	0.529	0.575	0.662	0.693
20	0.299	0.378	0.444	0.516	0.562	0.648	0.679
25	0.265	0.337	0.396	0.462	0.505	0.588	0.618
30	0.241	0.306	0.361	0.423	0.463	0.542	0.570
35	0.222	0.283	0.334	0.392	0.430	0.505	0.532
40	0.207	0.264	0.312	0.367	0.403	0.474	0.501
45	0.195	0.248	0.294	0.346	0.380	0.449	0.474
50	0.184	0.235	0.279	0.328	0.361	0.427	0.451
55	0.176	0.224	0.266	0.313	0.345	0.408	0.432
60	0.168	0.214	0.254	0.300	0.330	0.391	0.414
65	0.161	0.206	0.244	0.288	0.317	0.376	0.399
70	0.155	0.198	0.235	0.278	0.306	0.363	0.385
75	0.150	0.191	0.227	0.268	0.296	0.351	0.372
80	0.145	0.185	0.220	0.260	0.286	0.341	0.361
85	0.140	0.180	0.213	0.252	0.278	0.331	0.351
90	0.136	0.175	0.207	0.245	0.270	0.322	0.341
95	0.133	0.170	0.202	0.238	0.263	0.313	0.332
100	0.129	0.165	0.197	0.232	0.257	0.305	0.324

If the calculated Pearson's correlation coefficient is greater than the critical value from the table, then reject the null hypothesis

Table B.9 Binomial probability distribution table

n	x	$p = 0.05$	0.10	0.15	0.20	0.25	0.30	0.40	0.50
2	0	0.9025	0.8100	0.7225	0.6400	0.5625	0.4900	0.3600	0.2500
	1	0.9975	0.9900	0.9775	0.9600	0.9375	0.9100	0.8400	0.7500
3	0	0.8574	0.7290	0.6141	0.5120	0.4219	0.3430	0.2160	0.1250
	1	0.9928	0.9720	0.9393	0.8960	0.8438	0.7840	0.6480	0.5000
	2	0.9999	0.9990	0.9966	0.9920	0.9844	0.9730	0.9360	0.8750
4	0	0.8145	0.6561	0.5220	0.4096	0.3164	0.2401	0.1296	0.0625
	1	0.9860	0.9477	0.8905	0.8192	0.7383	0.6517	0.4752	0.3125
	2	0.9995	0.9963	0.9880	0.9728	0.9492	0.9163	0.8208	0.6875
	3	1.0000	0.9999	0.9995	0.9984	0.9961	0.9919	0.9744	0.9375
5	0	0.7738	0.5905	0.4437	0.3277	0.2373	0.1681	0.0778	0.0313
	1	0.9774	0.9185	0.8352	0.7373	0.6328	0.5282	0.3370	0.1875
	2	0.9988	0.9914	0.9734	0.9421	0.8965	0.8369	0.6826	0.5000
	3	1.0000	0.9995	0.9978	0.9933	0.9844	0.9692	0.9130	0.8125
	4	1.0000	1.0000	0.9999	0.9997	0.9990	0.9976	0.9898	0.9688
6	0	0.7351	0.5314	0.3771	0.2621	0.1780	0.1177	0.0467	0.0156
	1	0.9672	0.8857	0.7765	0.6554	0.5339	0.4202	0.2333	0.1094
	2	0.9978	0.9841	0.9527	0.9011	0.8306	0.7443	0.5443	0.3438
	3	0.9999	0.9987	0.9941	0.9830	0.9624	0.9295	0.8208	0.6563
	4	1.0000	1.0000	0.9996	0.9984	0.9954	0.9891	0.9590	0.8906
	5	1.0000	1.0000	1.0000	0.9999	0.9998	0.9993	0.9959	0.9844
7	0	0.6983	0.4783	0.3206	0.2097	0.1335	0.0824	0.0280	0.0078
	1	0.9556	0.8503	0.7166	0.5767	0.4450	0.3294	0.1586	0.0625
	2	0.9962	0.9743	0.9262	0.8520	0.7564	0.6471	0.4199	0.2266
	3	0.9998	0.9973	0.9879	0.9667	0.9294	0.8740	0.7102	0.5000
	4	1.0000	0.9998	0.9988	0.9953	0.9871	0.9712	0.9037	0.7734
	5	1.0000	1.0000	0.9999	0.9996	0.9987	0.9962	0.9812	0.9375
	6	1.0000	1.0000	1.0000	1.0000	0.9999	0.9998	0.9984	0.9922
8	0	0.6634	0.4305	0.2725	0.1678	0.1001	0.0576	0.0168	0.0039
	1	0.9428	0.8131	0.6572	0.5033	0.3671	0.2553	0.1064	0.0352
	2	0.9942	0.9619	0.8948	0.7969	0.6785	0.5518	0.3154	0.1445
	3	0.9996	0.9950	0.9787	0.9437	0.8862	0.8059	0.5941	0.3633
	4	1.0000	0.9996	0.9971	0.9896	0.9727	0.9420	0.8263	0.6367
	5	1.0000	1.0000	0.9998	0.9988	0.9958	0.9887	0.9502	0.8555

(continued)

Table B.9 (continued)

n	x	$p = 0.05$	0.10	0.15	0.20	0.25	0.30	0.40	0.50
	6	1.0000	1.0000	1.0000	0.9999	0.9996	0.9987	0.9915	0.9648
	7	1.0000	1.0000	1.0000	1.0000	1.0000	0.9999	0.9993	0.9961
9	0	0.6302	0.3874	0.2316	0.1342	0.0751	0.0403	0.0101	0.0019
	1	0.9288	0.7748	0.5995	0.4362	0.3003	0.1960	0.0705	0.0195
	2	0.9916	0.9470	0.8591	0.7382	0.6007	0.4628	0.2318	0.0898
	3	0.9994	0.9917	0.9661	0.9144	0.8343	0.7297	0.4826	0.2539
	4	1.0000	0.9991	0.9944	0.9804	0.9511	0.9012	0.7334	0.5000
	5	1.0000	0.9999	0.9994	0.9969	0.9900	0.9747	0.9006	0.7461
	6	1.0000	1.0000	1.0000	0.9997	0.9987	0.9957	0.9750	0.9102
	7	1.0000	1.0000	1.0000	1.0000	0.9999	0.9996	0.9962	0.9805
	8	1.0000	1.0000	1.0000	1.0000	1.0000	1.0000	0.9997	0.9980
10	0	0.5987	0.3487	0.1969	0.1074	0.0563	0.0283	0.0060	0.0010
	1	0.9139	0.7361	0.5443	0.3758	0.2440	0.1493	0.0464	0.0107
	2	0.9885	0.9298	0.8202	0.6778	0.5256	0.3828	0.1673	0.0547
	3	0.9990	0.9872	0.9500	0.8791	0.7759	0.6496	0.3823	0.1719
	4	0.9999	0.9984	0.9901	0.9672	0.9219	0.8497	0.6331	0.3770
	5	1.0000	0.9999	0.9986	0.9936	0.9803	0.9526	0.8338	0.6230
	6	1.0000	1.0000	0.9999	0.9991	0.9965	0.9894	0.9452	0.8281
	7	1.0000	1.0000	1.0000	0.9999	0.9996	0.9984	0.9877	0.9453
	8	1.0000	1.0000	1.0000	1.0000	1.0000	0.9999	0.9983	0.9893
	9	1.0000	1.0000	1.0000	1.0000	1.0000	1.0000	0.9999	0.9990
11	0	0.5688	0.3138	0.1673	0.0859	0.0422	0.0198	0.0036	0.0005
	1	0.8981	0.6974	0.4922	0.3221	0.1971	0.1130	0.0302	0.0059
	2	0.9848	0.9104	0.7788	0.6174	0.4552	0.3127	0.1189	0.0327
	3	0.9984	0.9815	0.9306	0.8389	0.7133	0.5696	0.2963	0.1133
	4	0.9999	0.9972	0.9841	0.9496	0.8854	0.7897	0.5328	0.2744
	5	1.0000	0.9997	0.9973	0.9883	0.9657	0.9218	0.7535	0.5000
	6	1.0000	1.0000	0.9997	0.9980	0.9924	0.9784	0.9006	0.7256
	7	1.0000	1.0000	1.0000	0.9998	0.9988	0.9957	0.9707	0.8867
	8	1.0000	1.0000	1.0000	1.0000	0.9999	0.9994	0.9941	0.9673
	9	1.0000	1.0000	1.0000	1.0000	1.0000	1.0000	0.9993	0.9941
	10	1.0000	1.0000	1.0000	1.0000	1.0000	1.0000	1.0000	0.9995
12	0	0.5404	0.2824	0.1422	0.0687	0.0317	0.0138	0.0022	0.0002
	1	0.8816	0.6590	0.4435	0.2749	0.1584	0.0850	0.0196	0.0032
	2	0.9804	0.8891	0.7358	0.5584	0.3907	0.2528	0.0834	0.0193

(continued)

Table B.9 (continued)

	3	0.9978	0.9744	0.9078	0.7946	0.6488	0.4925	0.2253	0.0730
	4	0.9998	0.9957	0.9761	0.9274	0.8424	0.7237	0.4382	0.1938
	5	1.0000	0.9995	0.9954	0.9806	0.9456	0.8821	0.6652	0.3872
	6	1.0000	1.0000	0.9993	0.9961	0.9858	0.9614	0.8418	0.6128
	7	1.0000	1.0000	0.9999	0.9994	0.9972	0.9905	0.9427	0.8062
	8	1.0000	1.0000	1.0000	0.9999	0.9996	0.9983	0.9847	0.9270
	9	1.0000	1.0000	1.0000	1.0000	1.0000	0.9998	0.9972	0.9807
	10	1.0000	1.0000	1.0000	1.0000	1.0000	1.0000	0.9997	0.9968
	11	1.0000	1.0000	1.0000	1.0000	1.0000	1.0000	1.0000	0.9998
13	0	0.5133	0.2542	0.1209	0.0550	0.0238	0.0097	0.0013	0.0001
	1	0.8646	0.6213	0.3983	0.2336	0.1267	0.0637	0.0126	0.0017
	2	0.9755	0.8661	0.6920	0.5017	0.3326	0.2025	0.0579	0.0112
	3	0.9969	0.9658	0.8820	0.7473	0.5843	0.4206	0.1686	0.0461
	4	0.9997	0.9935	0.9658	0.9009	0.7940	0.6543	0.3530	0.1334
	5	1.0000	0.9991	0.9925	0.9700	0.9198	0.8346	0.5744	0.2905
	6	1.0000	0.9999	0.9987	0.9930	0.9757	0.9376	0.7712	0.5000
	7	1.0000	1.0000	0.9998	0.9988	0.9943	0.9818	0.9023	0.7095
	8	1.0000	1.0000	1.0000	0.9998	0.9990	0.9960	0.9679	0.8666
	9	1.0000	1.0000	1.0000	1.0000	0.9999	0.9993	0.9922	0.9539
	10	1.0000	1.0000	1.0000	1.0000	1.0000	0.9999	0.9987	0.9888
	11	1.0000	1.0000	1.0000	1.0000	1.0000	1.0000	0.9999	0.9983
	12	1.0000	1.0000	1.0000	1.0000	1.0000	1.0000	1.0000	0.9999
14	0	0.4877	0.2288	0.1028	0.0440	0.0178	0.0068	0.0008	0.0001
	1	0.8470	0.5846	0.3567	0.1979	0.1010	0.0475	0.0081	0.0009
	2	0.9699	0.8416	0.6479	0.4481	0.2811	0.1608	0.0398	0.0065
	3	0.9958	0.9559	0.8535	0.6982	0.5213	0.3552	0.1243	0.0287
	4	0.9996	0.9908	0.9533	0.8702	0.7415	0.5842	0.2793	0.0898
	5	1.0000	0.9985	0.9885	0.9562	0.8883	0.7805	0.4859	0.2120
	6	1.0000	0.9998	0.9978	0.9884	0.9617	0.9067	0.6925	0.3953
	7	1.0000	1.0000	0.9997	0.9976	0.9897	0.9685	0.8499	0.6047
	8	1.0000	1.0000	1.0000	0.9996	0.9979	0.9917	0.9417	0.7880
	9	1.0000	1.0000	1.0000	1.0000	0.9997	0.9983	0.9825	0.9102
	10	1.0000	1.0000	1.0000	1.0000	1.0000	0.9998	0.9961	0.9713
	11	1.0000	1.0000	1.0000	1.0000	1.0000	1.0000	0.9994	0.9935
	12	1.0000	1.0000	1.0000	1.0000	1.0000	1.0000	0.9999	0.9991

(continued)

Table B.9 (continued)

	13	1.0000	1.0000	1.0000	1.0000	1.0000	1.0000	1.0000	0.9999
15	0	0.4633	0.2059	0.0873	0.0352	0.0134	0.0047	0.0005	0.0000
	1	0.8290	0.5490	0.3186	0.1671	0.0802	0.0353	0.0052	0.0005
	2	0.9638	0.8159	0.6042	0.3980	0.2361	0.1268	0.0271	0.0037
	3	0.9945	0.9444	0.8227	0.6482	0.4613	0.2969	0.0905	0.0176
	4	0.9994	0.9873	0.9383	0.8358	0.6865	0.5155	0.2173	0.0592
	5	1.0000	0.9978	0.9832	0.9389	0.8516	0.7216	0.4032	0.1509
	6	1.0000	0.9997	0.9964	0.9819	0.9434	0.8689	0.6098	0.3036
	7	1.0000	1.0000	0.9994	0.9958	0.9827	0.9500	0.7869	0.5000
	8	1.0000	1.0000	0.9999	0.9992	0.9958	0.9848	0.9050	0.6964
	9	1.0000	1.0000	1.0000	0.9999	0.9992	0.9963	0.9662	0.8491
	10	1.0000	1.0000	1.0000	1.0000	0.9999	0.9993	0.9907	0.9408
	11	1.0000	1.0000	1.0000	1.0000	1.0000	0.9999	0.9981	0.9824
	12	1.0000	1.0000	1.0000	1.0000	1.0000	1.0000	0.9997	0.9963
	13	1.0000	1.0000	1.0000	1.0000	1.0000	1.0000	1.0000	0.9995
	14	1.0000	1.0000	1.0000	1.0000	1.0000	1.0000	1.0000	1.0000
16	0	0.4401	0.1853	0.0742	0.0282	0.0100	0.0033	0.0003	0.0000
	1	0.8108	0.5147	0.2839	0.1407	0.0635	0.0261	0.0033	0.0003
	2	0.9571	0.7893	0.5614	0.3518	0.1971	0.0994	0.0183	0.0021
	3	0.9930	0.9316	0.7899	0.5981	0.4050	0.2459	0.0651	0.0106
	4	0.9991	0.9830	0.9210	0.7983	0.6302	0.4499	0.1666	0.0384
	5	0.9999	0.9967	0.9765	0.9183	0.8104	0.6598	0.3288	0.1051
	6	1.0000	0.9995	0.9944	0.9733	0.9204	0.8247	0.5272	0.2273
	7	1.0000	0.9999	0.9989	0.9930	0.9729	0.9256	0.7161	0.4018
	8	1.0000	1.0000	0.9998	0.9985	0.9925	0.9743	0.8577	0.5982
	9	1.0000	1.0000	1.0000	0.9998	0.9984	0.9929	0.9417	0.7728
	10	1.0000	1.0000	1.0000	1.0000	0.9997	0.9984	0.9809	0.8949
	11	1.0000	1.0000	1.0000	1.0000	1.0000	0.9997	0.9951	0.9616
	12	1.0000	1.0000	1.0000	1.0000	1.0000	1.0000	0.9991	0.9894
	13	1.0000	1.0000	1.0000	1.0000	1.0000	1.0000	0.9999	0.9979
	14	1.0000	1.0000	1.0000	1.0000	1.0000	1.0000	1.0000	0.9997
	15	1.0000	1.0000	1.0000	1.0000	1.0000	1.0000	1.0000	1.0000

Table B.10 Poisson probability distribution table

x	λ									
	0.1	0.2	0.3	0.4	0.5	0.6	0.7	0.8	0.9	1.0
0	0.9048	0.8187	0.7408	0.6703	**0.6065**	0.5488	0.4966	0.4493	0.4066	0.3679
1	0.0905	0.1637	0.2222	0.2681	0.3033	0.3293	0.3476	0.3595	0.3659	0.3679
2	0.0045	0.0164	0.0333	0.0536	0.0758	0.0988	0.1217	0.1438	0.1647	0.1839
3	0.0002	0.0011	0.0033	0.0072	0.0126	0.0198	0.0284	0.0383	0.0494	0.0613
4	0.0000	0.0001	0.0003	0.0007	0.0016	0.0030	0.0050	0.0077	0.0111	0.0153
5	0.0000	0.0000	0.0000	0.0001	0.0002	0.0004	0.0007	0.0012	0.0020	0.0031
6	0.0000	0.0000	0.0000	0.0000	0.0000	0.0000	0.0001	0.0002	0.0003	0.0005
7	0.0000	0.0000	0.0000	0.0000	0.0000	0.0000	0.0000	0.0000	0.0000	0.0001

x	λ									
	1.1	1.2	1.3	1.4	1.5	1.6	1.7	1.8	1.9	2.0
0	0.3329	0.3012	0.2725	0.2466	0.2231	0.2019	0.1827	0.1653	0.1496	0.1353
1	0.3662	0.3614	0.3543	0.3452	0.3347	0.3230	0.3106	0.2975	0.2842	0.2707
2	0.2014	0.2169	0.2303	0.2417	0.2510	0.2584	0.2640	0.2678	0.2700	0.2707
3	0.0738	0.0867	0.0998	0.1128	0.1255	0.1378	0.1496	0.1607	0.1710	0.1804
4	0.0203	0.0260	0.0324	0.0395	0.0471	0.0551	0.0636	0.0723	0.0812	0.0902
5	0.0045	0.0062	0.0084	0.0111	0.0141	0.0176	0.0216	0.0260	0.0309	0.0361
6	0.0008	0.0012	0.0018	0.0026	0.0035	0.0047	0.0061	0.0078	0.0098	0.0120
7	0.0001	0.0002	0.0003	0.0005	0.0008	0.0011	0.0015	0.0020	0.0027	0.0034
8	0.0000	0.0000	0.0001	0.0001	0.0001	0.0002	0.0003	0.0005	0.0006	0.0009
9	0.0000	0.0000	0.0000	0.0000	0.0000	0.0000	0.0001	0.0001	0.0001	0.0002

(continued)

Table B.10 (continued)

x	λ									
	2.1	2.2	2.3	2.4	2.5	2.6	2.7	2.8	2.9	3.0
0	0.1225	0.1108	0.1003	0.0907	0.0821	0.0743	0.0672	0.0608	0.0550	0.0498
1	0.2572	0.2438	0.2306	0.2177	0.2052	0.1931	0.1815	0.1703	0.1596	0.1494
2	0.2700	0.2681	0.2652	0.2613	0.2565	0.2510	0.2450	0.2384	0.2314	0.2240
3	0.1890	0.1966	0.2033	0.2090	0.2138	0.2176	0.2205	0.2225	0.2237	0.2240
4	0.0992	0.1082	0.1169	0.1254	0.1336	0.1414	0.1488	0.1557	0.1622	0.1680
5	0.0417	0.0476	0.0538	0.0602	0.0668	0.0735	0.0804	0.0872	0.0940	0.1008
6	0.0146	0.0174	0.0206	0.0241	0.0278	0.0319	0.0362	0.0407	0.0455	0.0504
7	0.0044	0.0055	0.0068	0.0083	0.0099	0.0118	0.0139	0.0163	0.0188	0.0216
8	0.0011	0.0015	0.0019	0.0025	0.0031	0.0038	0.0047	0.0057	0.0068	0.0081
9	0.0003	0.0004	0.0005	0.0007	0.0009	0.0011	0.0014	0.0018	0.0022	0.0027
10	0.0001	0.0001	0.0001	0.0002	0.0002	0.0003	0.0004	0.0005	0.0006	0.0008
11	0.0000	0.0000	0.0000	0.0000	0.0000	0.0001	0.0001	0.0001	0.0002	0.0002
12	0.0000	0.0000	0.0000	0.0000	0.0000	0.0000	0.0000	0.0000	0.0000	0.0001

x	λ									
	3.1	3.2	3.3	3.4	3.5	3.6	3.7	3.8	3.9	4.0
0	0.0450	0.0408	0.0369	0.0334	0.0302	0.0273	0.0247	0.0224	0.0202	0.0183
1	0.1397	0.1304	0.1217	0.1135	0.1057	0.0984	0.0915	0.0850	0.0789	0.0733
2	0.2165	0.2087	0.2008	0.1929	0.1850	0.1771	0.1692	0.1615	0.1539	0.1465
3	0.2237	0.2226	0.2209	0.2186	0.2158	0.2125	0.2087	0.2046	0.2001	0.1954

(continued)

Table B.10 (continued)

x	4.1	4.2	4.3	4.4	4.5	4.6	4.7	4.8	4.9	5.0
4	0.1734	0.1781	0.1823	0.1858	0.1888	0.1912	0.1931	0.1944	0.1951	0.1954
5	0.1075	0.1140	0.1203	0.1264	0.1322	0.1377	0.1429	0.1477	0.1522	0.1563
6	0.0555	0.0608	0.0662	0.0716	0.0771	0.0826	0.0881	0.0936	0.0989	0.1042
7	0.0246	0.0278	0.0312	0.0348	0.0385	0.0425	0.0466	0.0508	0.0551	0.0595
8	0.0095	0.0111	0.0129	0.0148	0.0169	0.0191	0.0215	0.0241	0.0269	0.0298
9	0.0033	0.0040	0.0047	0.0056	0.0066	0.0076	0.0089	0.0102	0.0116	0.0132
10	0.0010	0.0013	0.0016	0.0019	0.0023	0.0028	0.0033	0.0039	0.0045	0.0053
11	0.0003	0.0004	0.0005	0.0006	0.0007	0.0009	0.0011	0.0013	0.0016	0.0019
12	0.0001	0.0001	0.0001	0.0002	0.0002	0.0003	0.0003	0.0004	0.0005	0.0006
13	0.0000	0.0000	0.0000	0.0000	0.0001	0.0001	0.0001	0.0001	0.0002	0.0002
14	0.0000	0.0000	0.0000	0.0000	0.0000	0.0000	0.0000	0.0000	0.0000	0.0001

x	λ									
	4.1	4.2	4.3	4.4	4.5	4.6	4.7	4.8	4.9	5.0
0	0.0166	0.0150	0.0136	0.0123	0.0111	0.0101	0.0091	0.0082	0.0074	0.0067
1	0.0679	0.0630	0.0583	0.0540	0.0500	0.0462	0.0427	0.0395	0.0365	0.0337
2	0.1393	0.1323	0.1254	0.1188	0.1125	0.1063	0.1005	0.0948	0.0894	0.0842
3	0.1904	0.1852	0.1798	0.1743	0.1687	0.1631	0.1574	0.1517	0.1460	0.1404
4	0.1951	0.1944	0.1933	0.1917	0.1898	0.1875	0.1849	0.1820	0.1789	0.1755
5	0.1600	0.1633	0.1662	0.1687	0.1708	0.1725	0.1738	0.1747	0.1753	0.1755
6	0.1093	0.1143	0.1191	0.1237	0.1281	0.1323	0.1362	0.1398	0.1432	0.1462

(continued)

Table B.10 (continued)

7	0.0640	0.0686	0.0732	0.0778	0.0824	0.0869	0.0914	0.0959	0.1002	0.1044
8	0.0328	0.0360	0.0393	0.0428	0.0463	0.0500	0.0537	0.0575	0.0614	0.0653
9	0.0150	0.0168	0.0188	0.0209	0.0232	0.0255	0.0280	0.0307	0.0334	0.0363
10	0.0061	0.0071	0.0081	0.0092	0.0104	0.0118	0.0132	0.0147	0.0164	0.0181
11	0.0023	0.0027	0.0032	0.0037	0.0043	0.0049	0.0056	0.0064	0.0073	0.0082
12	0.0008	0.0009	0.0011	0.0014	0.0016	0.0019	0.0022	0.0026	0.0030	0.0034
13	0.0002	0.0003	0.0004	0.0005	0.0006	0.0007	0.0008	0.0009	0.0011	0.0013
14	0.0001	0.0001	0.0001	0.0001	0.0002	0.0002	0.0003	0.0003	0.0004	0.0005
15	0.0000	0.0000	0.0000	0.0000	0.0001	0.0001	0.0001	0.0001	0.0001	0.0002

x	λ									
	5.1	5.2	5.3	5.4	5.5	5.6	5.7	5.8	5.9	6.0
0	0.0061	0.0055	0.0050	0.0045	0.0041	0.0037	0.0033	0.0030	0.0027	0.0025
1	0.0311	0.0287	0.0265	0.0244	0.0225	0.0207	0.0191	0.0176	0.0162	0.0149
2	0.0793	0.0746	0.0701	0.0659	0.0618	0.0580	0.0544	0.0509	0.0477	0.0446
3	0.1348	0.1293	0.1239	0.1185	0.1133	0.1082	0.1033	0.0985	0.0938	0.0892
4	0.1719	0.1681	0.1641	0.1600	0.1558	0.1515	0.1472	0.1428	0.1383	0.1339
5	0.1753	0.1748	0.1740	0.1728	0.1714	0.1697	0.1678	0.1656	0.1632	0.1606
6	0.1490	0.1515	0.1537	0.1555	0.1571	0.1584	0.1594	0.1601	0.1605	0.1606
7	0.1086	0.1125	0.1163	0.1200	0.1234	0.1267	0.1298	0.1326	0.1353	0.1377
8	0.0692	0.0731	0.0771	0.0810	0.0849	0.0887	0.0925	0.0962	0.0998	0.1033

(continued)

Table B.10 (continued)

x										
9	0.0392	0.0423	0.0454	0.0486	0.0519	0.0552	0.0586	0.0620	0.0654	0.0688
10	0.0200	0.0220	0.0241	0.0262	0.0285	0.0309	0.0334	0.0359	0.0386	0.0413
11	0.0093	0.0104	0.0116	0.0129	0.0143	0.0157	0.0173	0.0190	0.0207	0.0225
12	0.0039	0.0045	0.0051	0.0058	0.0065	0.0073	0.0082	0.0092	0.0102	0.0113
13	0.0015	0.0018	0.0021	0.0024	0.0028	0.0032	0.0036	0.0041	0.0046	0.0052
14	0.0006	0.0007	0.0008	0.0009	0.0011	0.0013	0.0015	0.0017	0.0019	0.0022
15	0.0002	0.0002	0.0003	0.0003	0.0004	0.0005	0.0006	0.0007	0.0008	0.0009
16	0.0001	0.0001	0.0001	0.0001	0.0001	0.0002	0.0002	0.0002	0.0003	0.0003
17	0.0000	0.0000	0.0000	0.0000	0.0000	0.0000	0.0001	0.0001	0.0001	0.0001

x	λ									
	6.1	6.2	6.3	6.4	6.5	6.6	6.7	6.8	6.9	7.0
0	0.0022	0.0020	0.0018	0.0017	0.0015	0.0014	0.0012	0.0011	0.0010	0.0009
1	0.0137	0.0126	0.0116	0.0106	0.0098	0.0090	0.0082	0.0076	0.0070	0.0064
2	0.0417	0.0390	0.0364	0.0340	0.0318	0.0296	0.0276	0.0258	0.0240	0.0223
3	0.0848	0.0806	0.0765	0.0726	0.0688	0.0652	0.0617	0.0584	0.0552	0.0521
4	0.1294	0.1249	0.1205	0.1162	0.1118	0.1076	0.1034	0.0992	0.0952	0.0912
5	0.1579	0.1549	0.1519	0.1487	0.1454	0.1420	0.1385	0.1349	0.1314	0.1277
6	0.1605	0.1601	0.1595	0.1586	0.1575	0.1562	0.1546	0.1529	0.1511	0.1490
7	0.1399	0.1418	0.1435	0.1450	0.1462	0.1472	0.1480	0.1486	0.1489	0.1490
8	0.1066	0.1099	0.1130	0.1160	0.1188	0.1215	0.1240	0.1263	0.1284	0.1304

(continued)

Table B.10 (continued)

x	8.1	8.2	8.3	8.4	8.5	8.6	8.7	8.8	8.9	9.0
9	0.0723	0.0757	0.0791	0.0825	0.0858	0.0891	0.0923	0.0954	0.0985	0.1014
10	0.0441	0.0469	0.0498	0.0528	0.0558	0.0588	0.0618	0.0649	0.0679	0.0710
11	0.0245	0.0265	0.0285	0.0307	0.0330	0.0353	0.0377	0.0401	0.0426	0.0452
12	0.0124	0.0137	0.0150	0.0164	0.0179	0.0194	0.0210	0.0227	0.0245	0.0264
13	0.0058	0.0065	0.0073	0.0081	0.0089	0.0098	0.0108	0.0119	0.0130	0.0142
14	0.0025	0.0029	0.0033	0.0037	0.0041	0.0046	0.0052	0.0058	0.0064	0.0071
15	0.0010	0.0012	0.0014	0.0016	0.0018	0.0020	0.0023	0.0026	0.0029	0.0033
16	0.0004	0.0005	0.0005	0.0006	0.0007	0.0008	0.0010	0.0011	0.0013	0.0014
17	0.0001	0.0002	0.0002	0.0002	0.0003	0.0003	0.0004	0.0004	0.0005	0.0006
18	0.0000	0.0001	0.0001	0.0001	0.0001	0.0001	0.0001	0.0002	0.0002	0.0002
19	0.0000	0.0000	0.0000	0.0000	0.0000	0.0000	0.0000	0.0001	0.0001	0.0001

λ

x	8.1	8.2	8.3	8.4	8.5	8.6	8.7	8.8	8.9	9.0
0	0.0003	0.0003	0.0002	0.0002	0.0002	0.0002	0.0002	0.0002	0.0001	0.0001
1	0.0025	0.0023	0.0021	0.0019	0.0017	0.0016	0.0014	0.0013	0.0012	0.0011
2	0.0100	0.0092	0.0086	0.0079	0.0074	0.0068	0.0063	0.0058	0.0054	0.0050
3	0.0269	0.0252	0.0237	0.0222	0.0208	0.0195	0.0183	0.0171	0.0160	0.0150
4	0.0544	0.0517	0.0491	0.0466	0.0443	0.0420	0.0398	0.0377	0.0357	0.0337
5	0.0882	0.0849	0.0816	0.0784	0.0752	0.0722	0.0692	0.0663	0.0635	0.0607
6	0.1191	0.1160	0.1128	0.1097	0.1066	0.1034	0.1003	0.0972	0.0941	0.0911

(continued)

x	9.1	9.2	9.3	9.4	9.5	9.6	9.7	9.8	9.9	10.0
7	0.1378	0.1358	0.1338	0.1317	0.1294	0.1271	0.1247	0.1222	0.1197	0.1171
8	0.1395	0.1392	0.1388	0.1382	0.1375	0.1366	0.1356	0.1344	0.1332	0.1318
9	0.1256	0.1269	0.1280	0.1290	0.1299	0.1306	0.1311	0.1315	0.1317	0.1318
10	0.1017	0.1040	0.1063	0.1084	0.1104	0.1123	0.1140	0.1157	0.1172	0.1186
11	0.0749	0.0776	0.0802	0.0828	0.0853	0.0878	0.0902	0.0925	0.0948	0.0970
12	0.0505	0.0530	0.0555	0.0579	0.0604	0.0629	0.0654	0.0679	0.0703	0.0728
13	0.0315	0.0334	0.0354	0.0374	0.0395	0.0416	0.0438	0.0459	0.0481	0.0504
14	0.0182	0.0196	0.0210	0.0225	0.0240	0.0256	0.0272	0.0289	0.0306	0.0324
15	0.0098	0.0107	0.0116	0.0126	0.0136	0.0147	0.0158	0.0169	0.0182	0.0194
16	0.0050	0.0055	0.0060	0.0066	0.0072	0.0079	0.0086	0.0093	0.0101	0.0109
17	0.0024	0.0026	0.0029	0.0033	0.0036	0.0040	0.0044	0.0048	0.0053	0.0058
18	0.0011	0.0012	0.0014	0.0015	0.0017	0.0019	0.0021	0.0024	0.0026	0.0029
19	0.0005	0.0005	0.0006	0.0007	0.0008	0.0009	0.0010	0.0011	0.0012	0.0014
20	0.0002	0.0002	0.0002	0.0003	0.0003	0.0004	0.0004	0.0005	0.0005	0.0006
21	0.0001	0.0001	0.0001	0.0001	0.0001	0.0002	0.0002	0.0002	0.0002	0.0003
22	0.0000	0.0000	0.0000	0.0000	0.0001	0.0001	0.0001	0.0001	0.0001	0.0001

x	λ									
	9.1	9.2	9.3	9.4	9.5	9.6	9.7	9.8	9.9	10.0
0	0.0001	0.0001	0.0001	0.0001	0.0001	0.0001	0.0001	0.0001	0.0001	0.0000
1	0.0010	0.0009	0.0009	0.0008	0.0007	0.0007	0.0006	0.0005	0.0005	0.0005

(continued)

Table B.10 (continued)

2	0.0046	0.0043	0.0040	0.0037	0.0034	0.0031	0.0029	0.0027	0.0025	0.0023
3	0.0140	0.0131	0.0123	0.0115	0.0107	0.0100	0.0093	0.0087	0.0081	0.0076
4	0.0319	0.0302	0.0285	0.0269	0.0254	0.0240	0.0226	0.0213	0.0201	0.0189
5	0.0581	0.0555	0.0530	0.0506	0.0483	0.0460	0.0439	0.0418	0.0398	0.0378
6	0.0881	0.0851	0.0822	0.0793	0.0764	0.0736	0.0709	0.0682	0.0656	0.0631
7	0.1145	0.1118	0.1091	0.1064	0.1037	0.1010	0.0982	0.0955	0.0928	0.0901
8	0.1302	0.1286	0.1269	0.1251	0.1232	0.1212	0.1191	0.1170	0.1148	0.1126
9	0.1317	0.1315	0.1311	0.1306	0.1300	0.1293	0.1284	0.1274	0.1263	0.1251
10	0.1198	0.1210	0.1219	0.1228	0.1235	0.1241	0.1245	0.1249	0.1250	0.1251
11	0.0991	0.1012	0.1031	0.1049	0.1067	0.1083	0.1098	0.1112	0.1125	0.1137
12	0.0752	0.0776	0.0799	0.0822	0.0844	0.0866	0.0888	0.0908	0.0928	0.0948
13	0.0526	0.0549	0.0572	0.0594	0.0617	0.0640	0.0662	0.0685	0.0707	0.0729
14	0.0342	0.0361	0.0380	0.0399	0.0419	0.0439	0.0459	0.0479	0.0500	0.0521
15	0.0208	0.0221	0.0235	0.0250	0.0265	0.0281	0.0297	0.0313	0.0330	0.0347
16	0.0118	0.0127	0.0137	0.0147	0.0157	0.0168	0.0180	0.0192	0.0204	0.0217
17	0.0063	0.0069	0.0075	0.0081	0.0088	0.0095	0.0103	0.0111	0.0119	0.0128
18	0.0032	0.0035	0.0039	0.0042	0.0046	0.0051	0.0055	0.0060	0.0065	0.0071
19	0.0015	0.0017	0.0019	0.0021	0.0023	0.0026	0.0028	0.0031	0.0034	0.0037
20	0.0007	0.0008	0.0009	0.0010	0.0011	0.0012	0.0014	0.0015	0.0017	0.0019
21	0.0003	0.0003	0.0004	0.0004	0.0005	0.0006	0.0006	0.0007	0.0008	0.0009

(continued)

Table B.10 (continued)

22	0.0001	0.0001	0.0002	0.0002	0.0002	0.0002	0.0003	0.0003	0.0004	0.0004
23	0.0000	0.0001	0.0001	0.0001	0.0001	0.0001	0.0001	0.0001	0.0002	0.0002
24	0.0000	0.0000	0.0000	0.0000	0.0000	0.0000	0.0000	0.0001	0.0001	0.0001

x	λ									
	11	12	13	14	15	16	17	18	19	20
0	0.0000	0.0000	0.0000	0.0000	0.0000	0.0000	0.0000	0.0000	0.0000	0.0000
1	0.0002	0.0001	0.0000	0.0000	0.0000	0.0000	0.0000	0.0000	0.0000	0.0000
2	0.0010	0.0004	0.0002	0.0001	0.0000	0.0000	0.0000	0.0000	0.0000	0.0000
3	0.0037	0.0018	0.0008	0.0004	0.0002	0.0001	0.0000	0.0000	0.0000	0.0000
4	0.0102	0.0053	0.0027	0.0013	0.0006	0.0003	0.0001	0.0001	0.0000	0.0000
5	0.0224	0.0127	0.0070	0.0037	0.0019	0.0010	0.0005	0.0002	0.0001	0.0001
6	0.0411	0.0255	0.0152	0.0087	0.0048	0.0026	0.0014	0.0007	0.0004	0.0002
7	0.0646	0.0437	0.0281	0.0174	0.0104	0.0060	0.0034	0.0018	0.0010	0.0005
8	0.0888	0.0655	0.0457	0.0304	0.0194	0.0120	0.0072	0.0042	0.0024	0.0013
9	0.1085	0.0874	0.0661	0.0473	0.0324	0.0213	0.0135	0.0083	0.0050	0.0029
10	0.1194	0.1048	0.0859	0.0663	0.0486	0.0341	0.0230	0.0150	0.0095	0.0058
11	0.1194	0.1144	0.1015	0.0844	0.0663	0.0496	0.0355	0.0245	0.0164	0.0106
12	0.1094	0.1144	0.1099	0.0984	0.0829	0.0661	0.0504	0.0368	0.0259	0.0176
13	0.0926	0.1056	0.1099	0.1060	0.0956	0.0814	0.0658	0.0509	0.0378	0.0271
14	0.0728	0.0905	0.1021	0.1060	0.1024	0.0930	0.0800	0.0655	0.0514	0.0387

(continued)

Table B.10 (continued)

15	0.0534	0.0724	0.0885	0.0989	0.1024	0.0992	0.0906	0.0786	0.0650	0.0516
16	0.0367	0.0543	0.0719	0.0866	0.0960	0.0992	0.0963	0.0884	0.0772	0.0646
17	0.0237	0.0383	0.0550	0.0713	0.0847	0.0934	0.0963	0.0936	0.0863	0.0760
18	0.0145	0.0256	0.0397	0.0554	0.0706	0.0830	0.0909	0.0936	0.0911	0.0844
19	0.0084	0.0161	0.0272	0.0409	0.0557	0.0699	0.0814	0.0887	0.0911	0.0888
20	0.0046	0.0097	0.0177	0.0286	0.0418	0.0559	0.0692	0.0798	0.0866	0.0888
21	0.0024	0.0055	0.0109	0.0191	0.0299	0.0426	0.0560	0.0684	0.0783	0.0846
22	0.0012	0.0030	0.0065	0.0121	0.0204	0.0310	0.0433	0.0560	0.0676	0.0769
23	0.0006	0.0016	0.0037	0.0074	0.0133	0.0216	0.0320	0.0438	0.0559	0.0669
24	0.0003	0.0008	0.0020	0.0043	0.0083	0.0144	0.0226	0.0328	0.0442	0.0557
25	0.0001	0.0004	0.0010	0.0024	0.0050	0.0092	0.0154	0.0237	0.0336	0.0446
26	0.0000	0.0002	0.0005	0.0013	0.0029	0.0057	0.0101	0.0164	0.0246	0.0343
27	0.0000	0.0001	0.0002	0.0007	0.0016	0.0034	0.0063	0.0109	0.0173	0.0254
28	0.0000	0.0000	0.0001	0.0003	0.0009	0.0019	0.0038	0.0070	0.0117	0.0181
29	0.0000	0.0000	0.0001	0.0002	0.0004	0.0011	0.0023	0.0044	0.0077	0.0125
30	0.0000	0.0000	0.0000	0.0001	0.0002	0.0006	0.0013	0.0026	0.0049	0.0083
31	0.0000	0.0000	0.0000	0.0000	0.0001	0.0003	0.0007	0.0015	0.0030	0.0054
32	0.0000	0.0000	0.0000	0.0000	0.0001	0.0001	0.0004	0.0009	0.0018	0.0034
33	0.0000	0.0000	0.0000	0.0000	0.0000	0.0001	0.0002	0.0005	0.0010	0.0020
34	0.0000	0.0000	0.0000	0.0000	0.0000	0.0000	0.0001	0.0002	0.0006	0.0012

(continued)

Table B.10 (continued)

35	0.0000	0.0000	0.0000	0.0000	0.0000	0.0000	0.0000	0.0001	0.0003	0.0007
36	0.0000	0.0000	0.0000	0.0000	0.0000	0.0000	0.0000	0.0001	0.0002	0.0004
37	0.0000	0.0000	0.0000	0.0000	0.0000	0.0000	0.0000	0.0000	0.0001	0.0002
38	0.0000	0.0000	0.0000	0.0000	0.0000	0.0000	0.0000	0.0000	0.0000	0.0001
39	0.0000	0.0000	0.0000	0.0000	0.0000	0.0000	0.0000	0.0000	0.0000	0.0001

Table B.11 Negative exponential distribution table

	0	1	2	3	4	5	6	7
0	1	0.3679	0.1353	0.0498	0.0183	0.0067	0.0025	0.0009
0.1	0.9048	0.3329	0.1224	0.0450	0.0166	0.0061	0.0022	0.0008
0.2	0.8187	0.3012	0.1108	0.0408	0.0150	0.0055	0.0020	0.0007
0.3	0.7408	0.2725	0.1003	0.0369	0.0136	0.0050	0.0018	0.0007
0.4	0.6703	0.2466	0.0907	0.0334	0.0123	0.0045	0.0017	0.0006
0.5	0.6065	0.2231	0.0821	0.0302	0.0111	0.0041	0.0015	0.0006
0.6	0.5488	0.2019	0.0743	0.0273	0.0100	0.0037	0.0014	0.0005
0.7	0.4966	0.1827	0.0672	0.0247	0.0091	0.0033	0.0012	0.0005
0.8	0.4493	0.1653	0.0608	0.0224	0.0082	0.0030	0.0011	0.0004
0.9	0.4066	0.1496	0.0550	0.0202	0.0076	0.0027	0.0010	0.0004

Table B.12 Exponential distribution table: The exponential distribution is often concerned with the amount of time until some specific event occurs

	0	1	2	3	4	5	6	7
0	1.00	2.72	7.39	20.09	54.60	148.41	403.43	1096.63
0.1	1.11	3.00	8.17	22.20	60.34	164.02	445.86	1211.97
0.2	1.22	3.32	9.07	24.53	66.69	181.27	492.75	1339.43
0.3	1.35	3.67	9.97	27.11	73.70	200.34	544.57	1480.30
0.4	1.49	4.06	11.02	29.96	81.45	221.41	601.85	1635.98
0.5	1.65	4.48	12.18	33.12	90.02	244.69	665.14	1808.04
0.6	1.82	4.95	13.46	36.6	99.48	270.43	735.10	1998.20
0.7	2.01	5.47	14.88	40.45	109.95	298.87	812.41	2208.35
0.8	2.23	6.05	16.44	44.70	121.51	330.30	897.85	2440.60
0.9	2.46	6.69	18.17	49.4	134.29	365.04	992.27	2697.28

Table B.13 Required sample size to estimate a proportion (alpha $= 0.05$)

Margin of error P	0.05	0.1	0.2	0.25	0.33	0.5
0.50	1537	384	97	62	35	16
0.45	1879	470	118	76	43	19
0.40	2305	577	145	93	52	24
0.35	2855	714	179	115	65	29
0.30	3586	897	225	144	81	36
0.25	410	1153	289	185	104	47
0.20	6147	1537	385	246	139	62

(continued)

Table B.13 (continued)

Margin of error P	0.05	0.1	0.2	0.25	0.33	0.5
0.15	8707	2177	545	349	196	88
0.10	13,830	3458	865	554	312	139
0.08	17,672	4418	1105	707	398	177
0.06	24,075	6019	1505	963	542	241
0.05	29,197	7299	1825	1168	657	292
0.03	49,685	12,422	3106	1988	1118	497
0.02	75,296	18,824	4706	3012	1695	753
0.01	152,128	38,032	9508	6086	3423	1522
0.008	190,544	47,636	11,909	7622	4288	1906
0.006	254,570	63,643	15,911	10,183	5728	2546
0.005	305,792	76,448	19,112	12,232	6881	3058
0.003	510,677	127,670	31,918	20,428	11,491	5107
0.002	766,784	191,696	47,924	30,672	17,253	7668
0.001	1,535,103	383,776	95,944	61,405	34,540	15,352

Table B.14 Required sample size to estimate a proportion (alpha $= 0.05$)

Margin of error P	0.05	0.10	0.20	0.25	0.33	0.50
0.50	1083	271	68	44	25	12
0.45	1324	331	83	54	31	14
0.40	1624	406	102	66	37	17
0.35	2011	503	126	81	46	21
0.30	2526	632	158	102	57	26
0.25	3248	812	203	131	74	34
0.20	4330	1083	271	174	98	44
0.15	6134	1534	384	246	139	62
0.10	9742	2436	609	391	220	98
0.08	12,449	3112	778	498	281	125
0.06	16,959	4240	1060	679	382	170
0.05	20,567	5142	1284	823	463	206
0.03	34,999	8750	2188	1400	788	351

(continued)

Table B.14 (continued)

Margin of error P	0.05	0.10	0.20	0.25	0.33	0.50
0.02	53,039	13,260	3315	212	1194	531
0.01	107,160	26,790	6698	4287	2412	1073
0.008	134,220	33,555	8389	5369	3021	1343
0.006	179,320	44,830	11,208	7173	4035	1794
0.005	215,400	53,850	13,463	8617	4847	2155
0.003	359,721	89,931	22,483	14,390	8095	3598
0.002	540,123	135,031	33,758	21,606	12,154	5402
0.001	1,081,328	270,332	67,583	43,254	24,330	10,814

Table B.15 Square of sum of $s\left(z_{\frac{\alpha}{2}}+z_{\frac{\beta}{2}}\right)$

	α	0/40	0/20	0/10	0/05	0/02	0/01	0/002	0/001	0/0001	0/00001
β	Z	0.84	1.28	1.64	1.96	2.33	2.58	3.09	3.29	3.89	4.42
0.40	0.84	2.82	4.5	7.84	7/84	1/020	11.67	15.44	17.06	22.38	27.64
0.2	1.28	4.5	6.57	10.51	1/510	1/023	14.88	19.11	20.91	26.76	32.48
0.1	1.64	6.17	8.57	12.99	1/992	1/775	17.82	22.42	24.36	30.65	36.75
0.05	1.96	7.84	10.51	15.37	1/375	1/378	20.57	25.5	27.57	34.23	40.67
0.02	2.33	10.02	13,02	18.37	1/378	2/641	24.03	29.33	31.55	38.65	45.47
0.01	2.58	11.67	14.88	20.57	2/570	2/034	26.54	32.1	34.42	41.82	48.9
0.002	3.09	15.45	19.11	25.5	2/505	2/339	32.1	38.19	40.72	48.73	56.35
0.001	3.29	17.06	20.91	27.57	2/577	3/551	34.42	40.72	43.32	51.58	59.41
0.0001	3.89	22.38	26.76	34.23	3/234	3/658	41.82	48.73	51.58	60.56	69.02

Answer to Exercises

In order to make sure that the reader and the student can complete the exercises and questions at the end of each chapter, we offer the final answer (not the solution) in this section, having chosen a portion of them at random.

Chapter 1	
1-A	Shoe sizes are both quantitative and interval; the difference between any two half sizes is approximately 5 mm, and the zero has no meaning. The sizes of clothing (..., XS, S, M, L, XL, and XXL, ...) is ordinal and qualitative in nature
1-E	Quantitative and continuous
1-G	Quantitative and interval
1-L	Qualitative and nominal
1-N	Quantitative and continuous
1-O	Quantitative and discrete
1-S	Qualitative and ordinal
1-T	Quantitative and discrete
1-Z	Quantitative and discrete
Chapter 2	
Q2	Pie diagram
Q4	Bar and pie diagram
Q7	Finding the lowest and highest observations ($X_{min} = 0$, $X_{max} = 12$) as well as the three quartiles ($Q_1 = 3$, $Q_2 = $ median $= 5$, $Q_3 = 8$) is necessary in order to create a boxplot. In this case, there isn't an outlier
Chapter 3	
Q3	Mean $= 1.725$, median $= 2$, $Q_3 = 3$, Var $= 1.310$, and SD $= 1.145$
Q5	Since the coefficient of variation takes into account both the population means and the standard deviation, it is more appropriate

(continued)

(continued)

Chapter 1	
Q8	$Q_1 = 244.14$, $Q_3 = 328.96$ and $p70 = 317.67$
Q10	Man $= 4.1$, median $= 4.05$, Var $= 0.514$, and Sd $= 0.717$

Chapter 4	
Q3	$p(x = 4) = 0.01234$ and $p(x = 0) = 0.1975$
Q6	$p = 68/200 = 0.32$ and $n = 7$ A: one every other means that $p(\text{YNYNYNY or NYNYNYN}) = 0.00329 + 0.007 = 0.1030$ B: Only the first three patients means that $p(\text{YYYNNNN}) = 0.007$ C: 0.245
Q8	$p(\text{old}) = 0.50$ $p(\text{Young}) = 0.50$ $p(H\|O) = 0.35$ $p(H\|Y) = 0.15$ $P(OH) = 0.35 \times 0.50 = 0.175,$ $P(Y\|\overline{H}) = 0.85 \times 0.50 = 0.425,$ $P(H) = (0.50 \times 0.35) + (0.50 \times 0.15) = 0.25$
Q10	$P^{\circledR} = 0.3333$, $P(Y) = 0.6666$ A: All three seedlings are red: $P(X = 3R) = 0.037$ B: Maximum of two yellow: $P(X \leq 2) = 0.704$ C: At least one yellow:$P(X \geq 1) = 0.963$
Q14	$P(A) = 0.6$ A: The match will be finished in three sets. $p(X = \text{ABA}) + p(\text{BAB}) = 0.24$ B: The match ends in two sets. $p(X = \text{AA}) + p(X = \text{BB}) = 0.58$
Q19	Since Step Down model and Step Up model are independent events, $P(A*B) = P(A)*P(B) = 0.05*0.03 = 0.0015$ and the probability at least one of the model is: $P(A + B) = P(A \text{ or } B) = 0.05 + 0.03 - 0.0015 = 0.0785$
Q24	$P(orbit) = \frac{1}{7}$. $n = 5$ $P(x \geq 3) = p(X = 3) + P(x = 4) + p(x = 5) = 0.0647$

Chapter 5	
Q1	$\mu = \lambda = \frac{1}{4}$ and $p(2) = 0.0042$
Q3	$P = 0.2$ and $n = 6$ A: $p(x = 2) = 0.246$ B: $p(x \leq 2) = p(x = 0) + p(x = 1) + p(X = 2) = 0.9011$ C: Maximum of 2 patients survive which means that minimum of 4 patients die $p(X \geq 2) = 0.01696$ D: $p(X = 1) = 0.3932$
Q5	$\mu = \lambda = 6000 \times \frac{1}{1000} = 6$ and $p(x \leq 2) = 0.062$
Q7	$p = \frac{1}{4} = 0.25$, $n = 10$ and $x = 6$ Using binomial distribution: $p(x = 6) = 0.01622$

(continued)

(continued)

Chapter 1	
Q8	$\mu = \lambda = 2000 \times 0.002 = 4$ and $p(x = 0) = 0.183$
Q12	$\mu = \lambda = 10$ and $p(x = 0) = 0.000045$

Chapter 6	
Q4	$X \sim N(100, 15)$ $P(X > 126) = 1 - P(X \le 126) = 0.0418$ or 4.18%
Q6	$X \sim N(120, 15)$ $C = 0.75 \rightarrow P(-0.75 \le Z \le +0.75) = 0.50$
Q9	$X \sim N(600, 100)$ $P(650 \le X \le 750) = 0.2417$
Q11	$X \sim N(30, 5.5)$ $P(X \ge 20) = 0.9656$ or 96.56%
Q14	Yes, because calculated $\chi^2 = 8.35 > \chi_2^2 = 5.99$

Chapter 7	
Q3	$n = 34$
Q5	$P = 0.4, d = 0.2 \times 0.4 = 0.08,$ and $n = 144$
Q7	$n = 385$
Q10	$P = 0.333$ and $1-p = 0.666$. Assuming 0.05 if significant level and the maximum difference of 10% lead to $n = 86$

Chapter 8	
Q2	$P = 0.40, n = 225,$ and CI $= 0.336$–0.464
Q4	For $n = 100$: A: $z_{0.975} = 1.96$ and CI $= 22.04$–24.36 B: $z_{0.995} = 2.575$ and CI $= 21.68$–24.72 C: For $n = 20$: $t_{19 \text{ and } 0.975} = 2.093$ and CI $= 20.44$–25.96 $t_{19 \text{ and } 0.995} = 2.539$ and CI $= 19.85$–26.55
Q6	$n = 64$, Sd $= 12$, and mean $= 105$ 95% confidence interval (102–108) includes the actual mean blood pressure of the population
Q8	$P = \frac{11}{80} = 0.1375, n = 30, t_{29} = 2.045,$ then CI $= 0.0588$–0.216

(continued)

(continued)

Chapter 1	
Chapter 9	
Q1	$OR = 1.3125$ ($a = 42$, $b = 496$, $c = 28$, and $d = 434$)
Q3	Sen $= \frac{11}{81} = 0.1375$, Spe $= \frac{2037}{2069} = 0.984$, PPV $= \frac{11}{43} = 0.256$, NPV $= \frac{2037}{2107} = 0.967$
Q5	$OR = 11.65$
Q7	$P_e = 0.577$, $P_0 = 0.94$ and $K = 0.858$ Se $= 0.02981$ and CI $= 0.80$–0.916
Chapter 11	
Q1	Because Sd is known (18 min), z criteria should be used. Calculated $Z = 1.23 < 1.96$ and we can't reject the factory claim
Q2	$\bar{x} = 23.7$, Sd $= 6.7364$, $\mu_0 = 20$, $n = 20$ and $t_{19} = 2.093$ Calculated $t = 2.456 > t_{19} = 2.093$ and the researcher claim is rejected
Q5	A: Se$(p_1 - p_2) = 0.1637$ CI95% $= (0.4000 - 0.3333) \pm 1.96 \times 0.1637 = -0.254$ to 0.3877 B: Calculated $Z = 0.407 < 1.96$ C: Using Chi-square $\chi^2 = 0.1624 < 3.841$
Q7	Using paired t-test, $t = 1.990 < t_{15} = 2.131$ indicates that there is no difference between 2 methods
Q14	$Sp = \sqrt{\frac{(37-1)\times 1.59^2 + (36-1)2.36^2}{37+36-2}} = 2.01$ $Se = 2.01 \times \sqrt{\frac{1}{37} + \frac{1}{36}} = 0.47$ $t = \frac{7.79 - 7.63}{0.47} = 0.34 < t_{71} = 1.994$
Chapter 12	
Q1	A: Correlation is significant $r = 0.765$ B: Pl $= 0.077 + 0.044$W C: If $w = 60$, Pl $= 2.72$ If $w = 65$, Pl $= 2.94$ If $w = 70$, Pl $= 3.16$ D: We are unable to estimate plasma for those individuals as this equation is only applicable to weights that fall between the minimum and maximum values
Q3	Correlation coefficient ($r = 0.689$) is strong and significant
Q5	Correlation coefficient for group 1 is ($r1 = -0.18$) which is not significant and for group 2 is ($r2 = -0.239$) which is significant

(continued)

(continued)

Chapter 1	
Q8	A: Correlation between height of children and father's height $= 0.604$ Correlation between height of children and mother's height $= 0.262$ To compare this correlation, we use Z criteria $Z_1 = \frac{1}{2}\ln\left(\frac{1+0.604}{1-0.604}\right) = 0.699$, $Z_2 = \frac{1}{2}\ln\left(\frac{1+0.262}{1-0.262}\right) = 0.268$ $Z(r_1 - r_2) = \dfrac{0.699-0.268}{\sqrt{\left(\frac{1}{40-3}+\frac{1}{40-3}\right)}} = 1.854 < 1.96$ The difference in correlation coefficient between the fathers and mothers is not significant B: Correlation between height of girls and mother's height $= 0.795$ Correlation between height of Boys and father's height $= 0.241$ To compare this correlation, we use Z criteria $Z_1 = \frac{1}{2}\ln\left(\frac{1+0.795}{1-0.795}\right) = 1.085$, $Z_2 = \frac{1}{2}\ln\left(\frac{1+0.241}{1-0.241}\right) = 0.246$ $Z(r_1 - r_2) = \dfrac{1.085-0.246}{\sqrt{\left(\frac{1}{20-3}+\frac{1}{20-3}\right)}} = 2.446 > 1.96$ The difference in correlation coefficient between the boys with fathers and girls with mothers is significant

Chapter 13	
Q1	There are more than two groups to compare, so the ANOVA method is appropriate. In this instance, the average pulse rate of these groups is the same ($F = 0.079, p = 0.924$)
Q3	Since there are more than two groups to compare, the ANOVA technique is appropriate, as evidenced by the $F = 2.823$ with $p = 0.082 > 0.05$. So these teaching strategies have the same average score
Q5	The ANOVA technique is appropriate for comparing four groups. In this case, $F = 3.795, p = 0.015 < 0.05$. Then, there's a difference in the average scores for these types of insurance

References

1. Altman, D. G. (1996). *Practical statistics for medical research*. Chapman and Hall.
2. Andrade, C. (2017). Age as a variable: Continuous or categorical? *Indian* Journal *of Psychiatry, 59*(4).
3. Andrade, C. (2017). Age is a number, not a group. *Indian Journal of Psychiatry, 59*(2).
4. Armitage, P., & Colton, T. (1998). *Encyclopedia of biostatistics* (2nd ed.). Oxford Medical Publications 2015.
5. Armitage, P., Berry, G., & Matthews, J. N. S. (2008). *Statistical methods in medical research*. John Wiley & Sons.
6. Bégaud, B., Martin, K., Abouelfath, A., Tubert-Bitter, P., Moore, N., & Moride, Y. (2005). An easy to use method to approximate Poisson confidence limits. *European Journal of Epidemiology, 20*(3), 213–216.
7. Bland, M. (2015). *An introduction to medical statistics*. Oxford University Press.
8. Brink, D. (2010). *Essentials of statistics*. Bookboon.
9. Campbell, M. J., Machin, D., & Walters, S. J. (2010). *Medical statistics: A textbook for the health sciences*. John Wiley & Sons.
10. Carter, R., & Lubinsky, J. (2015). *Rehabilitation research: principles and applications*. Elsevier Health Sciences.
11. Le, C. T., & Eberly, L. E. (2016). *Introductory biostatistics*. John Wiley & Sons.
12. Davis, C. S. (2002). *Statistical methods for the analysis of repeated measurements* (No. 04; QA278, D38.). Springer, New York.
13. Dixon, W. J., & Massey Jr, F. J. (1983). *Introduction to statistical analysis*. McGraw Hill.
14. Everitt, B. S., & Skrondal, A. (2010). *The Cambridge dictionary of statistics*.
15. Folks G. L. (1981). *Ideas of statistics*. John Wiley & Sons.
16. Heiman, G. (2013). *Basic statistics for the behavioral sciences*. Cengage Learning.
17. Pezzullo, J. C. (2013). *Biostatistics for dummies*. John Wiley & Sons.
18. Kirkwood, B. R., & Sterne, J. A. (2010). *Essential medical statistics*. John Wiley & Sons.
19. Madsen, B. (2011). *Statistics for non-statisticians*. Springer.
20. Matsunawa, T. (1982). Some strong ε-equivalence of random variables. *Annals of the Institute of Statistical Mathematics, 34*(2), 209–224.
21. McGill, R., Tukey, J. W., & Larsen, W. A. (1978). Variations of box plots. *The American Statistician, 32*(1), 12–16.
22. Miller, H. (1995). *Descriptive statistics*. Open Learning Foundation.
23. Milton, J. S., & Tsokos, J. O. (1983). Statistical methods in the biological and health sciences. In *Statistical methods in the biological and health sciences*. McGraw-Hill.

24. Mondal, H., & Mondal, S. (2017). Pretesting and cognitive interviewing are integral parts in translation of survey instrument. *Indian Journal of Psychiatry, 59*(2).
25. Navidi, W. C., & Monk, B. J. (2019). *Elementary statistics*. McGraw-Hill.
26. Neale, J. M., & Liebert, R. M. (1983). *Science and behavior. An introduction to methods of research.* Prentice-Hall International
27. Ravid, R. (2019). *Practical statistics for educators*. Rowman & Littlefield Publishers.
28. Sanders, D. H. (1990). *Statistics: A fresh approach* (4th ed.). Mc Graw Hill.
29. Sturges, H. A. (1926). The choice of a class interval. *Journal of the American Statistical Association, 21*(153), 65–66.
30. Urdan, T. C. (2011). *Statistics in plain English*. Routledge.